Math 1

God's Gift of Numbers

2nd Printing, 2024.

Printed in the United States of America.

ISBN: 978-1-954745-30-8

Cover Design: Justin Turley
Interior Design: Sarah Lee Bryant

Published by:
Generations
PO Box 1398
Elizabeth, Colorado 80107-1398

For more information on this and other titles from Generations, visit Generations.org or call (888) 389-9080.

Math 1
God's Gift of Numbers

Creators and Writers
Elliott Best, (BS, Mathematics; MS, Mathematics)
Kevin Swanson (BS, Mechanical Engineering; MDiv.)

Editors
Tammy Sechrist (BS, Biology)
Kayla White

Generations
PASSING ON THE FAITH

Contents

Your Math Toolbox

Get ready for your journey in math by gathering the following helpful tools.

Pencils

Paper or Notebook

Crayons or Colored Pencils

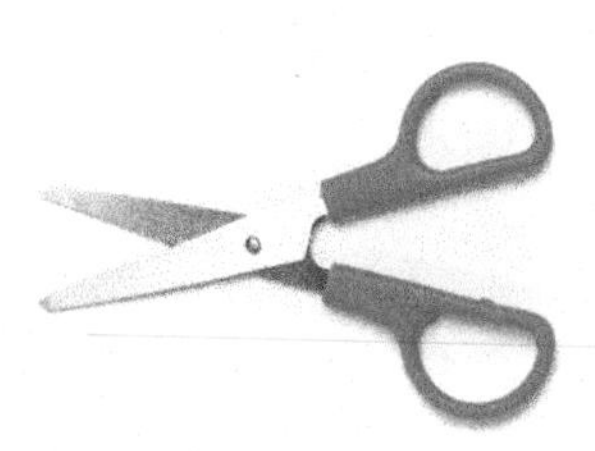

Scissors

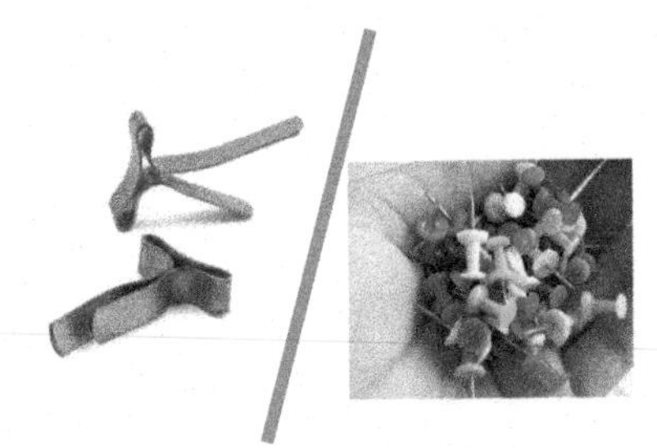

Brass Fasteners
or Push Pins

12 Small Stones

A Ruler (That measures
inches and centimeters)

Three Dice

Manipulatives

String (2-3 feet long)

A Tape Measure
(You can borrow the
family's when you need it!)

Flash Cards

To the Parent/Teacher

Welcome to the exciting world of mathematics!

This mathematics program has been produced by a team of math-lovers and homeschooling fans. We are a collection of experienced homeschool dads, moms, and graduates, with a love for math and a few advanced and undergraduate degrees amongst us. But mostly, we are "God-philes"—folks who love God's Word, God's works, and God's thoughts to think after Him. This is our motivation as we work to produce a new approach to math for the Christian family. We hope you and your little learners will be blessed by the unique features in this course.

Unique Features

God-Centered Focus. We will love to show you and your child math as the "pattern of God." Since math comes from God, we aim to keep the scriptures preeminent throughout this math course. Our goal is to help you implement the Deuteronomy 6:7-9 mandate, diligently teaching your children God's Word when you sit in your house, walk by the way, when you lie down, and when you rise up. Scripture is to be a "frontlet" before their eyes, integrated into every subject they study, and that includes this subject.

God-Glorifying Content. Proverbs 1:7 gives us the beginning of knowledge as the fear of God. The objective behind all learning should be to know God, and to learn more of His awesome power, wisdom, and goodness. Learning should never increase pride in the student's ability to know something. Instead, learning should humble us all and glorify God in our eyes.

Christian Worldview Built-in. Elements of a Christian worldview have been incorporated into the lessons—some of the principles may be more obvious than others. Family life, church life, honor for parents, honesty, and love for brothers and sisters are principles that are inculcated throughout.

Prayer and Faith Lessons. The time that you spend with your child will be habit forming: math habits, learning habits, and life habits. Beginning with prayer is a wonderful habit to impart: a simple confession of our dependence on God, a small acknowledgment of God's greatness, a sincere giving of thanks for the chance to learn, and a desire to see His kingdom come... even in the context of math.

Math Presented as a Tool. Throughout the course, students are encouraged to use math as a tool to steward God's creation, to take proper dominion in their household economy, and to serve others. Math can help them be "doers of the Word and not hearers only."

Math Demonstrated in Creation. Math shows God's wisdom, organization, and unchangeable nature—His patterns. We see God's attributes in all His creation. This course uses beautiful photos to point to God's design in creation whenever possible.

Use of the Concrete and the Abstract. This course uses concrete objects (like stones and manipulatives) to help with beginning math. These concrete objects are then tied to abstract numerals and equations. The abstract math is then reinforced with concrete applications like science and economics.

Math Connections Shown with Music and Games. Since there is a strong connection between music and math, we include exploration of the piano keyboard. We also explore the math and logic in games like sudoku, chess, and "Wonder Squares."

How to Use this Course

1. This course contains approximately 150 days of content for a 30 to 36-week school year. A suggested schedule follows this introduction but may be adapted to the student as necessary.
2. Typically, sequential math days will alternate between lessons of new content and the review of prior concepts. New lessons are varied and presented in an engaging, wonder-filled, enthusiastic way to immerse your child happily into the wonderful world of mathematics. Various lessons contain math exercises, thinking exercises, imagination exercises, consideration of God's creation, the use of math in scripture, the application of math to culture and science, drawing exercises, games, and opportunities to use math in everyday life. The "Practice" days use the time-tested idea of repetition to help the child retain a firm grasp on the old.
3. For your convenience everything you need is in one book. Special information for the parent/teacher is given within orange boxes. An answer key is also included in the back of the book. The simple, engaging text of the lessons is meant to be read aloud to the child so you don't need to take extra time to prepare.
4. Begin each day with prayer. We have suggested a prayer in each lesson, but feel free to adjust this as you find good and wise. Maybe some days could start with a hymn or psalm instead.
5. This course will generally be provided with flash cards. If that is not the case, ***be sure to purchase addition and subtraction flash cards as an essential supplement***.

God has given young children an amazing ability to memorize! Your diligence in helping your child practice the memory work (recommended in the blue boxes) will set your student up for a lifetime of success in math. Without this foundation, further math concepts will verge on futility and end in frustration. Scales are the foundation of music. Drills are the fundamentals for athletes. So too, there are basic ideas that should be memorized and rehearsed. This is essential for engaging in the art and science of mathematics.

6. Children are greatly helped in their math comprehension by the use of manipulatives, generally provided with this course. If these are not made available with the course, the parent/teacher is encouraged to purchase them separately.
7. Life application is the indication that real learning has taken place. The end of each chapter in this book contains suggestions of ways your child can apply math. Choose from several suggested family-life activities to give your child a tailor-made experience. Homeschoolers have wonderful access to learning opportunities that include grocery shopping, trips to grandparents, and all the counting and measuring that happens in their kitchen.
8. Some children will be naturally gifted in math and they may be able to begin this course at age four. Others may struggle a bit and not be ready until eight or nine years of age. We encourage you to allow for what we call "the principle of individuality" to operate. Allow the struggling child plenty of time with concrete manipulatives and don't worry about the schedule. For eager little mathematicians, we have included occasional "Extra Challenge" exercises. Most first graders will need one-on-one interaction from the parent/teacher for the first year or two. Once math facts are memorized and the student advances into second or third grade, he or she should be able to work more independently.

We Need Your Feedback

Feel free to offer any comments and suggestions to our director of publishing: *josh@generations.org*. We appreciate your input!

> And these words which I command you today shall be in your heart. You shall teach them diligently to your children, and shall talk of them when you sit in your house, when you walk by the way, when you lie down, and when you rise up. You shall bind them as a sign on your hand, and they shall be as frontlets between your eyes. You shall write them on the doorposts of your house and on your gates. (Deuteronomy 6:7-9)

Suggested Lesson Schedule

Planned Date	Day	Lessons & Practice	✓	Progress Notes
First Semester—First Quarter				
Week 1	Monday	Day 1 Lesson		
	Tuesday	Day 2 Lesson		
	Wednesday	Day 3 Practice		
	Thursday	Day 4 Lesson		
	Friday	Day 5 Lesson		
Week 2	Monday	Day 6 Practice		
	Tuesday	Day 7 Lesson		
	Wednesday	Day 8 Lesson		
	Thursday	Day 9 Practice		
	Friday	Day 10 Lesson		
Week 3	Monday	Day 11 Lesson		
	Tuesday	Day 12 Practice		
	Wednesday	Day 13 Lesson		
	Thursday	Day 14 Practice		
	Friday	Day 15 Lesson		
Week 4	Monday	Day 16 Practice		
	Tuesday	Day 17 Lesson		
	Wednesday	Day 18 Practice		
	Thursday	Day 19 Practice		
	Friday	Day 20 Lesson		
Week 5	Monday	Day 21 Lesson		
	Tuesday	Day 22 Practice		
	Wednesday	Day 23 Lesson		

Planned Date	Day	Lessons & Practice	✓	Progress Notes
	Thursday	Day 24 Practice		
	Friday	Day 25 Lesson		
Week 6	Monday	Memory Work		
	Tuesday	Day 26 Practice		
	Wednesday	Day 27 Lesson		
	Thursday	Day 28 Practice		
	Friday	Day 29 Lesson		
Week 7	Monday	Day 30 Practice		
	Tuesday	Day 31 Lesson		
	Wednesday	Day 32 Practice		
	Thursday	Day 33 Lesson		
	Friday	Day 34 Lesson		
Week 8	Monday	Memory Work		
	Tuesday	Day 35 Lesson		
	Wednesday	Day 36 Practice		
	Thursday	Day 37 Lesson		
	Friday	Day 38 Practice		
Week 9	Monday	Day 39 Lesson		
	Tuesday	Day 40 Practice		
	Wednesday	Day 41 Lesson		
	Thursday	Day 42 Practice		
	Friday	Day 43 Lesson		

First Semester—Second Quarter

Planned Date	Day	Lessons & Practice	✓	Progress Notes
Week 1	Monday	Memory Work		
	Tuesday	Day 44 Practice		
	Wednesday	Day 45 Lesson		
	Thursday	Day 46 Practice		

Planned Date	Day	Lessons & Practice	✓	Progress Notes
	Friday	Day 47 Lesson		
Week 2	Monday	Day 48 Practice		
	Tuesday	Day 49 Lesson		
	Wednesday	Day 50 Lesson		
	Thursday	Memory Work		
	Friday	Day 51 Practice		
Week 3	Monday	Day 52 Lesson		
	Tuesday	Day 53 Practice		
	Wednesday	Day 54 Lesson		
	Thursday	Memory Work		
	Friday	Day 55 Practice		
Week 4	Monday	Day 56 Lesson		
	Tuesday	Day 57 Lesson		
	Wednesday	Day 58 Practice		
	Thursday	Day 59 Lesson		
	Friday	Memory Work		
Week 5	Monday	Day 60 Practice		
	Tuesday	Day 61 Lesson		
	Wednesday	Day 62 Practice		
	Thursday	Day 63 Lesson		
	Friday	Day 64 Lesson		
Week 6	Monday	Day 65 Practice		
	Tuesday	Day 66 Lesson		
	Wednesday	Day 67 Practice		
	Thursday	Day 68 Practice		
	Friday	Day 69 Lesson		
Week 7	Monday	Day 70 Practice		
	Tuesday	Day 71 Lesson		

Planned Date	Day	Lessons & Practice	✓	Progress Notes
	Wednesday	Day 72 Practice		
	Thursday	Day 73 Practice		
	Friday	Day 74 Lesson		
Week 8	Monday	Day 75 Lesson		
	Tuesday	Day 76 Practice		
	Wednesday	Day 77 Lesson		
	Thursday	Memory Work		
	Friday	Day 78 Practice		
Week 9	Monday	Day 79 Lesson		
	Tuesday	Day 80 Practice		
	Wednesday	Day 81 Lesson		
	Thursday	Day 82 Practice		
	Friday	Day 83 Lesson		

Midterm Progress Notes	
Grade: ______	

Planned Date	Day	Lessons & Practice	✓	Progress Notes
Second Semester—Third Quarter				
Week 1	Monday	Day 84 Practice		
	Tuesday	Day 85 Lesson		
	Wednesday	Day 86 Practice		
	Thursday	Day 87 Lesson		
	Friday	Memory Work		
Week 2	Monday	Day 88 Lesson		
	Tuesday	Day 89 Practice		
	Wednesday	Day 90 Lesson		
	Thursday	Day 91 Practice		
	Friday	Day 92 Lesson		
Week 3	Monday	Day 93 Practice		
	Tuesday	Day 94 Lesson		
	Wednesday	Day 95 Practice		
	Thursday	Day 96 Lesson		
	Friday	Memory Work		
Week 4	Monday	Day 97 Practice		
	Tuesday	Day 98 Lesson		
	Wednesday	Day 99 Lesson		
	Thursday	Day 100 Practice		
	Friday	Day 101 Lesson		
Week 5	Monday	Day 102 Practice		
	Tuesday	Day 103 Lesson		
	Wednesday	Day 104 Practice		
	Thursday	Day 105 Lesson		
	Friday	Memory Work		
Week 6	Monday	Day 106 Practice		

Planned Date	Day	Lessons & Practice	✓	Progress Notes
	Tuesday	Day 107 Practice		
	Wednesday	Day 108 Lesson		
	Thursday	Day 109 Lesson		
	Friday	Memory Work		
Week 7	Monday	Day 110 Practice		
	Tuesday	Day 111 Lesson		
	Wednesday	Day 112 Practice		
	Thursday	Day 113 Lesson		
	Friday	Memory Work		
Week 8	Monday	Day 114 Practice		
	Tuesday	Day 115 Lesson		
	Wednesday	Day 116 Practice		
	Thursday	Day 117 Practice		
	Friday	Day 118 Lesson		
Week 9	Monday	Memory Work		
	Tuesday	Day 119 Practice		
	Wednesday	Day 120 Lesson		
	Thursday	Day 121 Lesson		
	Friday	Day 122 Practice		

Second Semester—Fourth Quarter

Planned Date	Day	Lessons & Practice	✓	Progress Notes
Week 1	Monday	Day 123 Lesson		
	Tuesday	Day 124 Practice		
	Wednesday	Day 125 Lesson		
	Thursday	Memory Work		
	Friday	Day 126 Practice		
Week 2	Monday	Day 127 Lesson		
	Tuesday	Day 128 Practice		

Planned Date	Day	Lessons & Practice	✓	Progress Notes
	Wednesday	Day 129 Lesson		
	Thursday	Memory Work		
	Friday	Day 130 Practice		
Week 3	Monday	Day 131 Lesson		
	Tuesday	Day 132 Lesson		
	Wednesday	Memory Work		
	Thursday	Day 133 Practice		
	Friday	Day 134 Lesson		
Week 4	Monday	Day 135 Practice		
	Tuesday	Day 136 Lesson		
	Wednesday	Memory Work		
	Thursday	Day 137 Practice		
	Friday	Day 138 Lesson		
Week 5	Monday	Memory Work		
	Tuesday	Day 139 Practice		
	Wednesday	Day 140 Lesson		
	Thursday	Day 141 Practice		
	Friday	Day 142 Lesson		
Week 6	Monday	Memory Work		
	Tuesday	Day 143 Practice		
	Wednesday	Day 144 Lesson		
	Thursday	Memory Work		
	Friday	Flex Day		
Week 7	Monday	Flex Day		
	Tuesday	Flex Day		
	Wednesday	Flex Day		
	Thursday	Flex Day		
	Friday	Flex Day		

Planned Date	Day	Lessons & Practice	✓	Progress Notes
Week 8	Monday	Flex Day		
	Tuesday	Flex Day		
	Wednesday	Flex Day		
	Thursday	Flex Day		
	Friday	Flex Day		
Week 9	Monday	Flex Day		
	Tuesday	Flex Day		
	Wednesday	Flex Day		
	Thursday	Flex Day		
	Friday	Flex Day		

Progress Notes	
Final Grade: _______	

God separated the
land from the sea!

CHAPTER 1

Making Separations

Introduction

God made the world, and that's when everything began for us. God made us too. He made humans like you and me. He gave us minds so we could study His amazing world, and He gave us math.

> In the beginning God created the heavens and the earth. The earth was without form, and void; and darkness was on the face of the deep. And the Spirit of God was hovering over the face of the waters. Then God said, "Let there be light"; and there was light. And God saw the light, that it was good; and God divided the light from the darkness. God called the light Day, and the darkness He called Night. So the evening and the morning were the first day. (Genesis 1:1-5)

Where do you think math came from? God thought of math first, before anybody else on earth learned about it. God made different kinds of things. When he made the plants, he made different kinds of plants. He made different kinds of animals. If God just made ducks, this would have been a boring world. God made many different kinds of animals.

When God made the world, he separated some things from other things. God separated the light from the darkness. He separated the heavens above from the earth below. He separated the dry land from the oceans and lakes. He separated the light part of the day from the dark part of the day. Can you see the darker area and the light in this picture?

Creating Sets DAY 1

This lesson explores the world God created. This will require about 20 minutes of instruction from the parent/teacher.

Prayer

Father, we love You. Teach us to see how You made so many different things. We want to learn more about Your creation so we can know You and worship You more. Amen.

Lesson

The first thing we will learn about math is the **set**. A set is a collection of things that look alike. It might be a collection of the same kind of animals, or the same kind of fruits, or some other kind of thing. Each piece of the set is called an **element** or **member**.

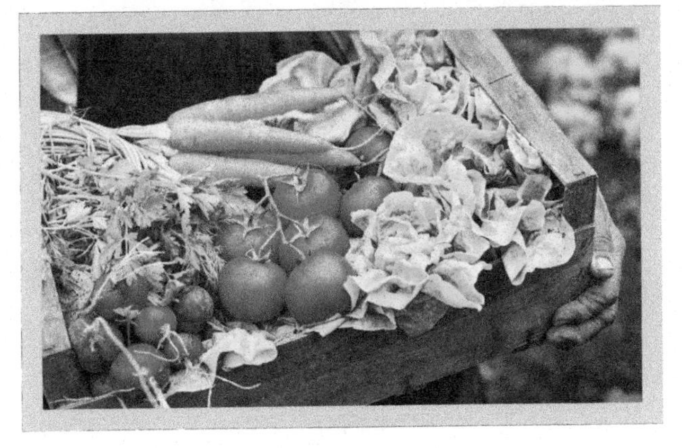

A set could also be a bunch of marbles in a bag, or some fruit in a basket.

When all the pieces of a set only look a little bit alike, the set is called a big set or (a **universal set**).

Think about your kitchen table at lunch time. Imagine everything that is put on the table. This is the big set or (the universal set). The bread and jam do not look like spoons and plates. These are all very different, but everything belongs on the table for lunchtime.

When we **partition** a set, we separate it into smaller sets with the pieces that look more like each other.

So, if we partition the things on the kitchen table into sets, what would we do? We would put all the food in one set, and the tableware (plates, cups, and spoons) in another set.

Or, let us look into the laundry basket. What do we see in there? There are so many different kinds of clothes in the basket. We can partition a laundry basket (the universal set) into three smaller sets — of shirts, pants, and socks.

The smallest set is called the **empty set**. This set doesn't have any member at all!

How many cows do you see in your bathroom? How many lions do you find in your living room? Are there any spiders crawling around in your shoes? There are none! These are empty sets!

Activity

Now, find a universal set in your house. You might use all the clean dishes in the drainer or the dishwasher. (See orange box on the next page for a few other suggested options for this activity). Now, make smaller sets out of the things that belong together. That is, separate the universal set into smaller sets.

1. Put all the cups together.
2. Put all the plates together.
3. Put all the silverware together.

There! Now that you have divided up the big set into three new sets, let's separate things some more. Now, think of the silverware as the big set or the universal set. Partition the silverware into three smaller sets.

1. Put all the forks together.
2. Put all the spoons together.
3. Put all the knives together.

Every time we wash and dry the dishes, we must separate the dishes into sets. We separate the forks, the spoons, and the knives in the silverware drawer. We put all the forks in one place, the spoons in another place, and the knives in still another place, so we can set the table later for another meal. Do you see how life is full of making sets?

Now, you can see how God created the world with sets of so many different things. When we organize things into sets like this, we are thinking a little bit like God thinks. God has organized His world by making things that are alike. And we organize things like spoons, forks, and knives in our kitchen too. Let us pray and thank God for the different sets—the different kinds of things He has made in this world.

Other Activities

Now, let's organize more things in the kitchen, school room, or your bedroom! You can separate out different kinds of toys in your toy box. You could organize your books on the bookshelf into different sets (according to size and color). Or you could organize a drawer full of office supplies into sets. You might separate by shapes, colors, or some other way.

Student Exercises

Draw a set of each of these types of things. There can be as many or as few members as you like!

Creating Shapes DAY 2

This is an imagination exploring lesson. The section includes a brief lesson, and two pages of exercises containing review and new material. This will require about 20 minutes of instruction and oversight from the parent/teacher.

Prayer

Our Father in heaven, we love You. Teach us to see Your patterns so that we can know and worship You more. Amen.

Lesson

The world is full of shapes and patterns. The simplest shape that God made is a **point**.

When a set of points are all put in a row real close to each other, we call this a **line**.

Some lines are **straight** and others are **curved**.

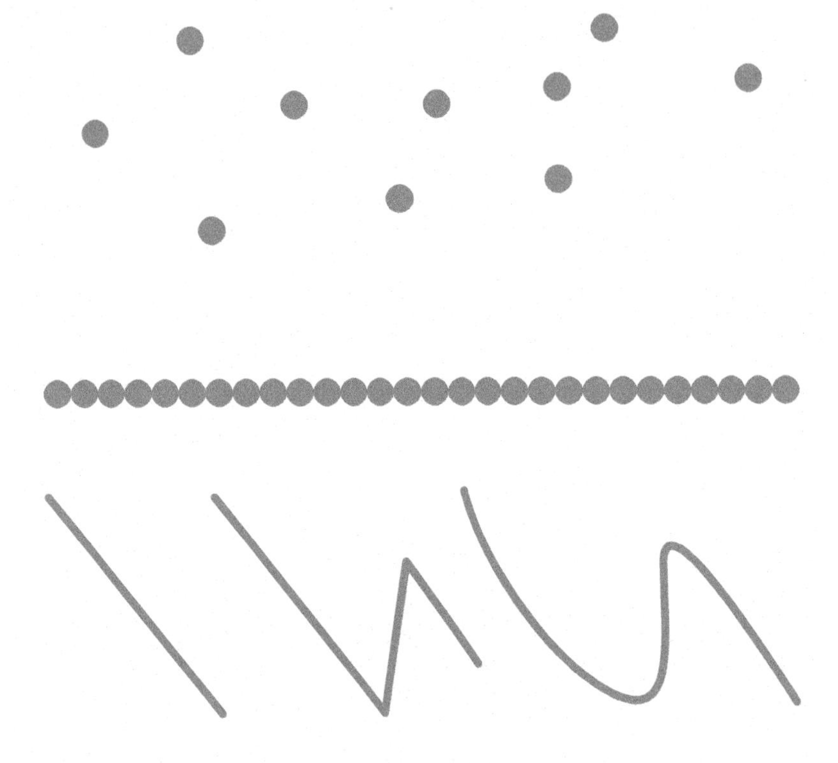

When you connect lines together you can make neat shapes like **triangles**, **squares**, **circles**, **stars** ... and so, so many more. Can you find some shapes in these pictures? You should find a star, a pyramid, and a hexagon. The pyramid was made by men. God created bees to make the cells of their honeycomb in the shape of a hexagon. A hexagon is a shape with 6 sides. God made the starfish too.

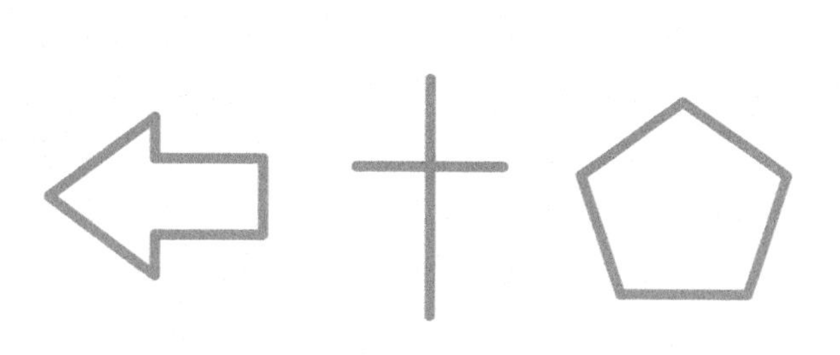

Student Exercises

In each of the boxes below, you will find three shapes. Two shapes will look alike, and they could be in a set together. Circle the shape that does not look like the other two. This one can be in a set by itself. Think about how God made the one shape different from the others. Do you know the name of each of these shapes? (The parent/teacher may need to help with this.)

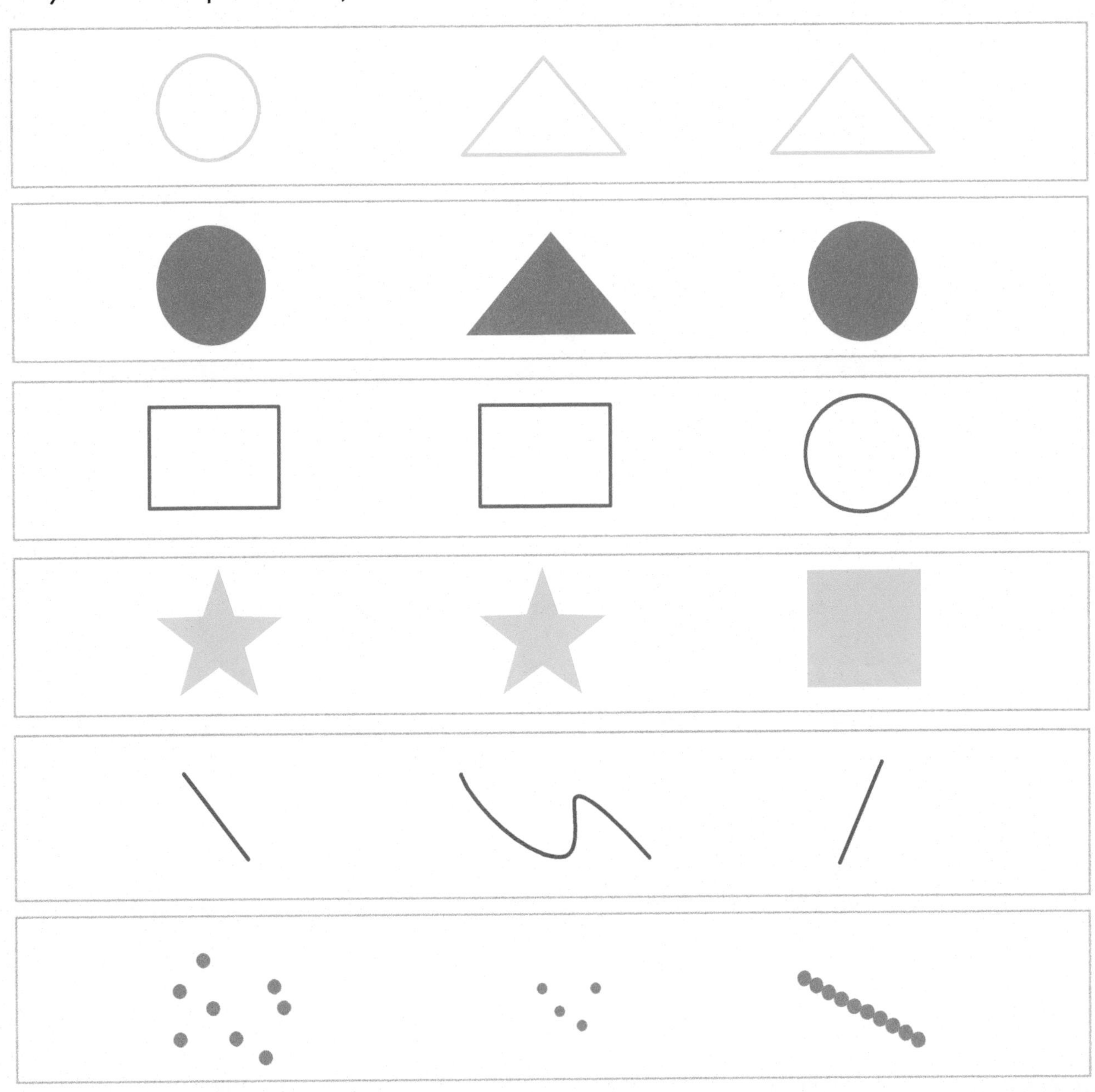

Trace these triangles, squares, circles, and other shapes too!

Practice DAY 3

Student Exercises

Show the child how to draw a straight line using a straight edge.

Now, draw some straight lines in the box below. Draw some curved ones too. You can draw nice curvy lines using a cup or a bottle lid. God's world is full of different kinds of lines. Can you be creative and draw different kinds of lines? Draw an interesting picture by drawing straight lines and curved lines. Connect the lines too.

Trace these shapes and lines! Add some color too!

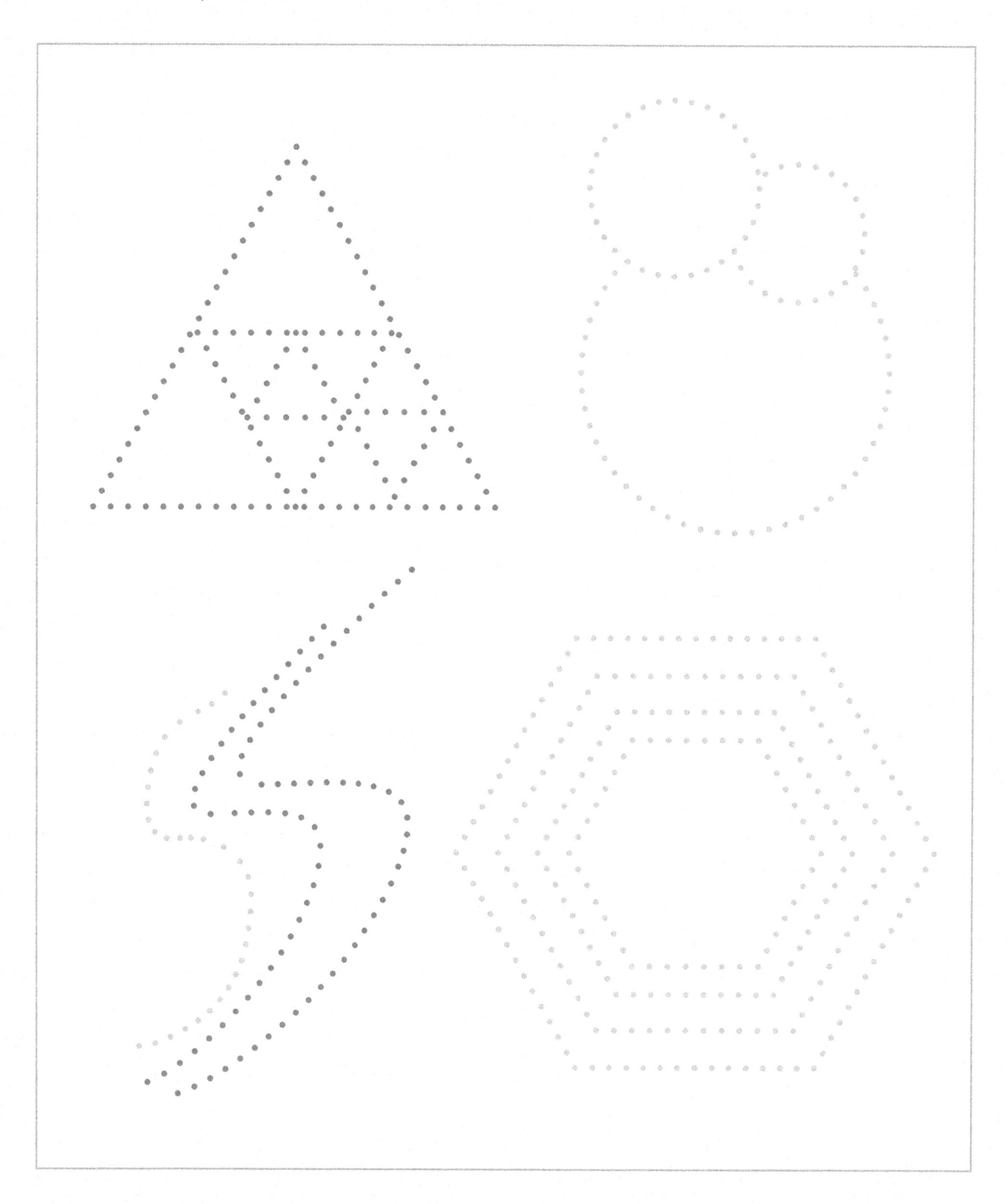

God's Separations DAY 4

This is a Scripture lesson. The section is related to the Scripture reading, followed by a page of exercises. This will require about 20 minutes of instruction from the parent/teacher.

Prayer

Father, we love You. Help us to see the beautiful patterns You made. We want to know You better and worship You more, God. Amen.

Activity

Let's read Genesis 1:1-25. (Parent/teacher reads aloud.) Where does God divide things (or make partitions)? God makes partitions when He separates a big set into smaller sets. Hint: He did it on each creation day!

God made so many kinds of animals. All of these animals are part of a very big set. We call it a "universal set." The universal set is a collection of things that only have a little bit in common. In this case, the universal set is simply all the animals God created. Animals are not like plants. That's because animals move around and have babies. The set of animals includes creatures like frogs, cows, birds, fish, butterflies, dogs, cats, and aardvarks. Let's divide up this big set of animals. Separate the animals according to the places we find them. Some animals live on the land. Some animals swim in the sea. Using the picture on the next page, draw some animals where you would find them—whether in the sky, the land, or the water.

The child could also paste pictures of animals on to the landscape drawing, with adult supervision.

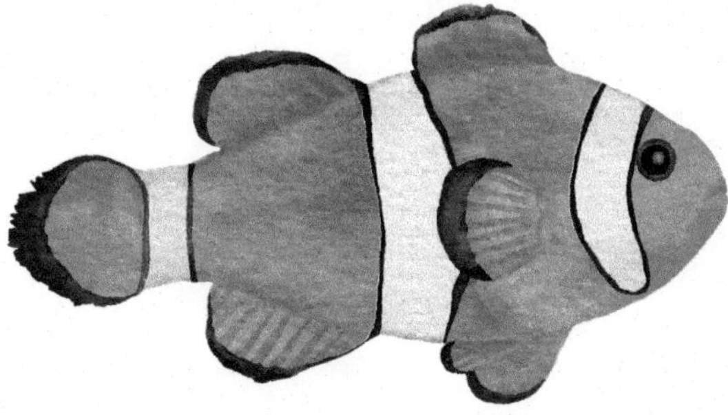

Student Exercises

What animals did God make for the sky? What animals did God make to live in the water, and on the land? Draw a few animals for each of these areas.

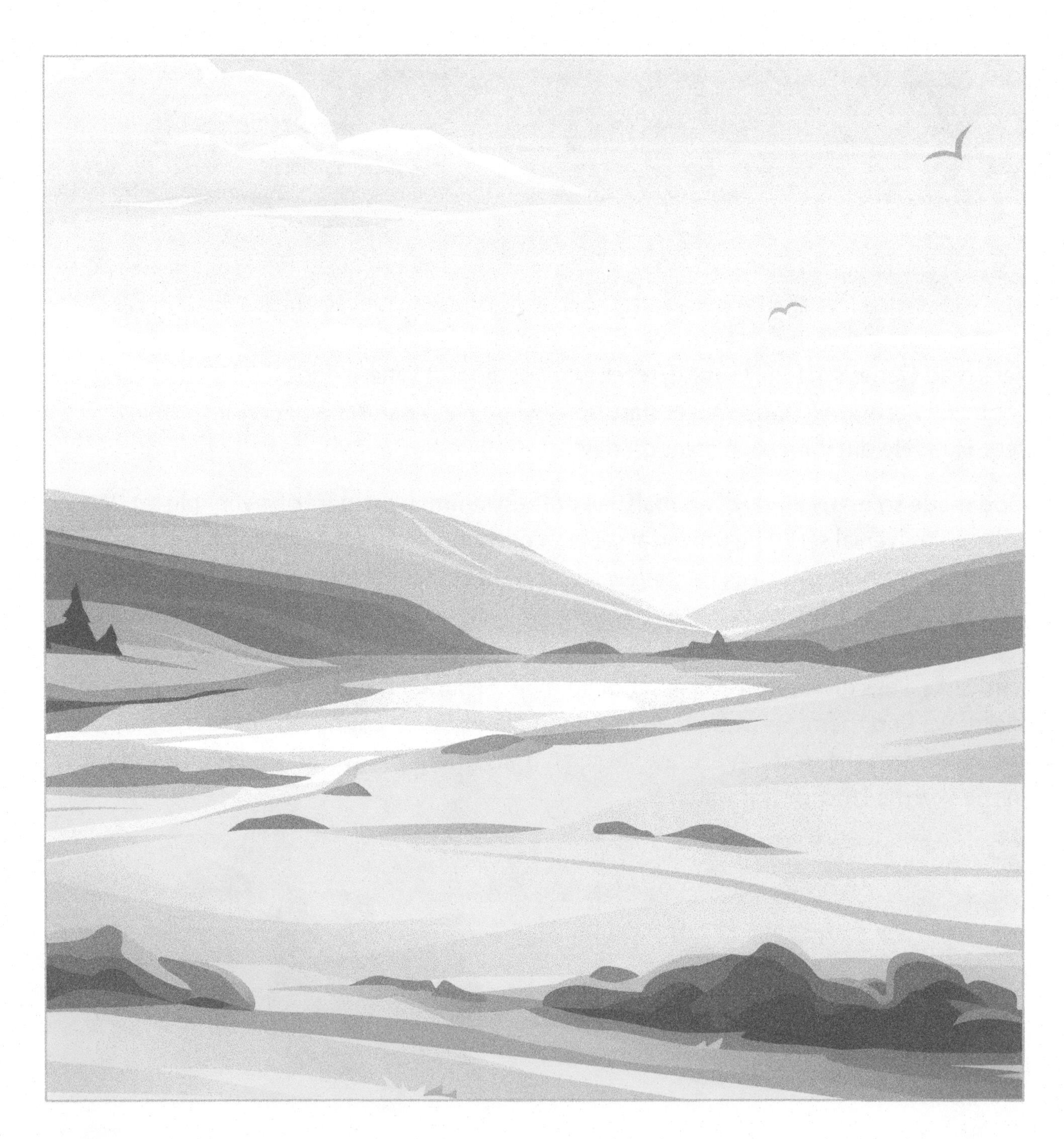

There are three different objects in each of the following sets. Circle the object that is not like the others. Why is that one object different from the others? How does God make things different from other things? (The parent/teacher may need to help the child with this.)

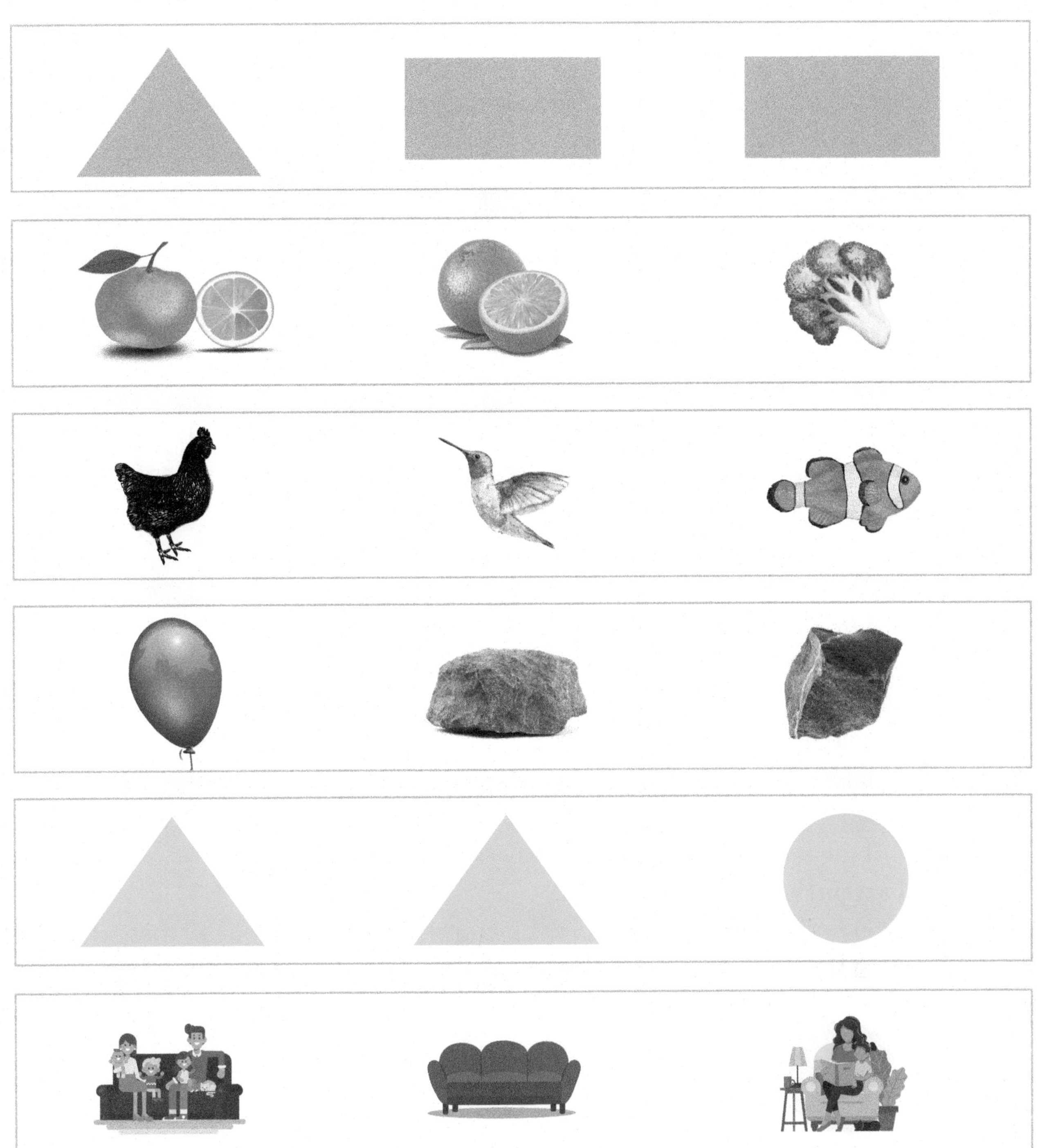

DAY 5 St. Paul's Cathedral

This lesson will draw in natural creation and human culture. The section invites the child to interact with the beautiful things made by God (and people), followed by a page of exercises. This will require about 15 minutes of instruction from the parent/teacher.

Prayer

Our Father in heaven, we love You. Help us to see Your patterns so that we can know You better and worship You more. Amen.

Activity

On the next page is a picture of St. Paul's Cathedral. It is a Christian church in London, England. What a beautiful building! What a beautiful rainbow! Of course, people made the building, but who made the rainbow?

Take a close look at the picture. Use your new understanding of shapes to think more deeply about this picture.

1. Do you see any straight lines in the picture?
2. Do you see any curved lines?
3. Find some squares or rectangles.
4. Find a circle.
5. What would you call the shape of the rainbow? What sort of shape would it make if the rainbow curved all the way around?
6. Find a triangle.
7. Can you find any other shapes in the picture? What would you call these shapes? Can you think of a good name for them?

> "The rainbow shall be in the cloud, and I will look on it to remember the everlasting covenant between God and every living creature of all flesh that is on the earth." (Genesis 9:16)

As you look at this picture, you can see the things that God makes and the things that people make. What is it that makes things so beautiful? God made people (like you and me), so that we can make things beautiful. We can make these shapes. We can make patterns, and we can create beauty. Can you make something beautiful too?

Student Exercises

Draw a set of members that you might find in a **garden**.

Draw a set of members that you might find in a **kitchen**.

Practice DAY 6

Student Exercises

There are three different objects in each of the following sets. Circle the object that is not like the others. Why is that one object different from the others? How does God make things different from other things? How do people make things different from other things? (The parent/teacher may need to help the child with this.)

Which animals live in the trees? Which animals sometimes like to crawl up into trees? Which animals usually crawl around on the ground? Draw lines from the tree to each animal that spends time in trees. Then, draw lines connecting those animals that crawl around on the ground to the grassy field in the picture.

Go Separate God's World DAY 7

This lesson connects math to real life, showing how we use math to take dominion of God's world. The section will take the child outside for an up-close-and-personal look at God's creation, to appreciate His wisdom, and to enjoy the patterns He has made. This is followed by two pages of exercises. This will require about 20 minutes of instruction from the parent/teacher.

Prayer

Father, we love You. Teach us to see Your patterns so that we can know and worship You more. Amen.

Activity

> "The Lord God formed every beast of the field and every fowl of the air; and brought them unto Adam to see what he would call them: and whatever Adam called every living creature, that was the name thereof." (Genesis 2:19)

In the beginning, God wanted Adam to do two things.

God told Adam to name the animals (by dividing the animals into sets). And then, He put Adam in the garden to take care of it.

Just like Adam, we can study God's world and learn about the animals God has made. Still today, people who study different kinds of animals will come up with new names for them. We are also called to take care of God's world. Today, we will go outside and enjoy God's creation. We want to study God's world and take good care of His creation in our own yard. We enjoy studying the various parts of God's creation, the sets of His plants and animals. And, we want to take good care of the garden or the yard where we live.

1. Go outside and look at the area around your home, either in the front or back. What are the things that are lying around? You might find some litter (paper and other garbage). You mind find sticks, leaves, and dead bugs. Separate what you find into piles.

2. Of the things you have collected, what can you learn about God's creation? What are the most interesting shapes and patterns? Search through the leaves for the most interesting shape. Take a moment to appreciate what God has made. Tell your parent/teacher about it.
3. Clean up the home a little bit. Sweep up the dirt and put that in one place. Remove the rocks from the pathway or the garden. Collect the trash and throw it away. Clean up the leaves and branches. Now, what have you done? You've taken a messy place and made it more orderly. You have enjoyed a part of God's beautiful creation. What will be the result? Maybe some plants will grow better. Your family will enjoy the beauty outside your home now.

Student Exercises

Trace these shapes. A shape is a pattern of points. Do you see how these shapes are patterns of points?

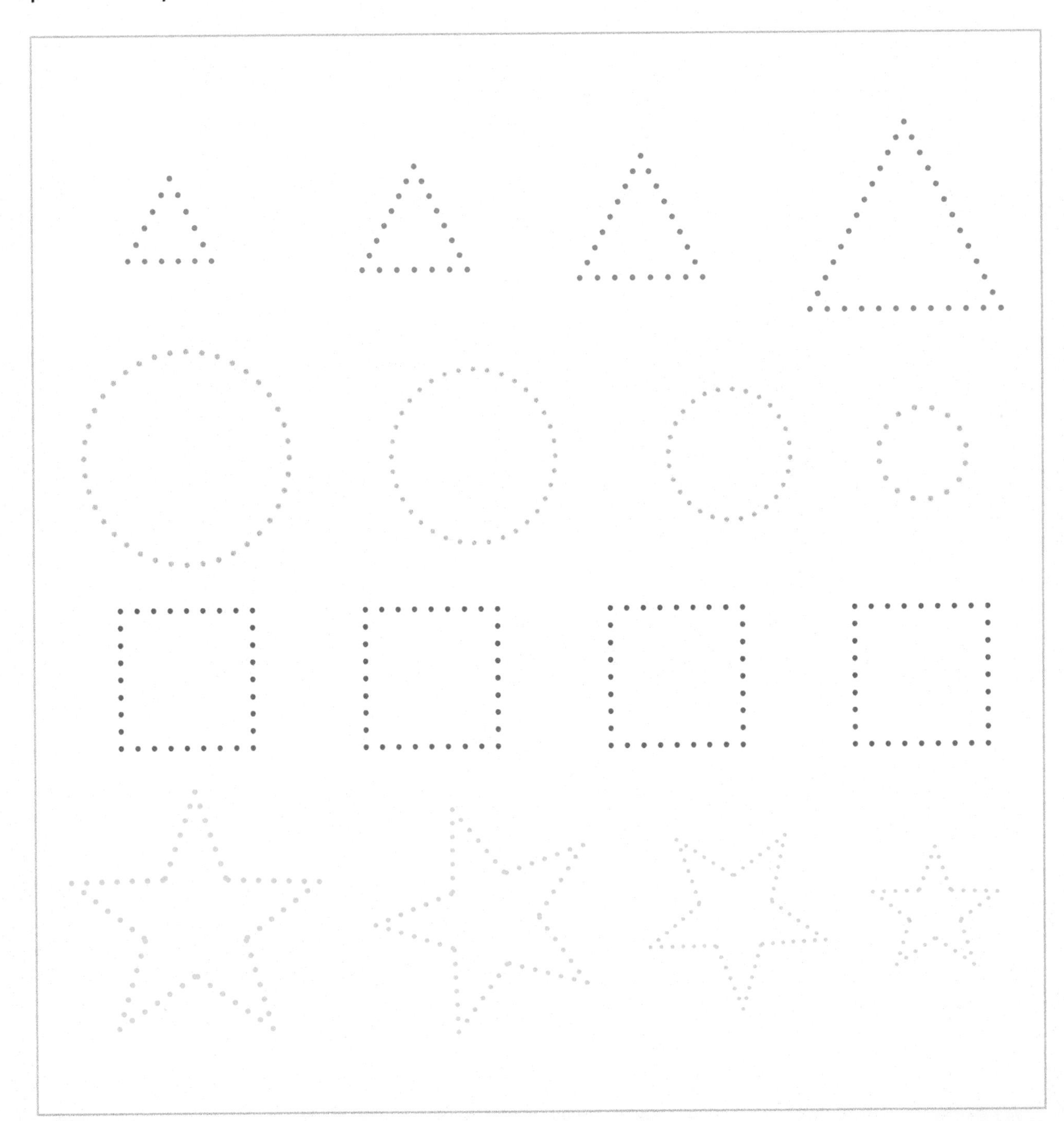

There are three different objects in each of the following sets. Circle the object that is not like the others. Why is that one object different from the others? How does God make things different from other things? (The parent/teacher may need to help the child with this.)

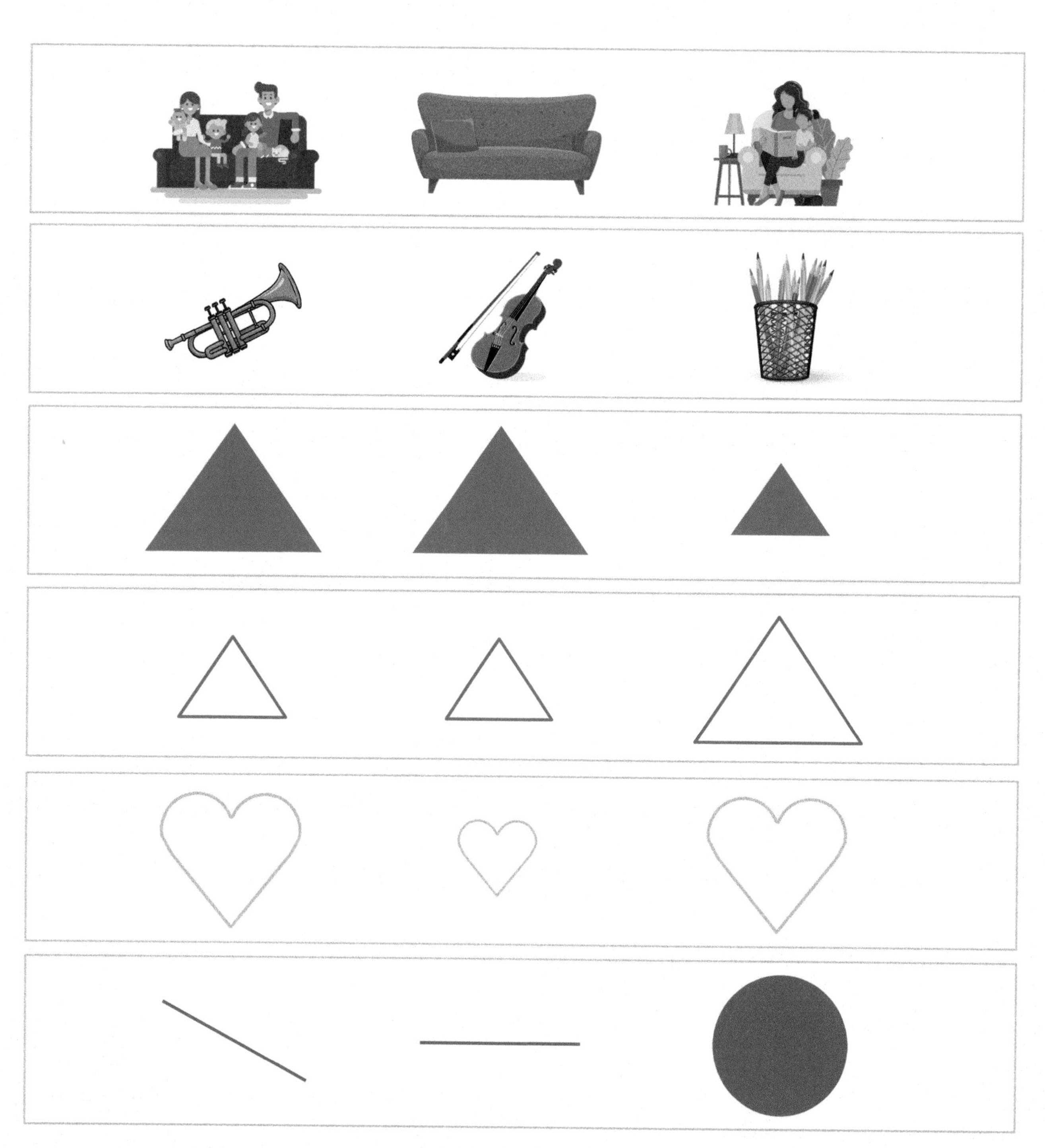

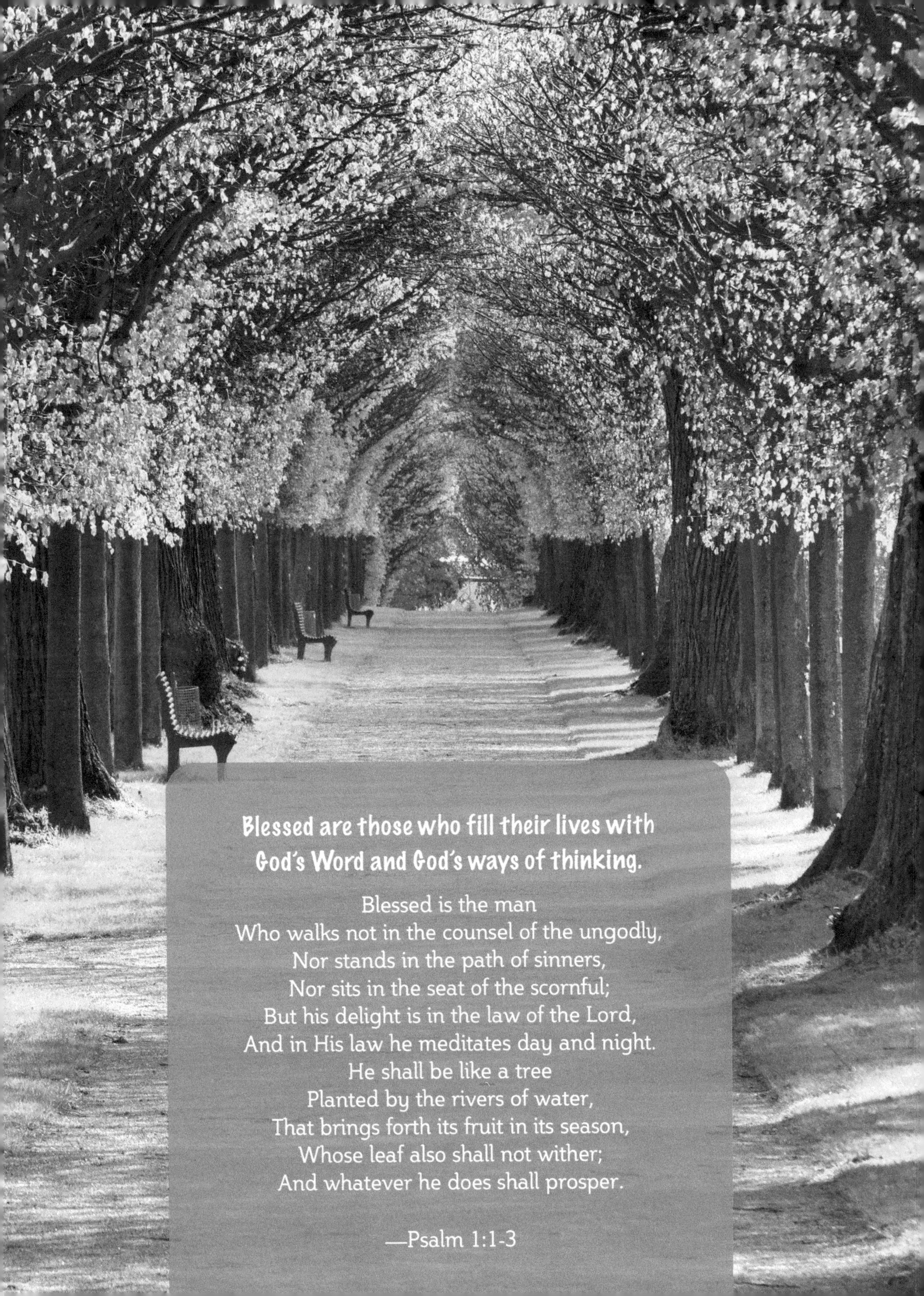
Blessed are those who fill their lives with God's Word and God's ways of thinking.
Blessed is the man
Who walks not in the counsel of the ungodly,
Nor stands in the path of sinners,
Nor sits in the seat of the scornful;
But his delight is in the law of the Lord,
And in His law he meditates day and night.
He shall be like a tree
Planted by the rivers of water,
That brings forth its fruit in its season,
Whose leaf also shall not wither;
And whatever he does shall prosper.
—Psalm 1:1-3

Each year God gives
this tree one new ring.

CHAPTER 2

Seeing Sizes

Introduction

In our first lessons, we learned about sets because God made a full world of sets. God also made a world of different-sized things. He made things big and small. For example, some animals are bigger than other animals.

The prophet Isaiah said:

> [The Lord] sits above the circle of the earth,
> And its inhabitants are like grasshoppers. (Isaiah 40:22)

As God looks down at people on the earth, they look as small as grasshoppers to Him. God is big. How big? God is bigger than the earth. He is bigger than the sun, the moon, and the stars. God is bigger than the whole creation!

Math helps us to tell the size of things. Math uses **numbers** to measure how big things are. We can use numbers to measure your height and your weight. We use math to compare big things and small things. Are you taller or shorter than your brother, or your sister, or your friend? The numbers will help you to compare big things with smaller things in God's creation. There is no number big enough to measure how big God is. But math can give us really big numbers! God is bigger than the biggest number anyone could ever think of. Can you see some big and small things in this picture?

Learning the Numbers DAY 8

This lesson explores Scripture. This will require about 20 minutes of instruction from the parent/teacher.

Prayer

Father God in Heaven, You are very great; no one is great like You. Thank You for showing us Your greatness with Your wonderful numbers.

Memory

Spend a few minutes repeating numbers 1 - 12 out loud.

Lesson

A **number** is a name that we give to the size of a set. Some sets are bigger than others, and big sets get big numbers.

Ten pieces of chocolate are more than 2 chocolates. A set of 0 chocolates is less than a set of 2 chocolates. There are different kinds of numbers. The simplest kind of number is what we use to count things. You can count 5 toes on 1 foot. Or you can count 2 pieces of chocolate.

Your Chocolate

Your Friend's Chocolate

We use math to measure the size of things. We also use math to count the number of things. Let's pretend the chocolates in the picture are for you and a friend to share. There are 2 pieces of chocolate on the right for your friend. And, there are 7 pieces of chocolate on the left for you. Now, what would happen if you broke the chocolate into same size chunks? Now, let's count the chunks. You would have 7 pieces, and your friend would have 8 pieces. Your friend's share would be bigger than yours. Giving your friend the bigger share is one way to show love. This is how we share the love of Jesus with others.

Today, let's learn the first numbers God has given to us. God uses these numbers to count things in His Word. We can use these numbers for counting all kinds of things. The first number is zero (or 0). This set has nothing in it. Then comes the number one (or 1), and then the number two (or 2).

God uses the number 1 when He speaks of Himself. There is only 1 God.

> Hear, O Israel: The Lord our God, the Lord is **one**! You shall love the Lord your God with all your heart, with all your soul, and with all your strength. (Deuteronomy 6:4-5)

God created **2 people** at the beginning — a man and a woman. Can you name them?

> And Noah begot **three** sons: Shem, Ham, and Japheth. (Genesis 6:10)

Two carts and **four** oxen [Moses] gave to the sons of Gershon. (Numbers 7:7)

Then [Jesus] took the **five** loaves and the **two** fish, and looking up to heaven, He blessed and broke them. (Luke 9:16)

"For in **six** days the LORD made the heavens and the earth, the sea, and all that is in them, and rested the **seventh** day." (Exodus 20:11)

Four carts and **eight** oxen [Moses] gave to the sons of Merari. (Numbers 7:8)

Hoshea became king of Israel, and he reigned **nine** years. (2 Kings 17:1)

Then the servant took **ten** of [Abraham's] camels and departed. (Genesis 24:10)

Then the **eleven** disciples went away into Galilee. (Matthew 28:16)

In the middle of its street, and on either side of the river, was the tree of life, which bore **twelve** fruits. (Revelation 22:2)

What does "1" look like? Everybody has 1 nose. Can you see 1 nose on somebody else's face? Can you feel "1" by touching your nose? How many eyes can you see on somebody else's face? How many ears can you see on somebody else's head? How many hands do you have? Can you see "2" when you look at your 2 hands?

You can use your fingers to count numbers 1, 2, 3, 4, 5, 6, 7, 8, 9, and 10. Try it now. Use your fingers to count the number of ducklings in the picture.

 Student Exercises

Practice writing the numbers 0 to 6.

0

1

2

3

4

5

6

Practice writing the numbers 7 to 12.

7

8

9

10

11

12

Counting with Numbers DAY 10

This lesson explores Scripture. It includes a brief lesson, an activity, and one page of exercises. This will require about 15 minutes of instruction from the parent/teacher.

Prayer

Thank God for something you have learned. Ask Him to help you as you do this lesson. OR Pray your own prayer of thanksgiving and praise to God. Pray for His help on this lesson.

Memory

Spend a few minutes repeating numbers 1 - 12 out loud.

Lesson

When you count the members of a set, you have found the size of the set.

How many fingers do you see on this hand? There are 5 members of this set of fingers. Now, 4 of these fingers can be divided into 12 segments, as you can see in the picture.

1 2 3
4 5 6
7 8 9
10 11 12

Now, you try counting to 12 using 4 fingers on one of your hands. Count each segment from 1 to 12. Don't use your thumb for this.

Activity

Go on a counting adventure around the house. Find sets that match each of the numbers you have learned (1 through 12).

Find 1 thing in the room. Find 2 things. Find 3 things. Touch each of these things as you count them. You might count 1 chair, 2 apples, 3 bananas, 4 legs, 5 bottles, 6 cups . . . What else can you count?

God made everything to be counted! Jesus's disciples counted the bread and fish in the little boy's lunch.

> One of Jesus' disciples, Andrew, Simon Peter's brother, said to Him, "There is a lad here who has five barley loaves and two small fish, but what are they among so many?" . . . And Jesus took the loaves, and when He had given thanks He distributed them to the disciples, and the disciples to those sitting down; and likewise of the fish, as much as they wanted. (John 6:8, 11)

How many loaves and how many fish did the little boy bring to Jesus? Later, we read that Jesus fed 5,000 people with this little bit of food. That was an amazing miracle, wasn't it?

Student Exercises

Copy these lists of numbers! Read each number out loud after you write it.

DAY 11 Counting with Shapes

This lesson explores shapes. It includes a brief lesson, an activity, and two pages of new exercises. This will require about 10 minutes of instruction from the parent/teacher.

Prayer

Thank God for something you have learned. Ask Him to help you as you do this lesson. OR Pray your own prayer of thanksgiving and praise to God.

Memory

Spend a few minutes repeating numbers 1- 12 out loud.

Lesson

Do you remember what a shape is? The simplest shape is a point. Many points all in a row make a line. Lines can be curved or straight. We can make many shapes with lines.

We see shapes all around us. Every line used to make a shape is called a **side**. The place where two sides touch is called a **corner** (or an **angle**).

A **triangle** is a shape with 3 sides and 3 corners (or angles).

This pink **rectangle** has 4 sides and 4 corners.

How many triangles and rectangles can you find on this barn? Point to the sides of one of the triangles. Now point to the corners of the triangle.

Student Exercises

How many sides can you count for each of these shapes? Write the number of sides you counted inside the shape.

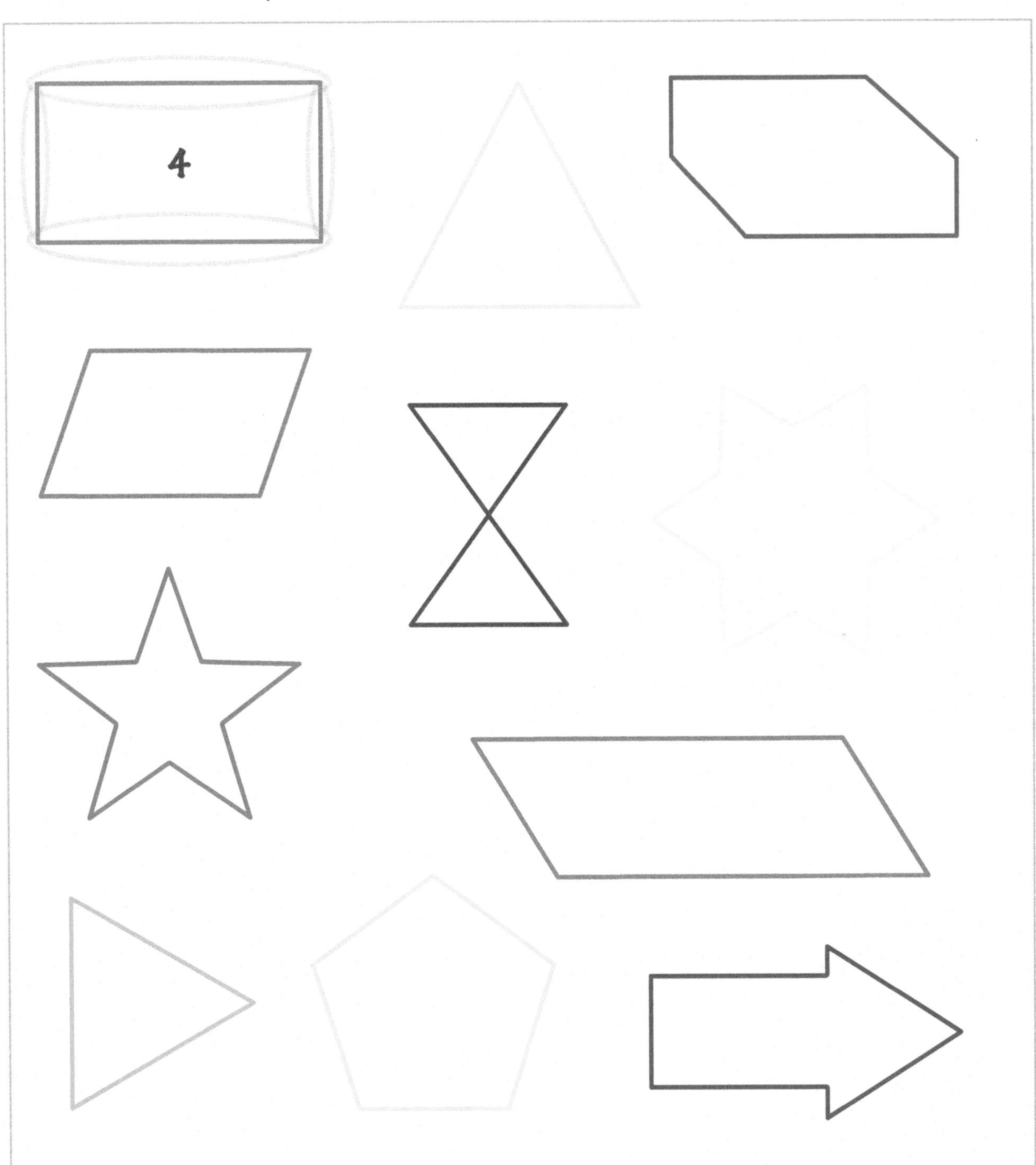

Student Exercises

How many corners (angles) can you count for each of these shapes? Write the number of corners you counted inside the shape.

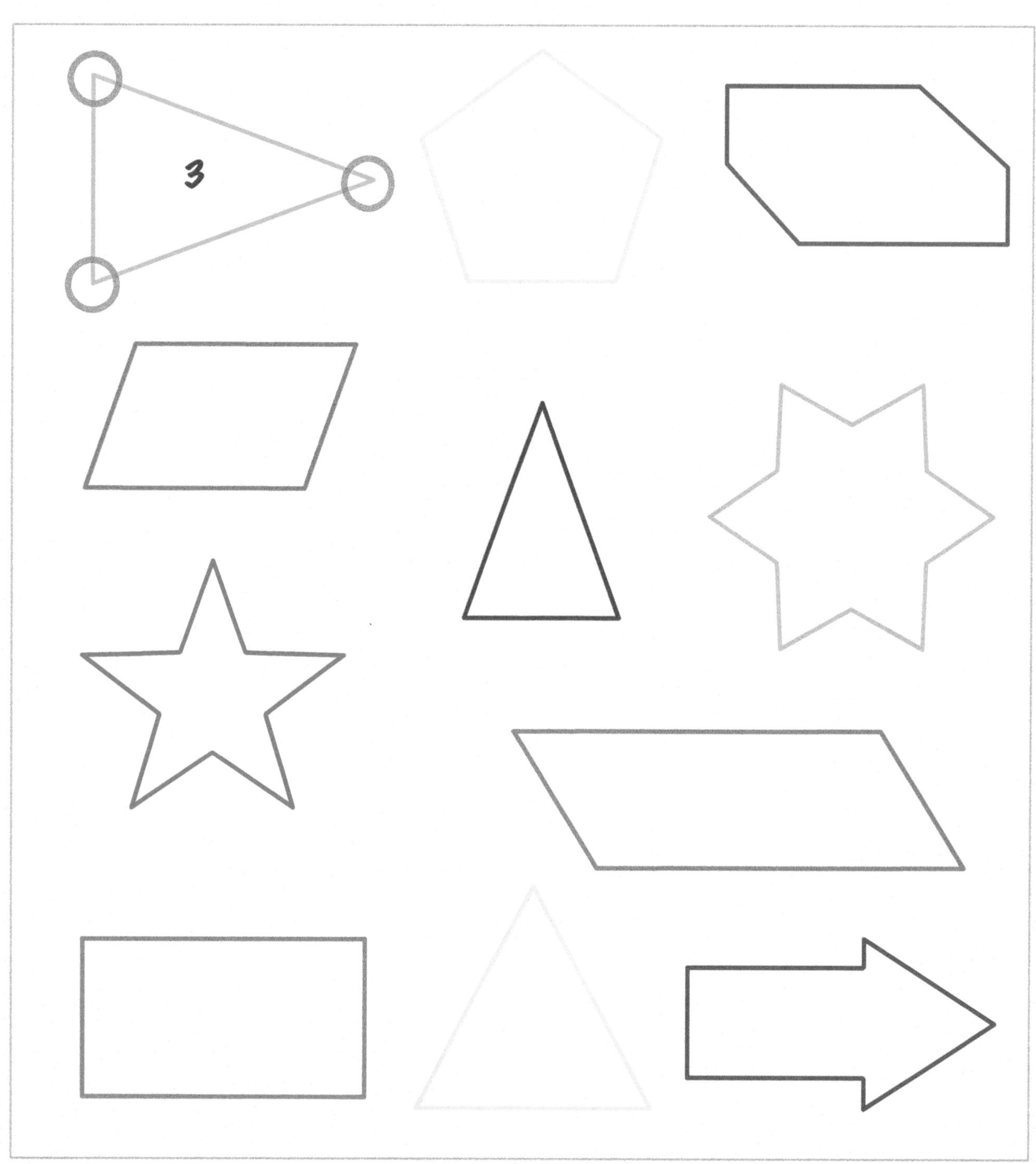

Student Exercises

Count the animals in each picture below. Write the number on the line. God's creatures are amazing!

Student Exercises

Count the number of shapes in each picture. Write the number on the blank line.

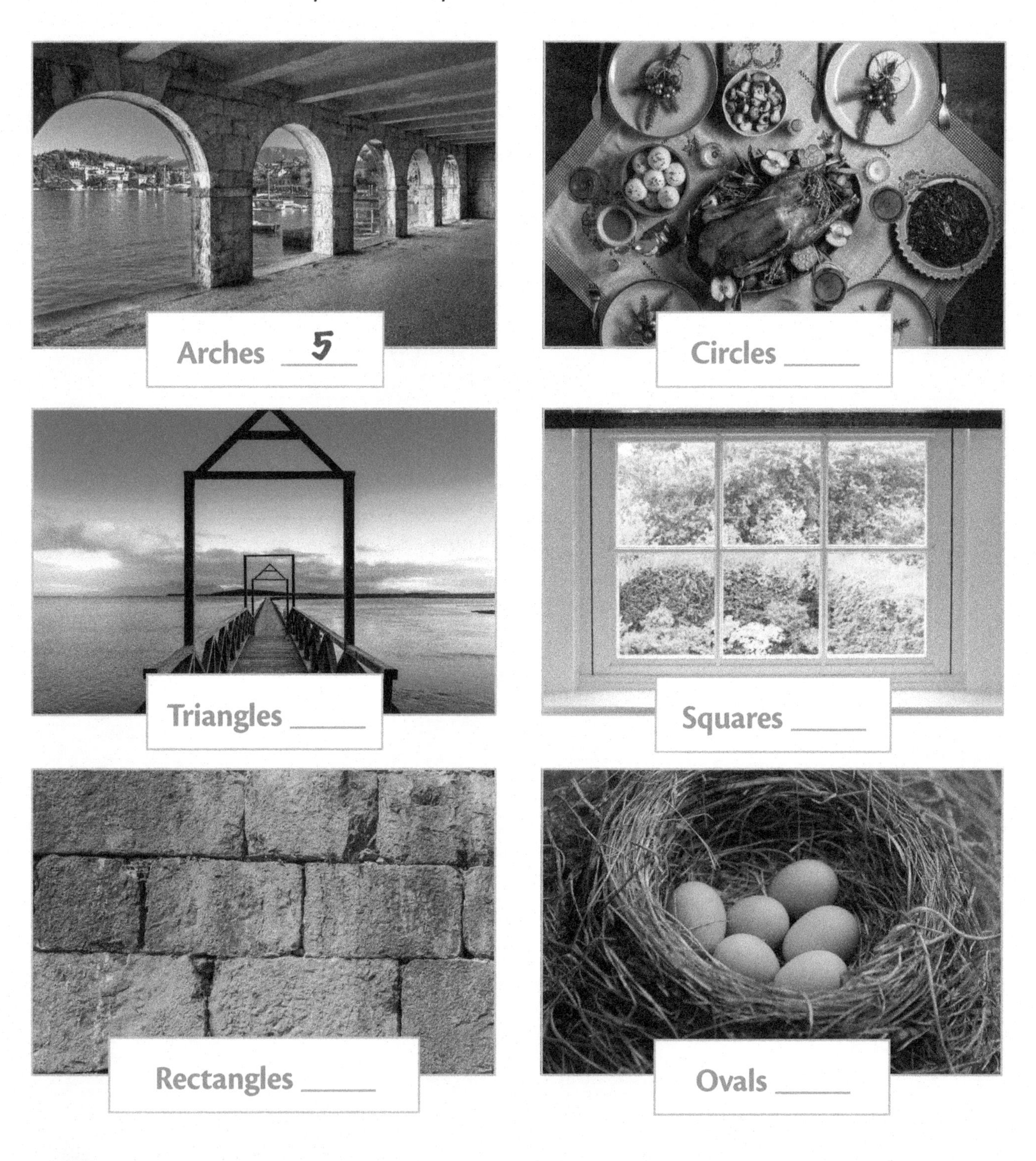

Number Patterns DAY 13

This lesson explores God's world by applying math to physical objects. It includes an activity and a page of review exercises. This will require about 20 minutes of instruction from the parent/teacher.

Prayer

Thank God for something you have learned. Ask Him to help you as you do this lesson. OR Pray your own prayer of thanksgiving and praise to God.

Memory

Spend a few minutes repeating numbers 1 - 12 out loud.

Lesson

Let's take a look at the 12 little stones in your math toolbox. God made more than 1 stone. He made more stones than we can count! Count your stones.

Make a set of 2 stones like the ones in the picture. Then, make another set of 2 stones. Keep the sets separated.

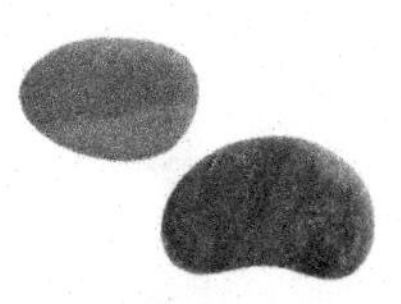

Now bring the 2 sets of stones together. How many stones are in this 1 set now? That's right: 4!

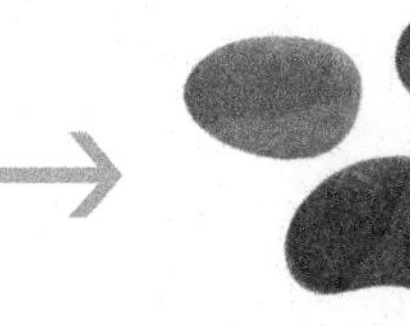

Take 1 member away from the set. How many stones are in your set now? That's right: 3! Take another stone away. Now how many are left in the set? 2!

We call this a pair. A pair is made of 2 things.

This pair can also be called a row. It is a row of 2 stones. Have you seen rows of chairs at your church? Maybe you have seen rows of shelves at the grocery store.

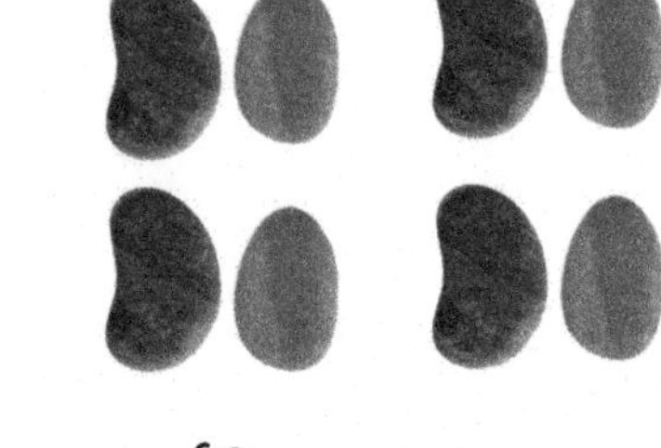

Now, make this pattern with your little stones. How many rows of 2 do you see in this set? What is the size of this set of stones? Hint: How many stones do you have altogether?

Set up three stones like the picture below. You can see a row of 2 stones. Can you make another row of 2? No! That's because we only have 1 stone left. We need 4 stones to make 2 rows of 2.

Look at your set of 3 stones again. You have a pair of 2 stones, right? They are friends. But 1 stone is all by itself. It doesn't have a friend.

If we arrange all the members in a set by twos, and every member has a friend, we say the set contains an even number of members. If one of the stones does not have a friend, we say the set has an odd number of members.

Let's look at sets of 1, 2, and 3 stones:

- The set of 1 stone: This stone does not have a friend. This is an **odd** number.
- The set of 2 stones: The stones are friends and nobody is left out. This is an **even** number.
- The set of 3 stones: One of the stones does not have a friend. This is an **odd** number.

What about 4 stones? Is this an **odd** number or an **even** number?

Let's find all the even numbers and odd numbers from 1 to 12. Draw a box around each of the even numbers. Circle each of the odd numbers.

1 2 3 4 5 6 7 8 9 10 11 12

There are more patterns to discover in God's world! Maybe you can make some rows of 3 using your collection of stones. Can you make some more neat patterns with the stones?

Student Exercises

God put numbers in order from 1 to 12. Write the numbers in the right order for each exercise. Sometimes you will have to count down (or backward) to find the number that comes before. And sometimes you will have to count up (forward) to find the number that comes next.

6, __7__, __8__

______, 10, ______

______, ______, 12

______, 2, ______, ______

______, ______, 7, ______

______, ______, 4

______, ______, ______, 8

______, 6, ______

DAY 14 Practice

Student Exercises

Make your own shapes. Use your ruler to draw straight lines for your shapes. The number in each box tells you how many sides to use for each of your shapes.

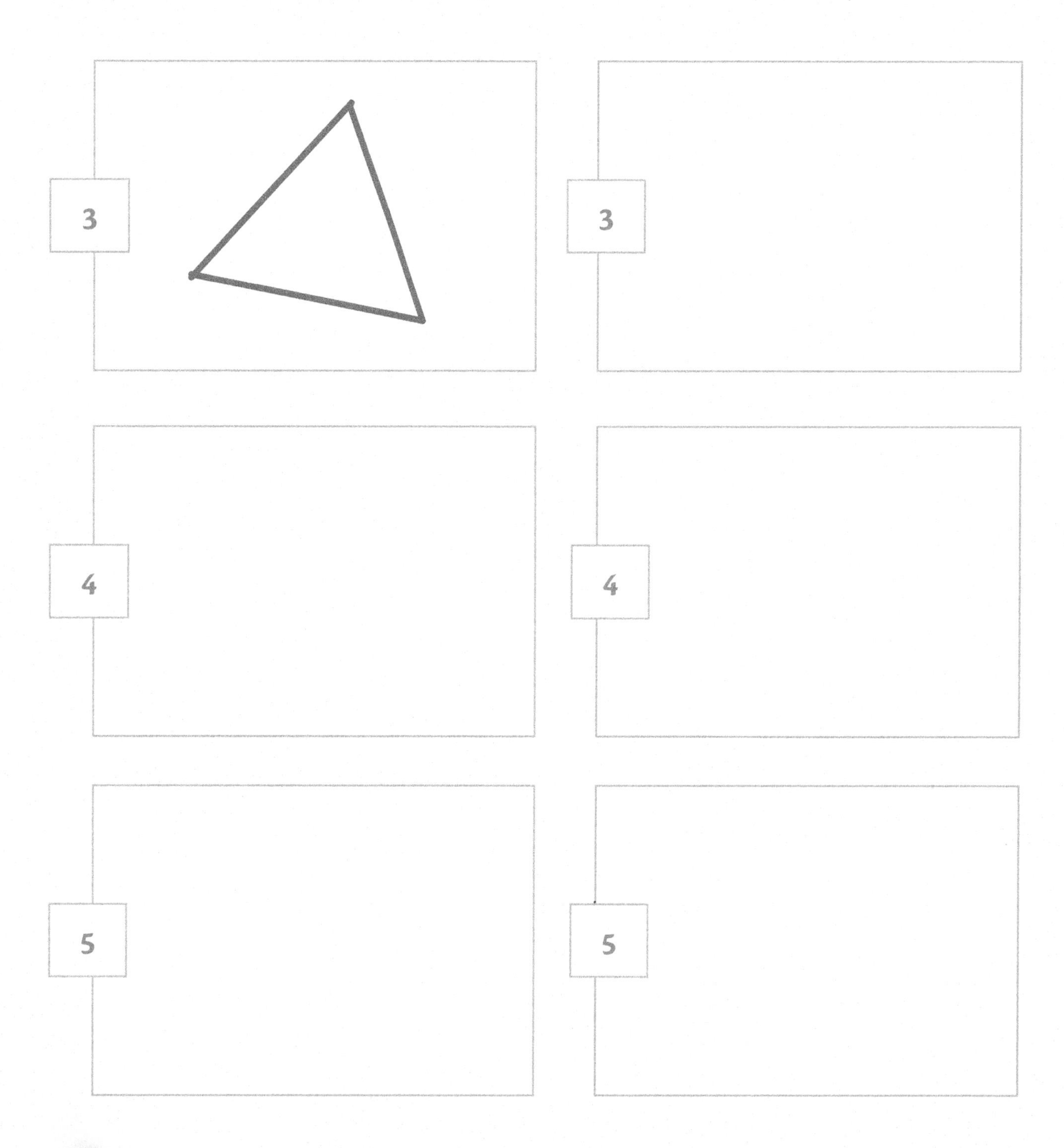

Use your ruler to turn the following shapes into triangles. Remember that a triangle has 3 sides and 3 corners (angles).

Hint: The best way to do this is to draw a straight line from one corner to another corner across from it. You may need to draw more than one line to turn all these shapes into triangles.

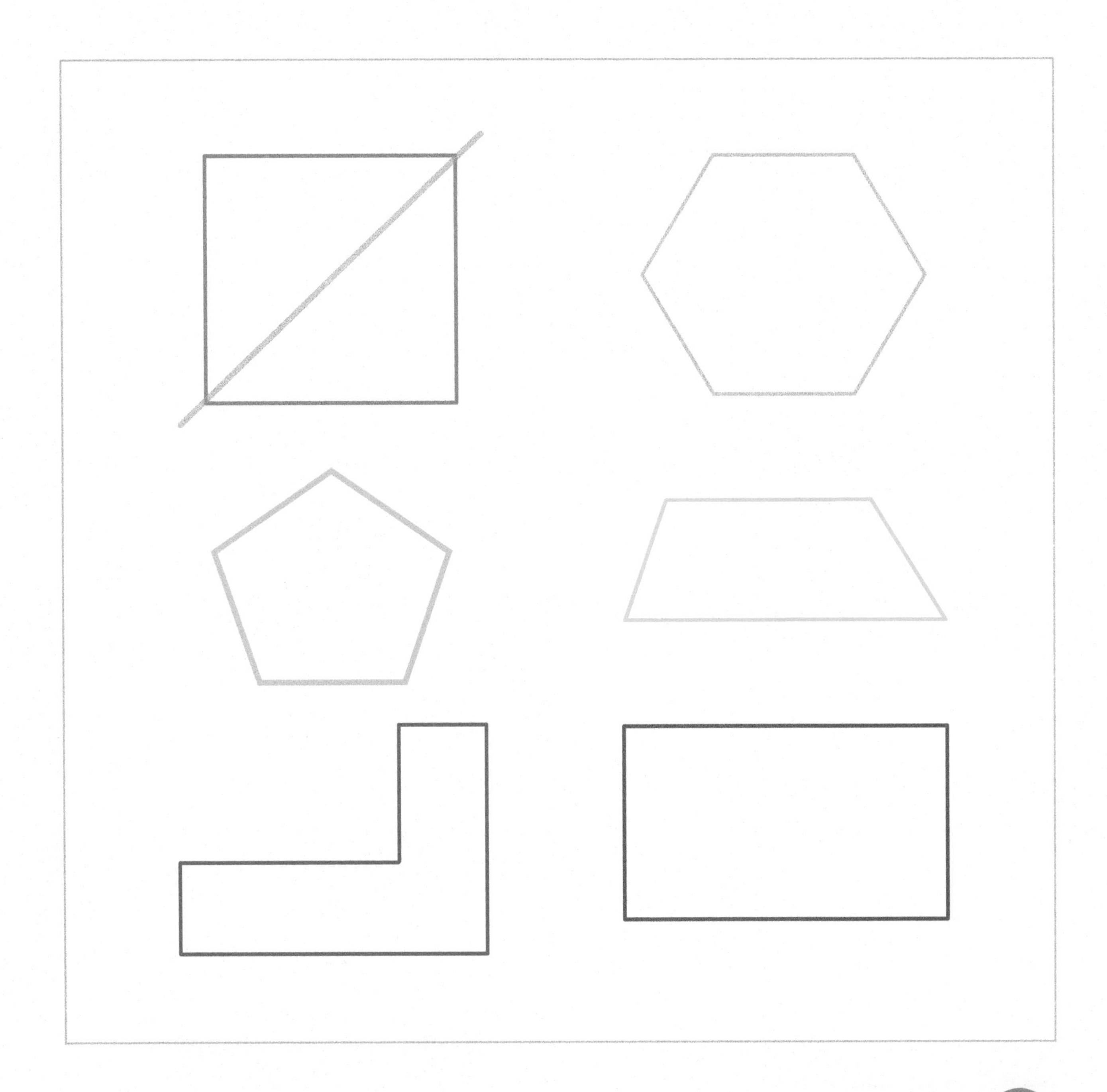

DAY 15 Combining Sets

This lesson introduces addition by combining sets, followed by two pages of exercises. This will require about 20 minutes of instruction from the parent/teacher.

Prayer

Thank God for something you have learned. Ask Him to help you as you do this lesson. OR Pray your own prayer of thanksgiving and praise to God.

Memory

Spend a few minutes repeating the odd and even numbers from 1 - 12 out loud.

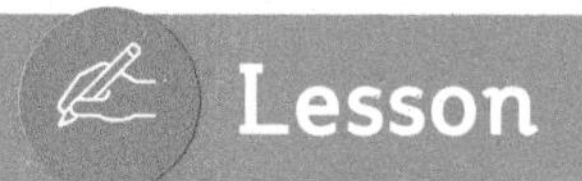

Lesson

Let's bring these sets together. Here are 2 sets.

Here is a set of parents.

Here is a set of children.

How many people are here? ________ How many people are here? ________

Let's bring the family together now into one bigger set of people. Here is what we have:

How many people are here now? __________

Now let's do this exercise with your blocks from your math tool box. You can use your own blocks, or just follow along here, using the pictures. Count the blocks in each set. Write the number of blocks you see on the blank line below each set. Then bring the two sets together.

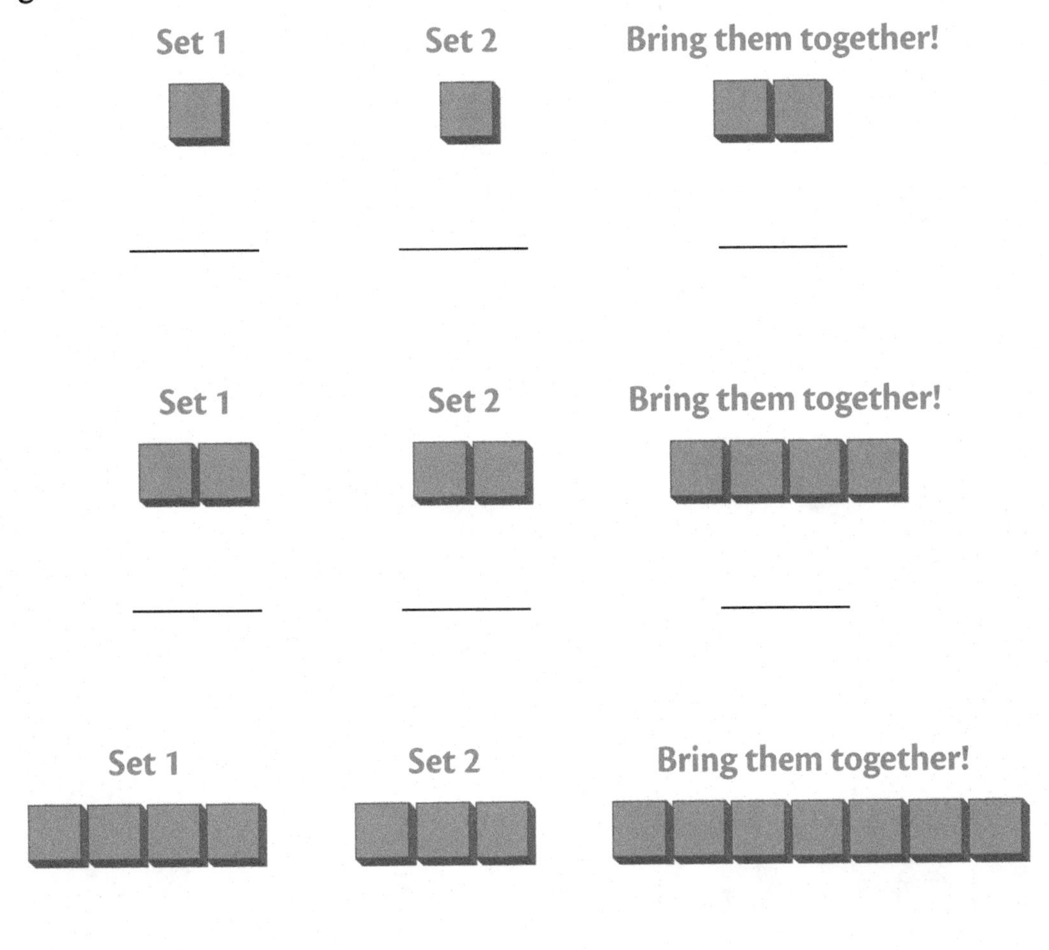

Student Exercises

How many do you see in these pictures?

Set 1	Set 2	Bring them together!
☐	☐	☐

Set 1	Set 2	Bring them together!
☐	☐	☐

Set 1	Set 2	Bring them together!
☐	☐	☐

Set 1	Set 2	Bring them together!
☐	☐	☐

THIS PAGE
INTENTIONALLY LEFT BLANK

DAY 16 **Practice**

Student Exercises

Write the numbers in the spaces provided in the order that God made for them. Sometimes you will have to count down (or backward) to find the correct number. Sometimes you will have to count up (or forward) to find the number that comes next.

2, ___3___, ___4___

________, 5, ________

________, ________, 2, ________

________, ________, 8

________, 10, ________, ________

________, ________, 9, ________

0, ________, ________, ________, ________

________, ________, ________, ________, 5

There are three different objects in each of the following sets. Circle the object that is not like the others. Why is that one object different from the others? How did God make it different? (The parent/teacher may need to help the child with this.)

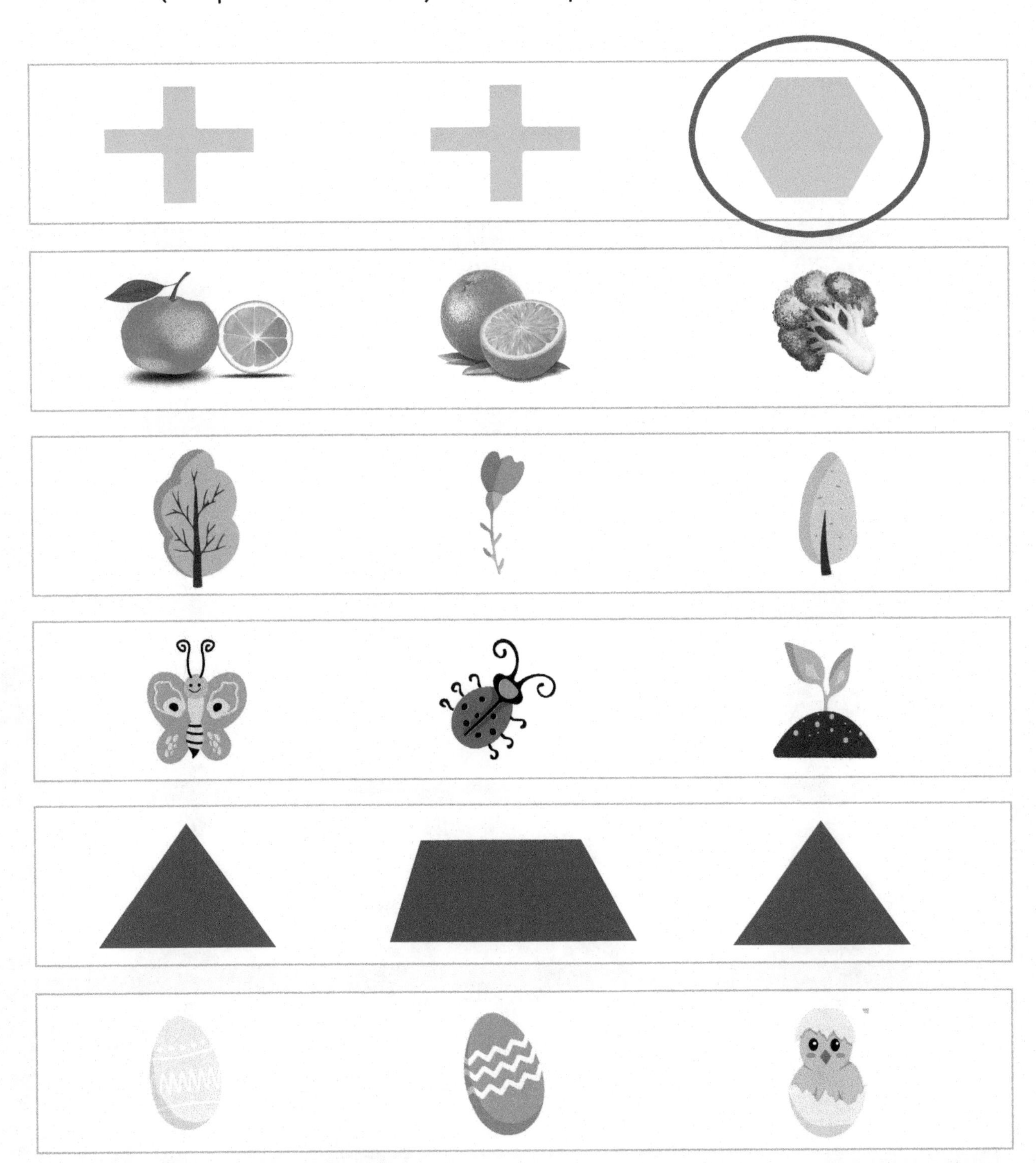

DAY 17 Counting in Yosemite

This lesson explores nature and culture. One page of exercises follows an activity that investigates a few of the beautiful things God and men have made. The parent/teacher is needed for about 15 minutes in this lesson.

Prayer

Thank God for something you have learned. Ask Him to help you as you do this lesson. OR Pray your own prayer of thanksgiving and praise to God.

Memory

Spend a few minutes repeating the odd and even numbers from 1 - 12 out loud.

Here is a photo of the Yosemite Valley in California. God didn't make any other place quite like this one.

As you study this picture, answer the following questions:

1. Do you see how God separated the sky from the ground?
2. Do you see how God separated the mountains from the valley?
3. Name ten different things you find in this picture.
4. How many of these things do you see in the picture?
    - Bears
    - Birds
    - People
    - Legs
    - Sides of the valley
    - Mountain peaks
5. Can you describe some of the shapes you see? What are the shapes of some of the rocks, the mountain peaks, or the trees?

> [God] sends the springs into the valleys;
> They flow among the hills.
> They give drink to every beast of the field;
> The wild donkeys quench their thirst.
> By them the birds of the heavens have their home;
> They sing among the branches. (Psalm 104:10-12)

Student Exercises

Now that you have seen God's picture, create your own picture using shapes. Use **5 squares**, **4 triangles**, **3 circles**, and **2 stars**. You may add extra shapes to make a better picture. Tell your parent/teacher about the picture.

Student Exercises

Use your ruler to make shapes using straight lines. The number in each box tells you how many sides to use for each shape. Be sure that every shape is different even if it has the same numbers of sides!

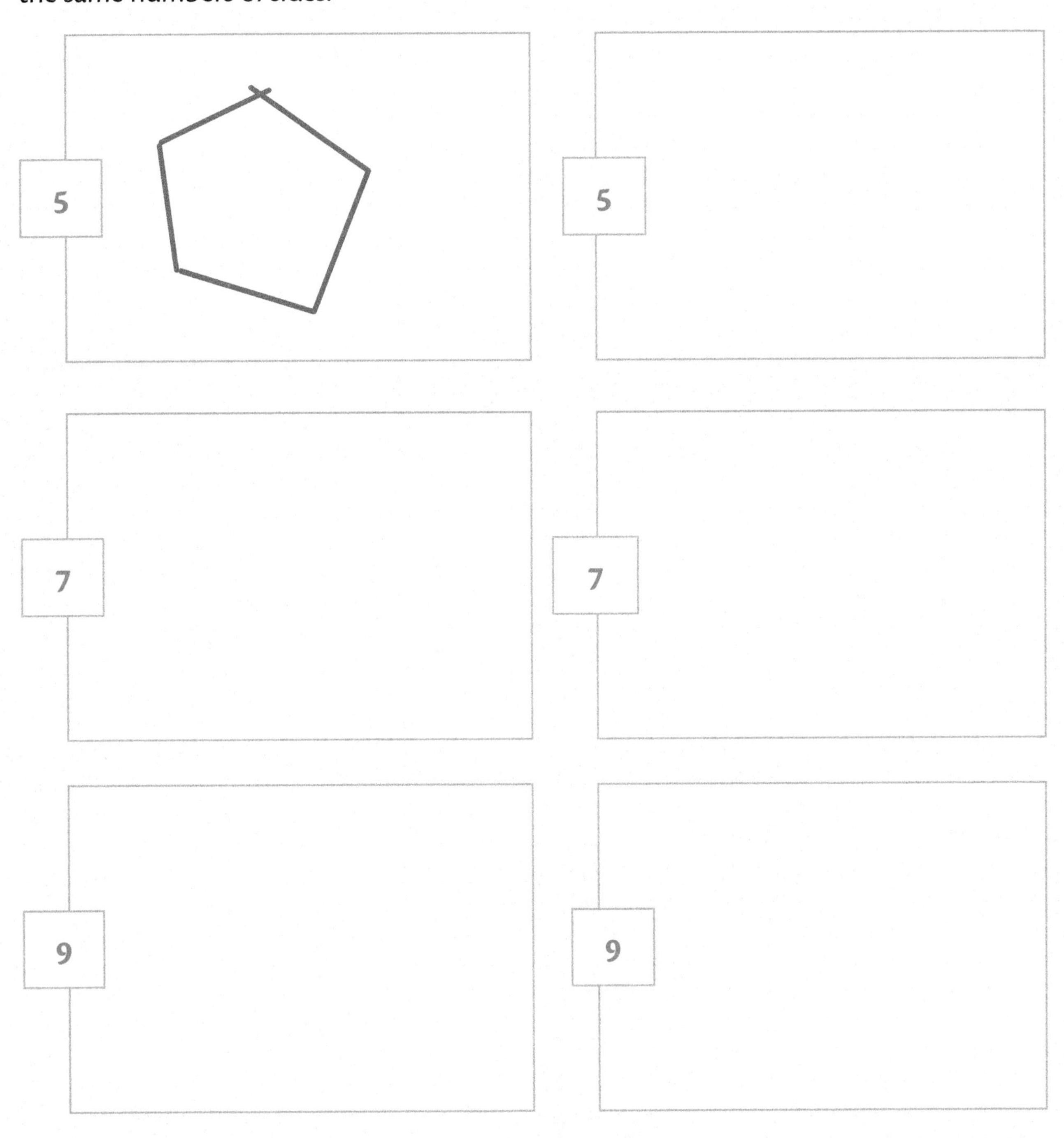

Write the numbers in the spaces provided in the order that God made for them. Sometimes you will have to count down (or backward) to find the correct number. Sometimes you will have to count up (or forward) to find the number that comes next.

3, __4__, __5__

________, ________, 10

________, ________, 3, ________

________, 9, ________, ________

________, ________, ________, 5

________, ________, 4, ________

5, ________, ________, ________, ________

________, ________, ________, ________, 12

 Student Exercises

Write the numbers in the spaces provided in the order that God for made them. Sometimes you will have to count down (or backward) to find the correct number. Sometimes you will have to count up (or forward) to find the number that comes next.

10, __11__, __12__

_____, 3, _____

_____, _____, 8

_____, 6, _____, _____

_____, _____, _____, 9

_____, _____, 4, _____

_____, _____, 6

8, _____, _____, _____, _____

Fill in the missing number. Write the number of animals there are in the first set. Next, write how many animals are joining them. Then, write how many there would be altogether.

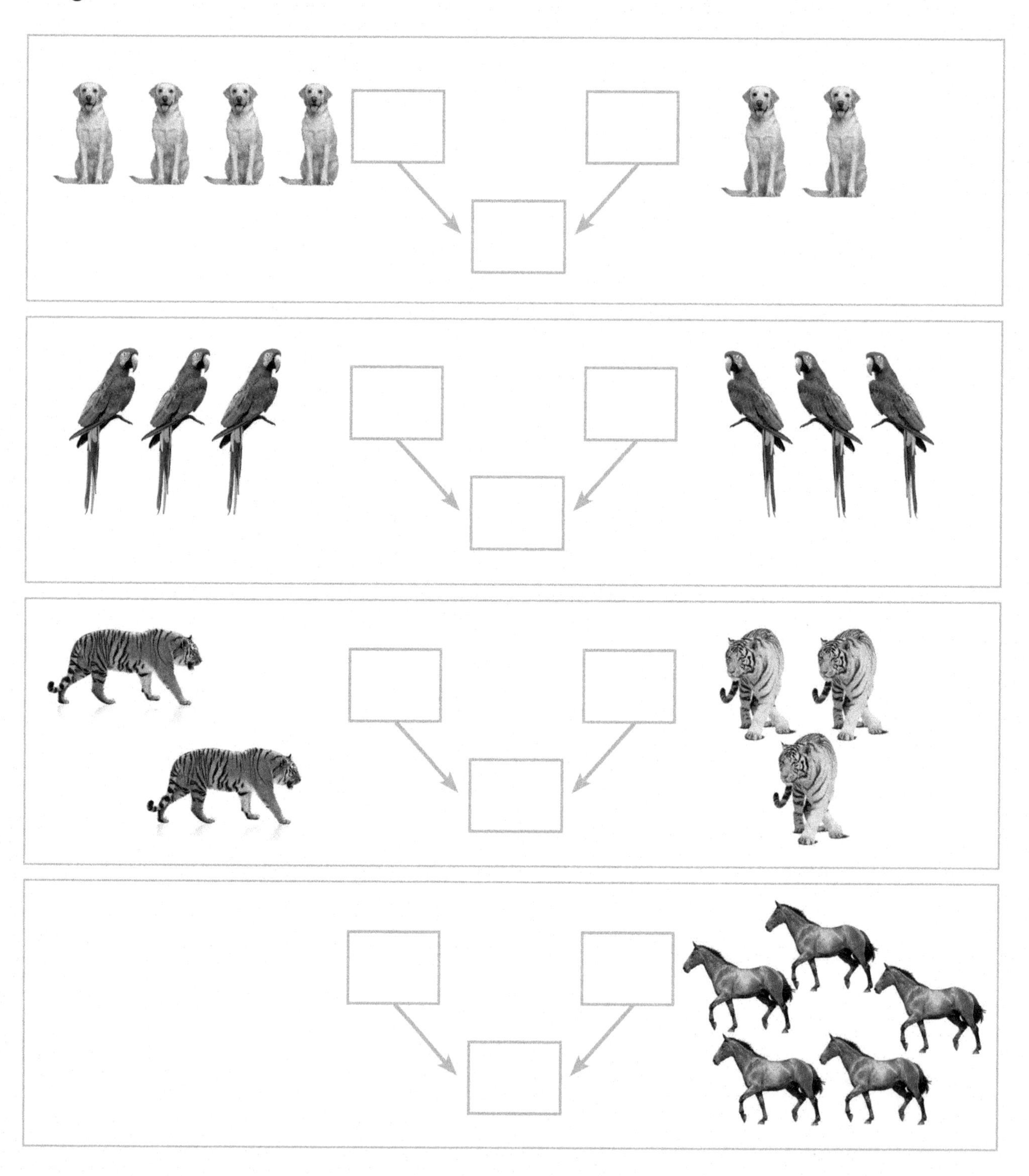

Go Count God's World! DAY 20

This lesson integrates math into everyday life. This is an essential element to learning. The child is encouraged to apply God's patterns and wisdom to life in the home and community. Take a break from memory work and academic exercises, and identify ways to make math part of your everyday life. The following are suggestions or examples, but other ideas may be added to the list.

Activity

Now that you have learned to count, you can help your family with your counting. What are all the ways you can love others and serve them by counting? Can you think of something else to count in your home?

- Having a snack. We need 10 grapes for each of us.
- Setting a table. We need 4 plates, 4 forks, and 4 spoons.
- Gardening. We need to dig 8 holes.
- Painting. We need to paint 10 fence posts.
- Shopping. We need to buy 5 boxes of that.
- Planning. We are going to Grandma's house in 4 days.
- Cooking dinner. We need 7 apples for this dish.
- Singing. How many hymnals do we need when we sing?
- ____

  ____
- ____

  ____

With a little help from your parents, choose one of these exercises and serve your family today by counting! Maybe you can set the table for lunch or dinner.

Kids get bigger and bigger. How about numbers?

CHAPTER 3

Making Bigger Things

Introduction

God made sets, and God made numbers. Do you remember what a set is? It's a group of similar things. What about a number? A number tells us how big a set is.

God made sets to come together with other sets. He also made numbers to come together with other numbers. This makes bigger numbers.

> Adam lived one hundred and thirty years, and begot a son in his own likeness, after his image, and named him Seth. After he begot Seth, the days of Adam were eight hundred years; and he had sons and daughters. (Genesis 5:3-4)

God adds things to our world in different ways. He adds rabbits to the world by making baby bunnies. He adds more plants by making seeds. He also adds people to the world in families. When a daddy and mommy come together and get married, they live together as two people. Jesus said they are one flesh because they have come together. Usually, when a daddy and a mommy get married, they will have a child. How many people do you see in these families?

If a daddy and the mommy have twins first, how many people will there be in the family?

Bringing Sets Together DAY 21

This lesson explores addition. It includes a brief chapter introduction, a lesson, and an activity. This will require about 20 minutes of instruction from the parent/teacher.

Prayer

Our Father in Heaven, teach us how You make bigger numbers by adding things together. Show us how You have made the world to change by adding things together. Now we have more reason to praise You! Amen.

Memory

Spend a few minutes counting backwards from 12 to 1.

Lesson

When two numbers come together to make something new, we call it **addition**. This is kind of like mixing things together. When you mix blue paint with yellow paint, what is the new color that you make? Green! Have you ever mixed flour, oil, and salt to make cookies or bread?

The **union** or addition of sets is something we do to bring more than 1 set together. This makes a bigger set. You can see what happens when three sets of plants and flowers are brought together. What a beautiful bouquet of flowers!

Today, you will learn to add numbers together. **Addition** is adding numbers together to make a bigger number. This bigger number is called a **sum**.

Do you know what this is?

$$2 + 3 = 5$$

This is called a math **equation**. An equation is a way of writing a sentence with numbers. In this equation, we are adding two numbers to find the sum. Here's how you read this: "Two plus three equals five." Can you repeat that a few times?

When we say two numbers added together are equal to another number, we mean that they are the same as each other. You can say it this way: 2 plus 3 is the same as 5. Read each of the addition equations below:

$2+4=6$ $7+1=8$ $3+3=6$

This is how we put two things together to get something bigger. Do you see the union of the 2 bananas? Write the number of bananas you see below each of the pictures.

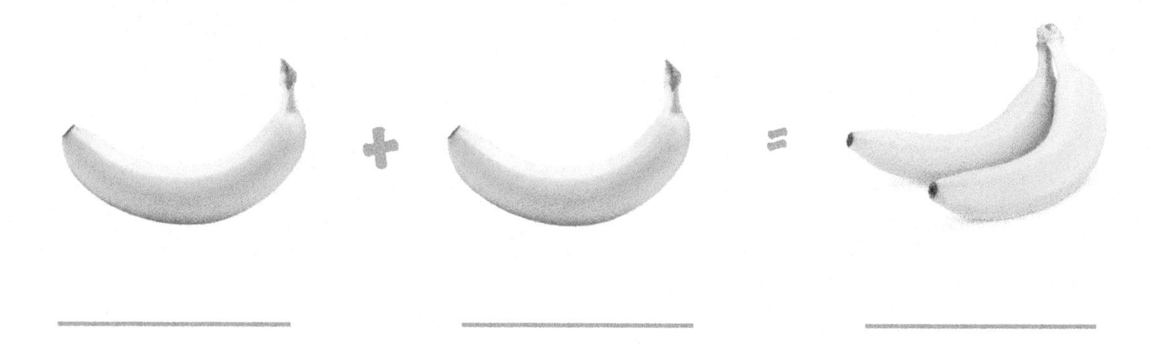

__________ __________ __________

While you look at the picture, say this out loud two times:

"1 banana plus 1 banana equals 2 bananas."

Look very carefully at these bananas again. 1 banana is not the same thing as 2 bananas. But if you add 1 banana to 1 banana, you get 2 bananas. You can see that the bananas are separated on one side of the equal sign. Each banana is sitting there all by itself over there on the other side of the equation. On the other side of the equation, the bananas are not separated anymore. There are 2 bananas on a stem. This is one set of 2 bananas.

Activity

Count the number of bananas in each exercise. First, write the number of bananas under each picture. Then recite the equation out loud. Here is an example: "1 banana plus 2 bananas equals 3 bananas."

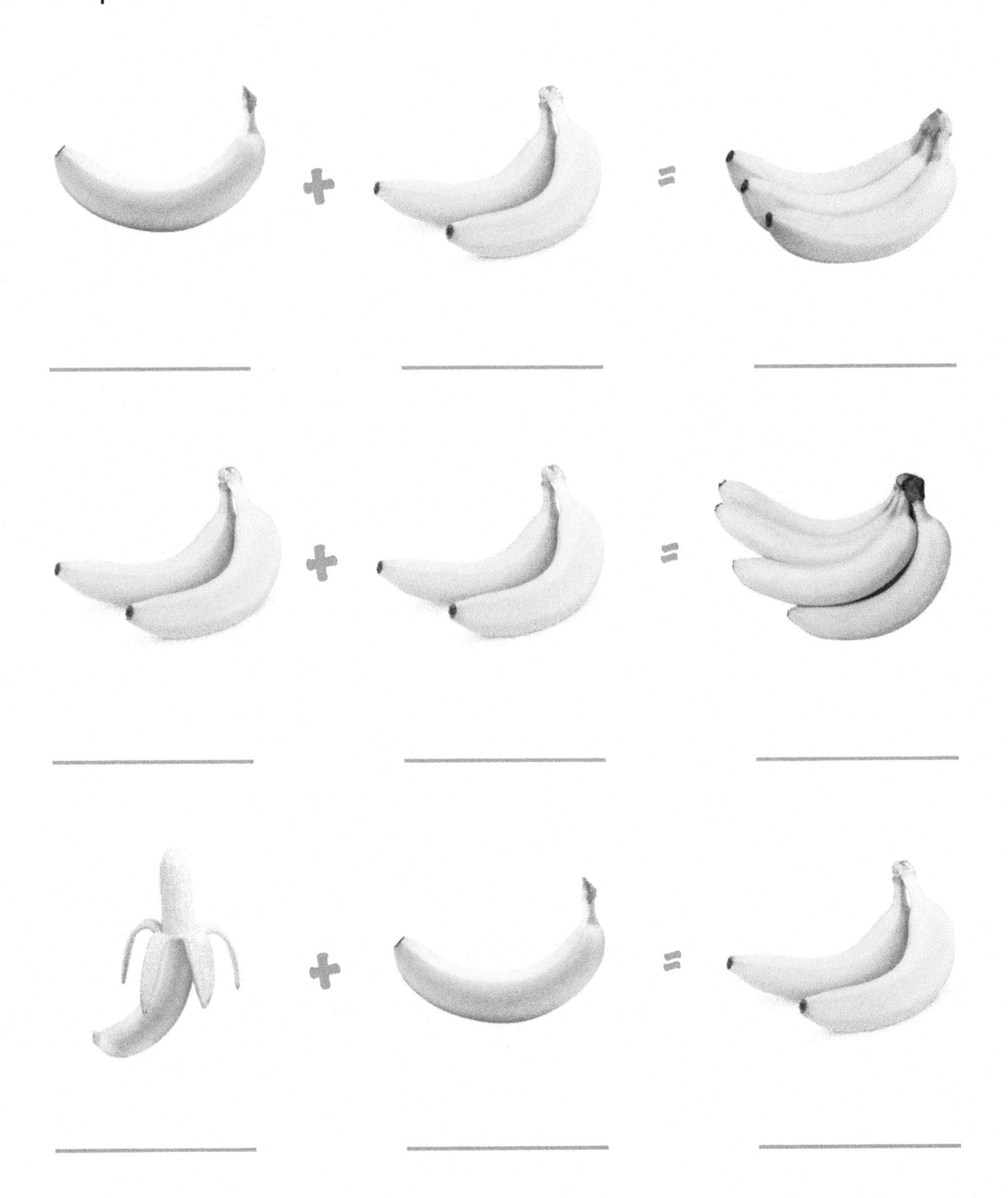

Get the small stones out of your math toolbox. Let's use these little objects God made to learn about addition. Here's how:

First, you will put two groups (or sets) of stones in front of you. Put 2 stones in one group. Put 3 stones in the other group. Now bring the two groups of stones together to make one bigger group.

__________ __________ __________

Count the number of stones in this new set. How many did you count? You have just done a simple addition exercise!

What does $2 + 3$ equal? Write the number of stones under each set. Then, say the equation out loud: $2 + 3 = 5$.

Find the answers to these addition equations. Use your stones to help you!

$6 + 1 =$ ____7____

$5 + 2 =$ __________

$3 + 0 =$ __________

$0 + 9 =$ __________

$5 + 0 =$ __________

Hint: Do you see a pattern here? Every time we add 0 to a number, the number does not change!

Practice DAY 22

Student Exercises

Count the number of cherries, bananas, or cookies in each picture and write the number under it. Now read the equation out loud. For the first one, you would say, "1 cherry plus 1 cherry equals 2 cherries."

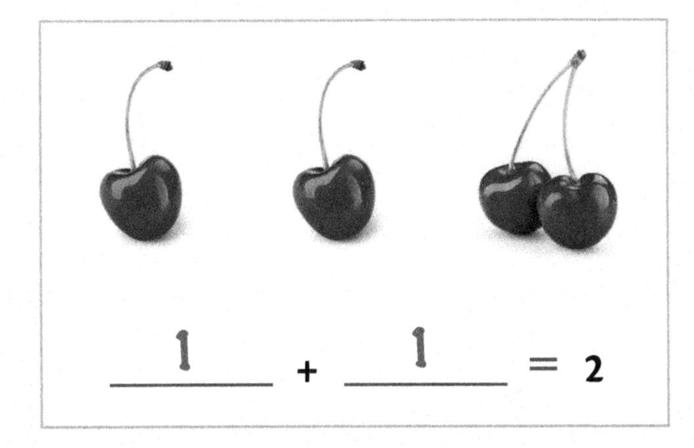

1 + 1 = 2

______ + ______ = 3

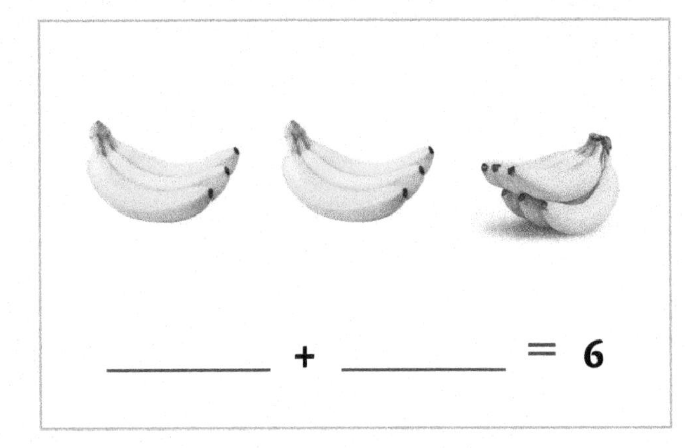

______ + ______ = 6

______ + 2 = ______

______ + ______ = 8

______ + 7 = ______

Find the answers to these addition exercises. Use your twelve little stones to help you. Write the sum (answer) on the line.

Hint: For the first exercise, make a set of 3 little stones. Then make another set of 4 little stones. Bring the two sets together. How many stones do you have now?

$3 + 4 =$ 7	$5 + 2 =$ ______
$6+6=$ ______	$7 + 5 =$ ______
$3 + 0 =$ ______	$1+9=$ ______
$7 + 5 =$ ______	$3 + 3 =$ ______
$5 + 4 =$ ______	$10+1=$ ______

Let's look at another way to write an addition exercise. Sometimes we write an addition equation on the same line like this:

$$2 + 3 = 5$$

Other times we write the addition exercise like this:

$$\begin{array}{r} 2 \\ +\,3 \\ \hline 5 \end{array}$$

Find the answers to these addition exercises. Use your twelve little stones to help you. Write the sum under the line.

$\begin{array}{r} 1 \\ +\,2 \\ \hline \end{array}$	$\begin{array}{r} 2 \\ +\,1 \\ \hline \end{array}$	$\begin{array}{r} 5 \\ +\,2 \\ \hline \end{array}$
$\begin{array}{r} 5 \\ +\,0 \\ \hline \end{array}$	$\begin{array}{r} 3 \\ +\,1 \\ \hline \end{array}$	$\begin{array}{r} 4 \\ +\,3 \\ \hline \end{array}$
$\begin{array}{r} 1 \\ +\,1 \\ \hline \end{array}$	$\begin{array}{r} 2 \\ +\,4 \\ \hline \end{array}$	$\begin{array}{r} 6 \\ +\,4 \\ \hline \end{array}$

DAY 23 Finding the Right Answer

This lesson explores several ways to add. It includes a lesson and one page of review exercises. This will require about 20 minutes of instruction from the parent/teacher.

Prayer

Thank God for something you have learned. Ask Him to help you as you do this lesson. OR Pray your own prayer of thanksgiving and praise to God.

Memory

Spend a few minutes with flash cards (0 + 1, + 2, + 3 . . . series).

> The Lord added to the church daily those who were being saved. (Acts 2:47)

> So the churches . . . increased in number daily. (Acts 16:5)

Lesson

Acts tells us that more people joined the church. This means the size of the church grew. The number of people who were part of the church got bigger and bigger. God was adding to the set of people He called "the Church."

Today we are going to think about adding things.

Imagine a grassy yard. Sometimes it helps to close your eyes. Can you describe the yard? Now imagine the yard has 2 trees. Then, a gardener comes along, and he plants 3 more trees. Can you use your imagination to count the number of trees in the yard now? The following picture might help.

$$1 + 1 = 2$$

1 plus 1 always equals 2. 1 plus 1 never equals 3. 1 plus 1 cannot equal 3. If you say that 1 plus 1 equals 3, you are wrong. You have the wrong answer. That's because math comes from God. Math is God's truth. God does not change. God's truth does not change. To say that 1 plus 1 equals 3 is to say something that is not true. It is a lie.

When we learn that 1 plus 1 equals 2, we learn something that is true. It is always true. For the rest of your life, 1 plus 1 will always equal 2. Truth does not change because all truth comes from God. God's truth does not change because God does not change.

How Can You Get the Right Answer?

There are different ways to add. God made each of us differently, so we will add numbers in different ways. But the answer should always be the same.

Here are four ways to get the right answer when you add:

1. **Memory**. You can memorize $1 + 1 = 2$. And $2 + 3 = 5$. That's why you will do memory practice using flash cards.
2. **Imagination**. At the beginning of the lesson, you imagined the trees or bushes in the yard. You can add like this too! Imagine 2 trees. Imagine 3 more trees. Then, imagine all 5 trees together.
3. **Drawings**. You can make little drawings of things like trees on paper.
4. **Stones**. You can use your small stones from your math toolbox.

Find the answers to these addition exercises using these four ways. Which way works the best for you?

$3 + 6 =$ ___________ $8 + 4 =$ ___________

$2 + 4 =$ ___________ $7 + 0 =$ ___________

$6 + 2 =$ ___________ $2 + 2 =$ ___________

Student Exercises

Count the number of cherries, bananas, or cookies in each picture and write the number under it. Now read the equation out loud. For the first exercise, you would say, "2 cherries plus 3 cherries equals 5 cherries."

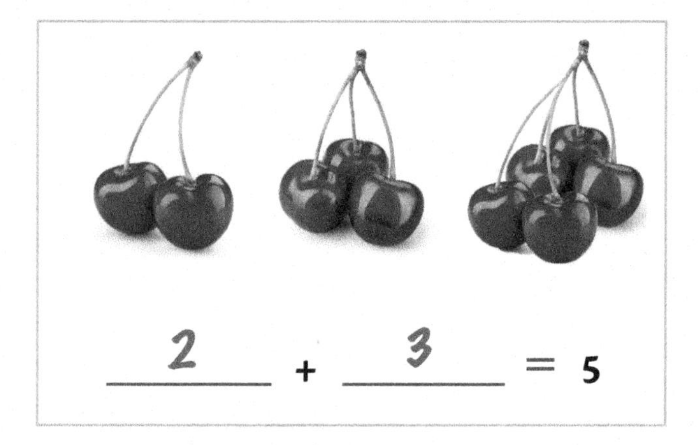

2 + 3 = 5

_____ + _____ = 4

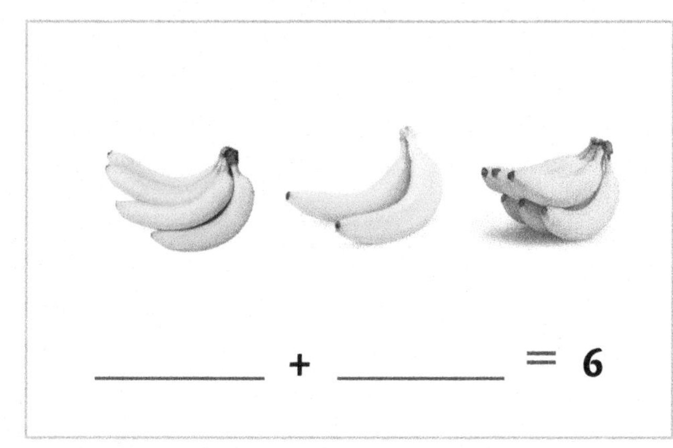

_____ + _____ = 6

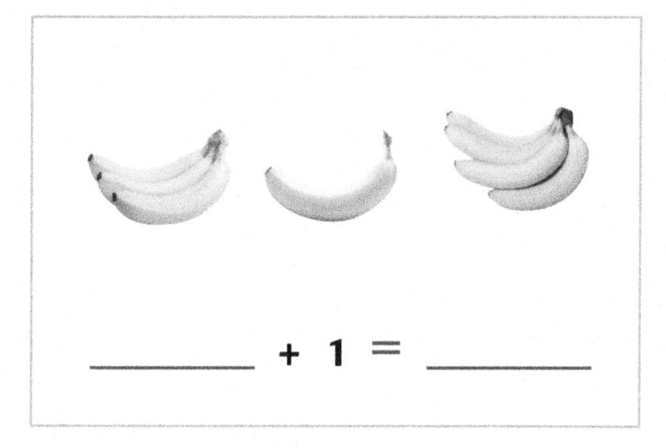

_____ + 1 = _____

_____ + _____ = 9

_____ + 2 = _____

Practice **DAY 24**

Student Exercises

Try these addition exercises from memory. Have you figured out a trick for when you add zero?

$1 + 0 =$ ________	$0 + 5 =$ ________
$12 + 0 =$ ________	$7 + 0 =$ ________
$2 + 0 =$ ________	$0 + 6 =$ ________
$0 + 0 =$ ________	$0 + 8 =$ ________
$3 + 0 =$ ________	$5+0 =$ ________

Use your imagination. Use your little stones. Use your memory. Make little drawings. Try using different ways to find the bigger number!

Write the answers for the first six exercises on the line. Write the answers for the last six exercises under the line.

8 + 4 = 12	7+3 = ______
1 + 4 = ______	7 + 5 = ______
11 + 0 = ______	4+2 = ______

$$\begin{array}{r} 5 \\ +5 \\ \hline \end{array} \qquad \begin{array}{r} 4 \\ +4 \\ \hline \end{array} \qquad \begin{array}{r} 3 \\ +3 \\ \hline \end{array}$$

$$\begin{array}{r} 6 \\ +0 \\ \hline \end{array} \qquad \begin{array}{r} 3 \\ +4 \\ \hline \end{array} \qquad \begin{array}{r} 11 \\ +1 \\ \hline \end{array}$$

This is an interesting (and important) exercise because numbers represent both the members of the set and its size. They are all numbers, but we are using them in two different ways.

Here are some sets of numbers. How many numbers are in each set? Find the size of each set and write it on the line below.

3, 4, 7 3	0, 1, 2, 3, 4, 5 ______	2, 4, 6, 8 ______
1, 2, 3, 4, 5, 6, 7, 8 ______	0, 3, 6, 9, 12 ______	12, 11, 10, 9, 8, 7, 6 ______
2, 3 ______	5, 10 ______	1 ______

DAY 25

Comparing Numbers: Bigger and Smaller Numbers

This lesson explores comparison. It includes a lesson and a page of new exercises. This will require about 20 minutes of instruction from the parent/teacher.

Prayer

Thank God for something you have learned. Ask Him to help you as you do this lesson. OR Pray your own prayer of thanksgiving and praise to God.

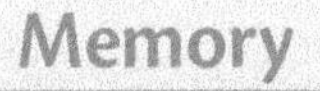

Memory

Spend a few minutes with flash cards (0 + 1, + 2, + 3 . . . series).

Lesson

> "The kingdom of heaven is like a mustard seed, which a man took and sowed in his field, which indeed is the least of all the seeds; but when it is grown it is greater than the herbs and becomes a tree, so that the birds of the air come and nest in its branches." (Matthew 13:31-32)

God made big things and small things. He made a little seed to grow into a big tree.

Some numbers are bigger than other numbers. We use symbols to compare numbers. These are the symbols we use:

A 6-year-old child is older (and bigger) than a 4-year-old child. 6 is bigger than 4. Here is the way we write this with numbers: $6 > 4$.

A 3-year-old child is younger (and smaller) than a 7-year-old child. 3 is smaller than 7. Here is the way we write this math sentence with numbers: $3 < 7$.

The little arrow points at the little number. The wide opening of the arrow points to the big number. You might also think about a baby shark: the small shark opens his mouth wide to eat the big number. You can pretend the bigger-than and smaller-than signs are the baby shark's open mouth!

What if two numbers are the same? Can we compare them? Yes. We say one number is "equal to" the other number! How would you read this math sentence?

$$12 = 12$$

That's right! 12 is equal to 12!

Now try writing a math sentence to compare two numbers!

Student Exercises

There are two numbers for each of these exercises. Which of these two numbers is smaller? Which number is bigger? You can use your stones to help you if you wish. Write the correct numbers on the blank lines. Remember the baby shark always eats the bigger number.

1	2, 7	7 (shark) 2
2	3, 1	______ < ______
3	4, 6	______ > ______
4	0, 12	______ < ______
5	5, 2	______ < ______
6	8, 4	______ > ______

Student Exercises

We're always thinking about "one more": one more hour of playing, one more story before bed, or one more prayer for Dad. Try these "one more" exercises from memory!

$10 + 1 =$ 11	$1 + 6 =$ ______
$1 + 1 =$ ______	$1 + 9 =$ ______
$2 + 1 =$ ______	$1 + 8 =$ ______
$3 + 1 =$ ______	$1 + 5 =$ ______
$11 + 1 =$ ______	$4 + 1 =$ ______

God made some numbers bigger than others. Which number is bigger? Which number is smaller? Fill in the correct symbol. Then read the math sentence. We read the first example like this: 1 is smaller than 3.

1 < 3	4 ______ 9	6 ______ 2
1 ______ 0	5 ______ 10	3 ______ 7
5 ______ 4	10 ______ 0	8 ______ 6
0 ______ 3	5 ______ 9	3 ______ 2

Use your imagination. Use your little stones. Use your memory. Or make little drawings. Try using different ways to add these numbers together.

Write the answers for the first six exercises on the line. Write the answers for the last six exercises under the line.

$1 + 2 =$ 3

$2+3 =$ ______

$3 + 4 =$ ______

$4 + 5 =$ ______

$5 + 6 =$ ______

$6+3 =$ ______

$$\begin{array}{r} 5 \\ +7 \\ \hline \end{array}$$

$$\begin{array}{r} 4 \\ +6 \\ \hline \end{array}$$

$$\begin{array}{r} 3 \\ +5 \\ \hline \end{array}$$

$$\begin{array}{r} 3 \\ +8 \\ \hline \end{array}$$

$$\begin{array}{r} 7 \\ +5 \\ \hline \end{array}$$

$$\begin{array}{r} 2 \\ +0 \\ \hline \end{array}$$

DAY 27 Adding Sounds

This lesson explores God's world by applying math to music and culture. It includes an activity and three pages of new and review exercises. This will require about 15 minutes of instruction from the parent/teacher.

Prayer

Father, teach us about change in math, and show us a changing world that is full of Your praise.

Memory

Spend a few minutes with flash cards (0 + 1, + 2, + 3 . . . series).

Optional Activity

> Praise the LORD with the harp;
> Make melody to Him with an instrument of ten strings.
> Sing to Him a new song;
> Play skillfully with a shout of joy. (Psalm 33:2-3)

Math is a little bit like music. God made music for us to use. God made music for us to enjoy. God made math for us to use. God also made math for us to enjoy.

Music has low sounds and high sounds. Music can be played faster or slower. Music can be played louder or softer. We can measure these things about music using math.

Take a look at the piano keys on your piano or in the picture. At the middle of the keyboard is a white key. If you push it, it plays a sound called a note. We call this note middle C. We've given it the number 1. Notice that there are two black keys on one side of middle C and three black keys on the other side.

Put your finger on middle C. Do you know what the next notes to the right are? D comes after middle C. Then comes E, F, G, A, B, and high C. We've given each of these notes numbers too: 1, 2, 3, 4, 5, 6, 7, 8. God made high C to have a higher sound than middle C.

If you are using a real piano, play the keys labeled 1, 2, 3, 4, 5, 6, 7, and 8. Can you hear the sounds going higher and higher? The sounds get higher as the numbers get bigger.

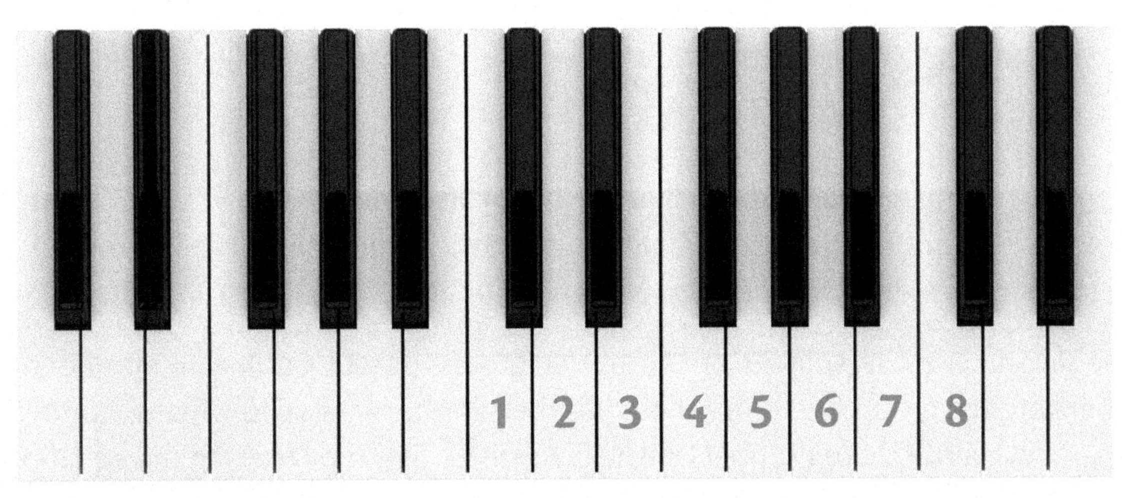

Now play the keys called 8, 5, 3, and 1 in this order. Can you hear the sounds going lower and lower as you go down?

Let's play two keys at once to add sounds together. The sound will be louder when you play two keys together. Play key 1 and key 2 at the same time. Then play 1 + 2, 1 + 3, 1 + 4, 1 + 5, 1 + 6, 1 + 7, and 1 + 8. Which sound is the loudest? Which sound is the most pleasant?

Let's play 2 keys at the same time again. But this time, play 1 + 3, 2 + 4, 3 + 5, 4 + 6, 5 + 7, and 6 + 8. Each set of keys is separated by 1 note. Do these make good sounds or bad sounds? How do you feel when you play these notes?

Now play keys that are right next to each other at the same time. Play 1 + 2, 2 + 3, 3 + 4 . . . Do these make good sounds or bad sounds? How do you feel when you play these notes?

Try playing any three keys at the same time. Try it more times with other sets of three notes. Which keys sound the best to you? Does it make a louder sound when you play more keys at the same time? Try putting a few other keys together. Which sounds are the best?

We have added sounds together. This is kind of like adding sets together. When you add different things together, you get something different. When you make different patterns with sounds, you may feel differently. Some sounds make you feel good or happy. Some sounds make you feel bad or sad.

God makes things to go together. God makes a daddy and a mommy for each other. God makes sounds to go together too. Some sounds do not fit together nicely with other sounds. Some sounds do fit nicely together. Two sounds are louder (or bigger) than one sound.

Student Exercises

Let's try something different. You will need your little stones to help you. We will make two smaller sets out of one big set. Count out 6 stones. This is your big set. You can split 6 into two smaller sets by making sets of 3 and 3, or 2 and 4, or 6 and 0.

For the second exercise, split 2 stones into two groups or sets. Count the number of members in each group, and write these numbers in the blanks. How many stones are in each new set? Write these numbers in the blank lines. Finish by reading the equation you made. (There will be several right answers for these exercises.)

6 = 3 + 3	2 = _____ + _____
3 = _____ + _____	10 = _____ + _____
5 = _____ + _____	11 = _____ + _____
7 = _____ + _____	12 = _____ + _____

Student Exercises

Adding 1 more is just like counting. 4 plus 1 is 1 more than 4. 4 plus 1 equals. . .? Think: 1, 2, 3, 4 . . . 5! 4 + 1 = 5. Can you figure out the answer in each of these addition exercises?

4 + 1 = 5	1 + 8 = ____
1 + 9 = ____	1 + 3 = ____
5 + 1 = ____	1 + 2 = ____
3 + 1 = ____	1 + 5 = ____
1 + 1 = ____	0 + 1 = ____

God made some numbers bigger than others. Which number is bigger? Which number is smaller? Fill in the correct symbol. Then read the math sentence.

Use your stones to help you if you need to. Remember, the arrow points at the smaller number. The bigger number goes on the side that opens wide. The baby shark eats the bigger number.

6 ______ 7	5 ______ 4	3 ______ 5
6 ______ 8	7 ______ 4	5 ______ 0
0 ______ 2	3 ______ 8	9 ______ 11
1 ______ 12	4 ______ 6	10 ______ 9

For these exercises, decide which is heavier, bigger, taller, longer, or more than the other. Fill in the box with the symbol < or >.

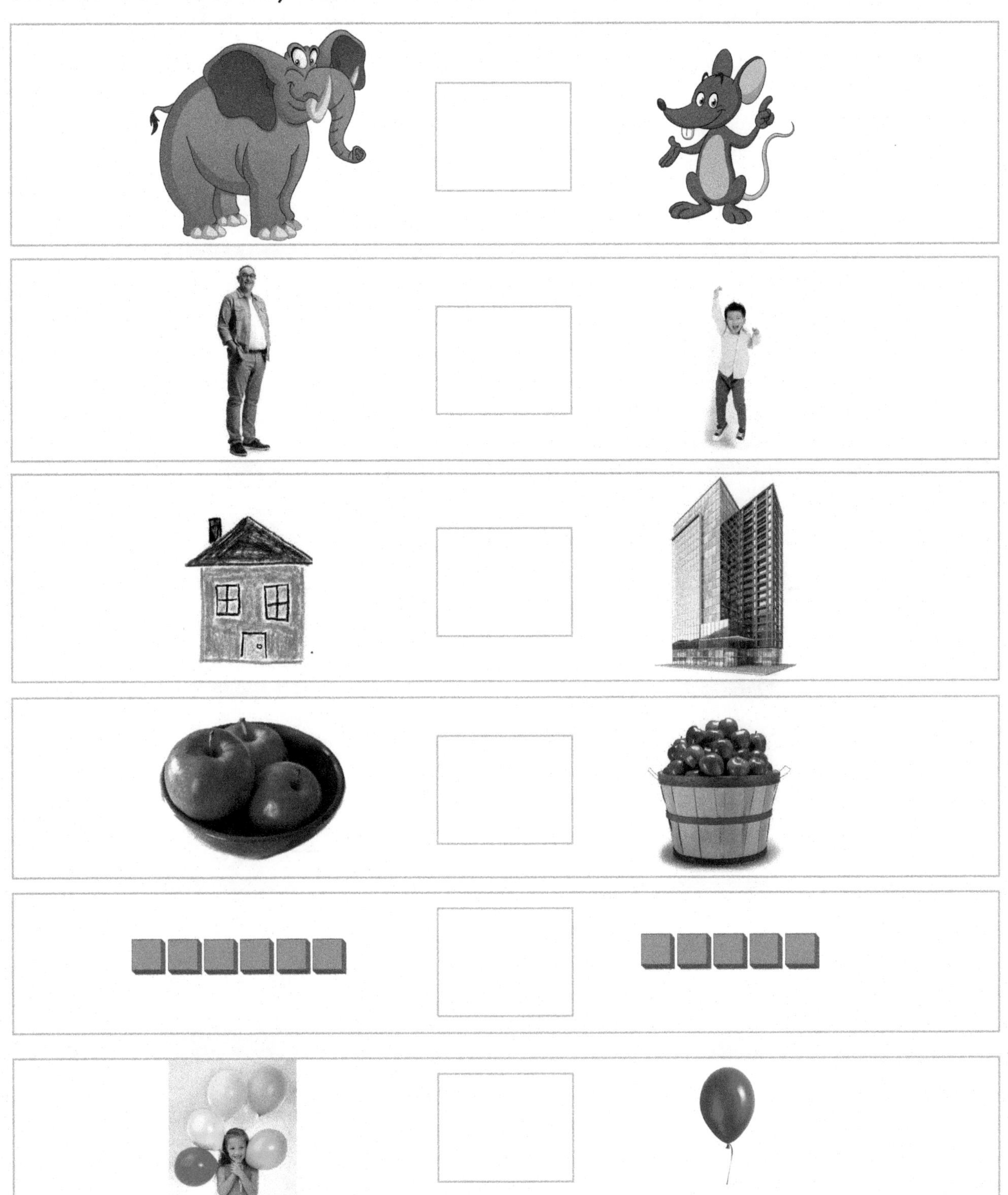

Let's make two smaller sets out of one big set. Use your little stones to create the two sets if needed. For the first exercise, take out 7 stones. Split the 7 stones into two groups or sets. Count the members of each group, and write these numbers in the blanks. How many stones are in each new set? For every exercise on this page, there will be several right answers.

7 = 3 + 4	1 = _____ + _____
3 = _____ + _____	9 = _____ + _____
4 = _____ + _____	10 = _____ + _____
6 = _____ + _____	12 = _____ + _____

Find the answers to these addition exercises. Use your 12 little stones to help you. Write the sum under the line.

1 + 0 ____	3 + 4 ____	1 +2 ____
4 + 5 ____	2 + 2 ____	5 +5 ____
4 + 2 ____	3 + 2 ____	4 +0 ____

DAY 29 Combining Patterns

This lesson explores math in day-to-day life. It includes a lesson and an activity. This will require about 20 minutes of instruction from the parent/teacher. You may refer back to Day 13 to review the lesson on even and odd numbers.

Prayer

Thank God for something you have learned. Ask Him to help you as you do this lesson. OR Pray your own prayer of thanksgiving and praise to God.

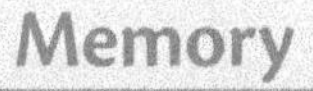

Spend a few minutes with flash cards (1 + 1, + 2, + 3 . . . series).

Lesson

> Two are better than one, . . .
> If two lie down together, they will keep warm;
> But how can one be warm alone? (Ecclesiastes 4:9, 11)

Today, we will use our little stones from our math toolbox to review even and odd numbers.

Remember that an even number of things (like stones) can be split into pairs. If every stone is paired up with another stone, the number is even. When you split an odd number of stones into pairs, one stone will be left out. One stone will be all by itself.

Now let's do this exercise together.

1. **Get out 2 stones.** Can you make a pair with them? Yes! 2 is an even number.

   Get out 6 more stones. How many pairs can you make with these 6 stones? Is 6 an even number? Yes!

   Now bring the two sets together. Do you still have an even number of stones? Put them all together in pairs just to make sure.

   Now lay out a set of 4 stones and a set of 6 stones. Are 4 and 6 even numbers or odd numbers? Combine the two sets of stones. Do you still have an even number? Put them all together in pairs just to make sure. We've found a pattern. What is this pattern?

Even + Even = ____________

2. **Get out 2 stones.** Can you make a pair with them? Yes! 2 is an even number. Now get out 7 more stones. Can you make pairs with these? No! Is 7 an even number or an odd one?

Now bring the stones together. Do you have an even number of stones or an odd number? If you aren't sure, try making pairs out of your new big set.

In this case, one stone is by itself. This is an odd number of stones.

Now try a set of 3 stones and a set of 4 stones. Is 3 an even number or an odd number? What about 4? Now combine the two sets. Is the big set an even number or an odd one?

We've just learned another pattern! What is this pattern?

Even + Odd = ____________

3. **Let's try one more.** Get out 5 stones. Can you make pairs with 5 stones? No! Is 5 an even or an odd number?

Get out 3 more stones. Can you make pairs with 3 stones? Is 3 an even or an odd number?

Now bring both sets of stones together. Do you get an even number or an odd number of stones? The 2 stones that were all by themselves in the first sets are teamed up with each other when the sets are brought together. Now every stone is part of a pair.

Here is our third pattern:

Odd + Odd = ____________

Activity

Let's apply our math to the world around us. God wants us to use our math every day. Use your math skills to fill in the blanks and find the answers to each of the exercises. You can use your stones or draw pictures if you need to.

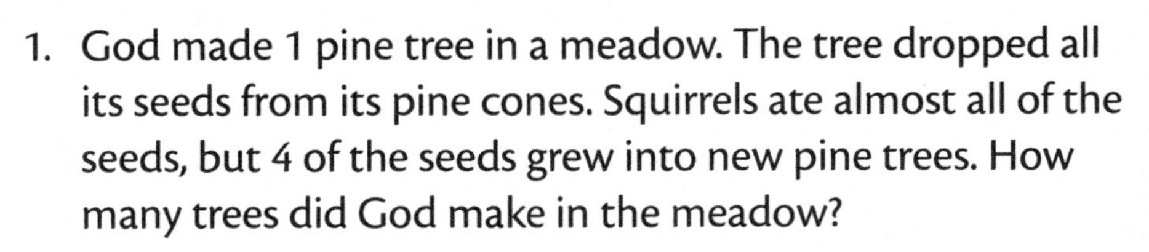

1. God made 1 pine tree in a meadow. The tree dropped all its seeds from its pine cones. Squirrels ate almost all of the seeds, but 4 of the seeds grew into new pine trees. How many trees did God make in the meadow?

______ + ______ = ______

2. Daniel prayed for his father and mother, and then he prayed for his 3 sisters. How many people did Daniel pray for?

______ + ______ + ______ = ______

3. Last year, a family was raising 4 goats. This year, 2 of the mother goats had 3 babies each. How many goats does the family take care of now?

______ + ______ + ______ = ______

4. Jack, Brooke, Chris, and Kimberly each bought 2 apples at the store. How many apples did they buy all together?

______ + ______ + ______ + ______ = ______

Practice DAY 30

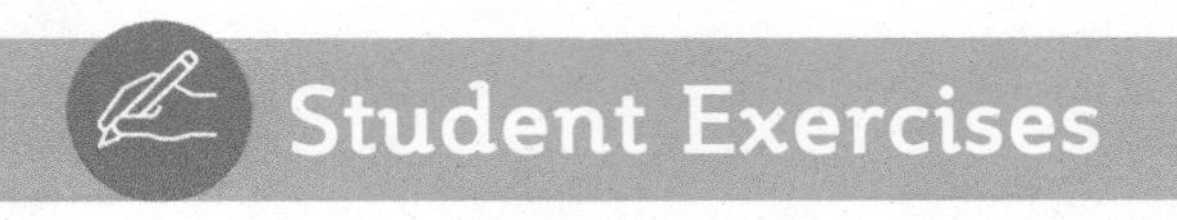

Draw four houses using different shapes. Think of each house as a set of shapes. Use triangles, circles, and rectangles. Here are some examples you might like to copy.

Try these addition exercises. What do they all have in common? What is the pattern? The pattern is that adding 1 is like counting to the next number. Does that help you figure them out? Even if you're not too comfortable with it, try these from memory! Someday these will be no exercise for you! Thank You, God.

$10 + 1 =$ 11	$1 + 1 =$ ______
$1 + 7 =$ ______	$1 + 2 =$ ______
$8 + 1 =$ ______	$1 + 11 =$ ______
$9 + 1 =$ ______	$1 + 3 =$ ______
$5 + 1 =$ ______	$4 + 1 =$ ______

Racing with Dice DAY 31

This lesson explores addition. The student will play a game and solve about 6 addition exercises for each lap around the track. The parent/teacher may take as much time as necessary to improve on the student's ability to add.

Prayer

Thank God for something you have learned. Ask Him to help you as you do this lesson. OR Pray your own prayer of thanksgiving and praise to God.

Memory

Spend a few minutes with flash cards (1 + 1, + 2, + 3 . . . series).

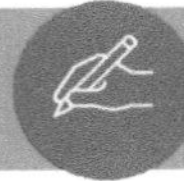

Lesson

Do you like games? Let's make our own game and play it together!

Here's what you'll need from your math toolbox:

- 40 coins
- Two dice
- One small object (like a button or token) for each player

Arrange your 40 coins, one after another, to make a loop on the table or the floor. This is your racetrack. Each coin will mark a space on the track.

Next, mark the starting position on the racetrack. Decide how many people will play the game, and give each person a token or button. Place all the tokens or buttons at the starting position on the track. Now you can go first.

Now roll the two dice and add the two numbers together. What number do you get? This tells you how many coins your race car can move down the track. Count the coins as you move your token. Take turns. The first person to make it around the whole track wins!

As you play, compare each new sum you get with the last one. You might write down each sum on a piece of paper. Was the new sum bigger or smaller than the last sum?

DAY 32 Practice

Student Exercises

God made some numbers bigger than others. Which number is bigger? Which number is smaller? Use the bigger than (>) and smaller than (<) symbols to show which numbers are smaller and which numbers are bigger in these exercises. Use the equal to (=) symbol if the numbers are equal. Then read the completed exercise out loud. For the first exercise, you would say, “1 is smaller than 3, and 3 is smaller than 5.”

1 __<__ 3 __<__ 5	1 ______ 4 ______ 9
6 ______ 2 ______ 1	2 ______ 1 ______ 0
5 ______ 10 ______ 11	3 ______ 7 ______ 11
5 ______ 4 ______ 2	10 ______ 4 ______ 0
1 ______ 11 ______ 12	8 ______ 9 ______ 10

Which of these four numbers are even? Which of these four numbers are odd? You may use your stones if you need to, dividing the number into pairs. Every even number can be paired into sets. Each stone will have a friend. Write the even numbers in the circle marked **Even**. Odd numbers will have one stone without a friend. Write the odd numbers in the circle marked **Odd**.

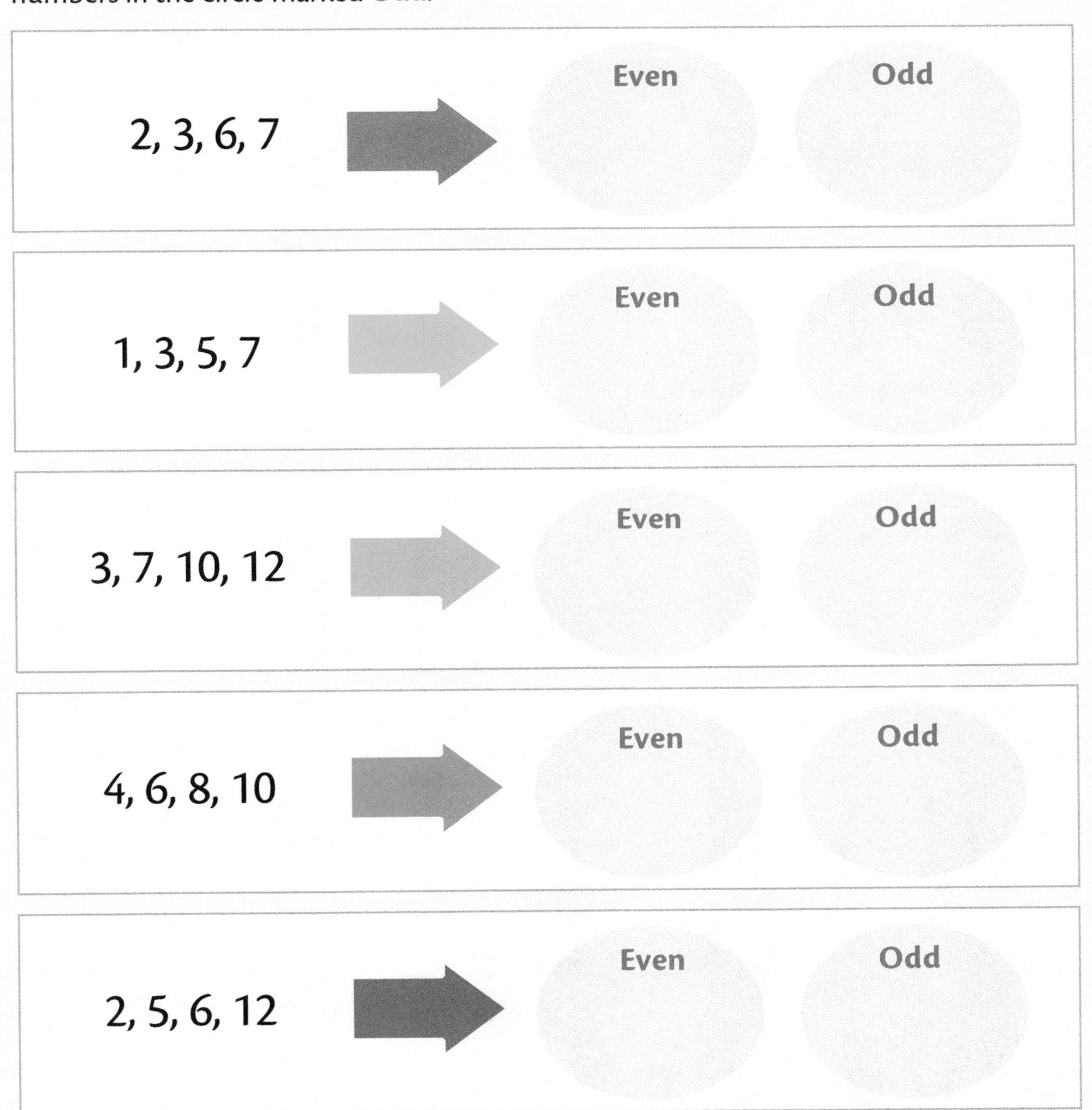

DAY 33 Joining Shapes

This lesson explores God's creation using shapes and numbers. It includes a lesson and an activity. This will require about 20 minutes of instruction from the parent/teacher.

Prayer

Thank God for something you have learned. Ask Him to help you as you do this lesson. OR Pray your own prayer of thanksgiving and praise to God.

Memory

Spend a few minutes with flash cards (1 + 1, + 2, + 3 . . . series)

Lesson

You have already learned how we can bring two little sets together to make one bigger set. We add small numbers together to get a bigger number. God also makes big sets out of little sets in His creation. This happens all around us all the time. Do you see the 3 birds flying separately in the picture? They are joining the larger flock of birds. How many birds are there now? Count as high as you can.

Do you see more fish joining the set (or school) of fish in the picture above? Praise God for creating so many fish! He is always adding more fish to His world. He is always adding more birds to His world.

We can create something bigger too when we combine shapes together. When we bring 2 shapes together to make 1 bigger shape, we call this **joining**.

Activity

Cut out the shapes on the next page. Then follow the directions to make new shapes. You can also trace the shapes on another piece of paper if you would like more shapes to work with.

1. Join some squares to create a rectangle.
2. Join two triangles to create a square.
3. Can you create a shape with exactly seven sides?
4. Join four squares to make one square.
5. Join a triangle and a square so that your shape is a square (overlapping is good).
6. Make a fun shape! How many sides does it have?
7. Make a serious shape! How many corners does it have?
8. Make a sad shape! How many shapes did you join?

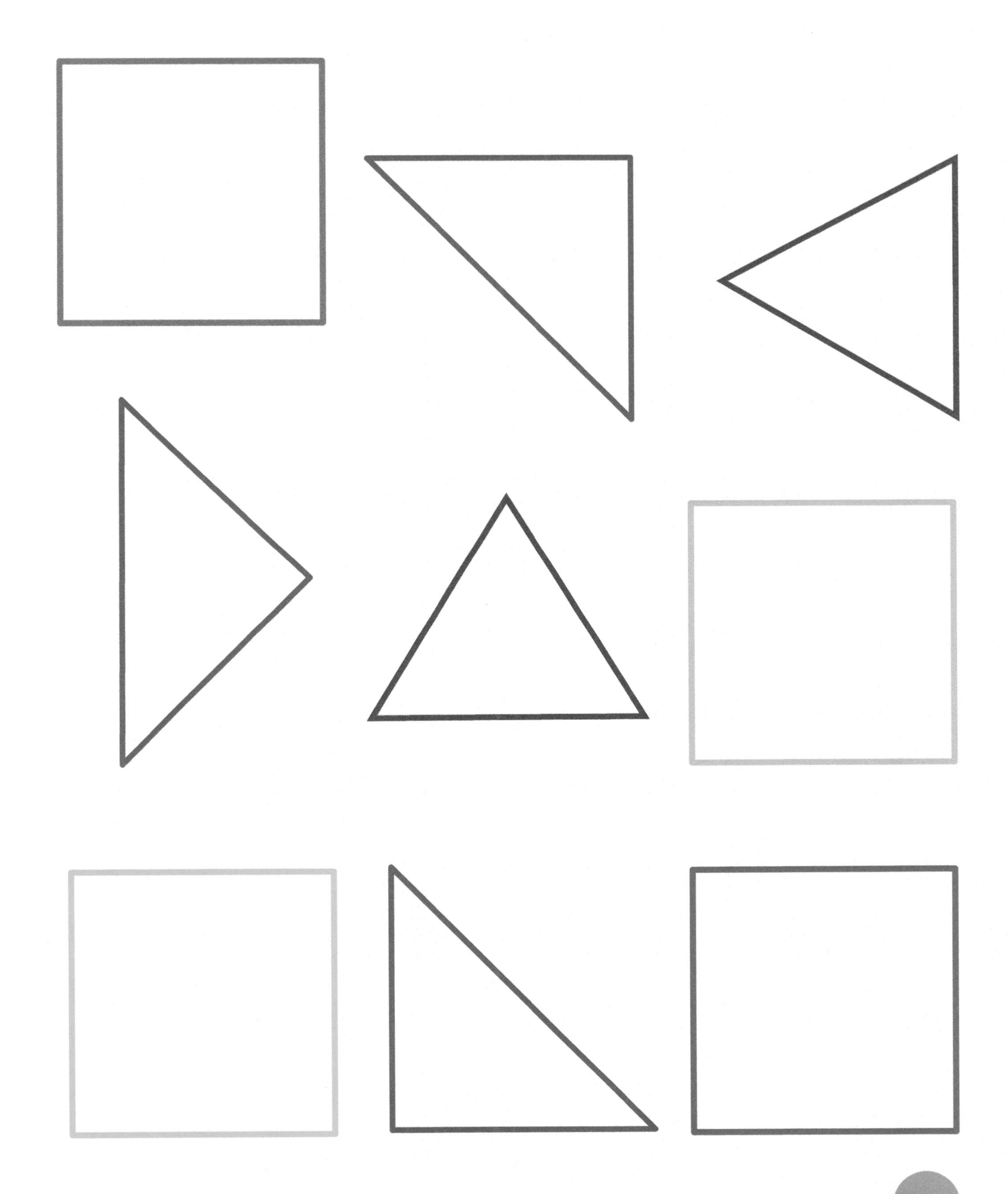

THIS PAGE
INTENTIONALLY LEFT BLANK

Go Help with Addition! DAY 34

This lesson integrates math into everyday life. This is an essential element to learning. The child is encouraged to apply God's patterns and wisdom to life in the home and community. Take a break from memory work and academic exercises, and identify ways to make math part of your everyday life. The following are suggestions or examples, but other ideas may be added to the list.

Activity

You've learned so much in this chapter! Great job! Now let's choose one of the following exercises or come up with your own idea to use math in everyday life. What are some ways you can use addition to love and serve others?

1. Buy Groceries

How many apples or bananas does your family eat in one day? How many will they eat in three days? Add these numbers together. Then go to the grocery store and count out the correct number of apples or bananas you need to buy.

2. Fix a Snack

Fix a snack for two or three children in your home. Ask your mom how many crackers, cookies, or chocolates each person gets. Add them together. Be sure to wash your hands before you touch the food. Count out the total number of snacks. Then divide them into sets for each of the children.

3. Make a Meal

Does your family enjoy eating pancakes for breakfast? How many pancakes will each person eat? Father? Mother? The children? Add these numbers together, and make a batch of pancakes to feed the whole family.

4. Drawing Pictures

Grandma and Grandpa always enjoy getting cards and pictures from their grandchildren. Make up six drawings or cards with a little help from the rest of your family.

When we give,
others gain.

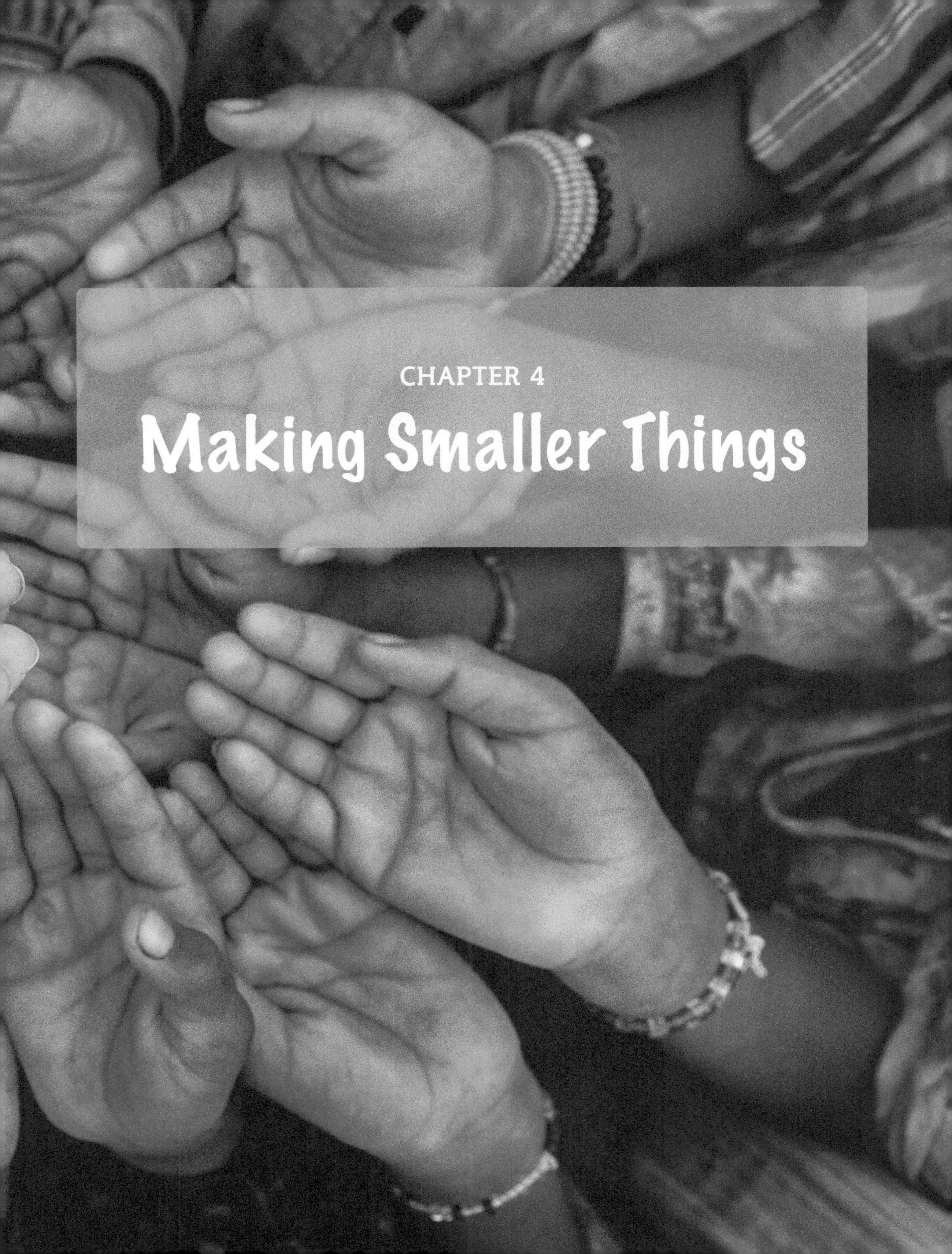

CHAPTER 4
Making Smaller Things

Introduction

This lesson introduces subtraction in God's world, and is followed by a page of review exercises. This will require about 20 minutes of instruction from the parent/teacher.

God's world is full of **opposites**. Can you try these opposite activities?

- Nod your head up and down.
- Shake your head from left to right.
- Sing a high note and a low note.
- Look backwards and forwards.

You can do opposite exercises with math too. We have learned to use addition to make bigger sets from smaller sets. Today, we will learn how to do the opposite. We will make smaller sets from bigger sets.

God made big things and small things. God counts big things and small things. He counts the hairs on our heads. This is what Jesus said:

> The very hairs of your head are all numbered." (Matthew 10:30)

God counts the big things too. He counts the stars in the sky. He names every one of them!

> He counts the number of the stars;
> He calls them all by name. (Psalm 147:4)

Can you see these small things God has made?

Can you see these big things God has made?

In the picture below, what is getting smaller (or less)? What are we taking away? What is getting bigger (or growing)? What are we adding?

We are adding water to the glass. Soon there will be more water in the glass than what we started with. At the same time, we are taking water out of the pitcher. There will be less water in the pitcher. Math explores these patterns. Math can tell you about patterns that grow using addition. Math can also tell you about the opposite. It can tell us how things are shrinking. It shows us how things become less and less. We call this **subtraction**.

DAY 35 Separating Sets

Prayer

Our Father God, we can't learn these things without You. We wouldn't want to learn without You and Your wisdom. Teach and help us today! Amen.

Memory

Spend a few minutes with addition flash cards (add only numbers 1 - 6 to each other).

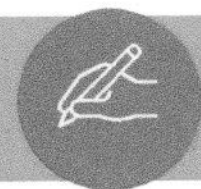

Lesson

Do you see the set of 4 fish in the picture? God made these fish. He made them so they can move! 2 fish are swimming away. 1 is swimming to the right and 1 is swimming to the left. Can you imagine how many will stay?

God made numbers to work together in lots of different ways! When we add sets or bring two small sets of things together, we get a bigger set. What happens when we take members away? We get a smaller set. This is subtraction.

How can we use numbers to tell us what is happening with the fish? This is what it looks like:

$$4 - 2 = 2$$

This is called a subtraction equation. This is how you would read it: "4 **minus** 2 equals 2." Can you say this? Do you see the minus sign? Look at the picture of the 4 fish one more time. First there were 4 fish. But 2 fish swam away. Now there are only 2 fish left. 4 **minus** 2 equals 2.

What do we call the fish that are left? They are called the **difference**. The answer to a subtraction equation is called the difference.

What is the difference between 7 and 3? We bought 7 apples. After we eat 3 of the apples, how many are left? 7 minus 3 equals 4. The difference is 4. There are only 4 apples left. That's a smaller set of apples, isn't it?

You can learn subtraction as you learned addition. Can you say "4 - 3?" Read these out loud:

5-2 10-7 12-6

Let's find the answer to 4 - 1. Take your little stones out of your math toolbox. Put a set of 4 stones on the table. Now take 1 away. How many stones are left? 3! That's the difference! That's the answer!

How many cats do you see in this picture? Take away the black cat. Now how many are left? 3!

Use subtraction to find the difference. You can use your stones if you need to. Use the first number to tell you how many stones to start with. Use the second number to tell you how many stones to take away. How many stones are left? Write that number in the blank.

2 – 1 = ___________ 3 – 2 = ___________

3 – 1 = ___________ 4 – 2 = ___________

Student Exercises

Let's practice addition! Add these numbers together to find a bigger number. Try to answer these from memory. You can use your little stones if you need help. Write the answers for the first six exercises **on** the line. Write the answers for the last six exercises **under** the line.

$1 + 2 =$ ___3___	$1 + 3 =$ ______
$1 + 4 =$ ______	$9 + 1 =$ ______
$8 + 1 =$ ______	$1 + 7 =$ ______

$5 + 0$	$4 + 0$	$3 + 0$
$12 + 0$	$7 + 5$	$3 + 9$

Student Exercises

Here are some subtraction exercises. Let's say your dog has 8 puppies. You keep them all. How many do you have left? Take away 0 puppies, and you still have 8 puppies! These exercises answer questions like that. Try these without any help. Just use the memory God has given you!

5 – 0 = 5	8 – 0 = ______
11 – 0 = ______	3 – 0 = ______
7 – 0 = ______	12 – 0 = ______
4 – 0 = ______	2 – 0 = ______
0 – 0 = ______	5 – 0 = ______

Let's subtract birds! In the exercises below, how many birds start out on the branch? Write that number on the first line. How many birds fly away? Write that number on the second line. How many birds are left perching on the branch? Write that number on the third line.

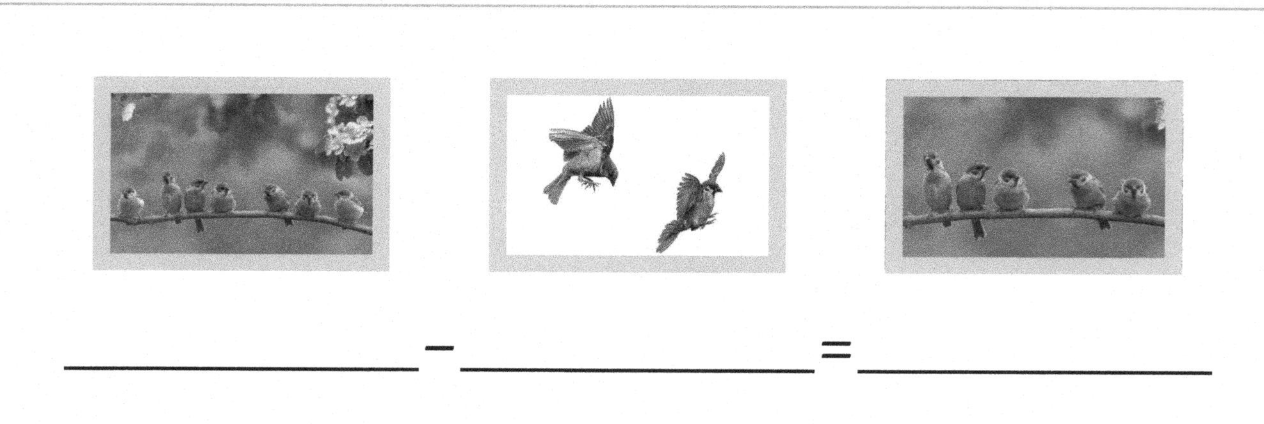

Subtracting Numbers DAY 37

This lesson explores subtraction using imagination. It is followed by one page of review exercises. This will require about 20 minutes of instruction from the parent/teacher.

Prayer

Thank God for something you have learned. Ask Him to help you as you do this lesson.

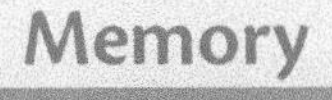

Memory

Spend a few minutes with addition flash cards (add only numbers 1 - 6 to each other).

Lesson

When you learned about addition, you imagined planting more trees in the yard. What will you do when you subtract? You will take things away!

Imagine 3 trees. What if somebody cuts a tree down? How many are left? Can you see the 2 trees in the picture? Where was the third tree? Now all you can see is a stump. Here is how we use math to say this:

$$3 - 1 = 2$$

Let's imagine the bananas God made. When we add bananas, we bring them together. We can join 1 banana with 1 banana to get 2 bananas. But now we are going to take bananas away. We start with 2 bananas. We take 1 away to find the difference.

Point to the pictures of the bananas above and say this out loud three times:

"2 bananas minus 1 banana equals 1 banana."

2 bananas is not the same as 1 banana. You can see that 1 banana has been taken away from 2 bananas. We started with 2 bananas. We took 1 banana away. We have 1 banana left. What if there are 2 bananas in the fruit basket? If you eat 1 banana (and throw away the banana peel), how many bananas are left in the fruit basket? Only 1!

In the picture below, you can see 2 children holding hands. One child walks away. How many children are left?

Point to the pictures above and say this out loud three times:

"2 children minus 1 child equals 1 child."

Let's practice subtraction! First, count the number of bananas in each picture. Write the number under the picture. Now read the answer out loud. You're doing subtraction!

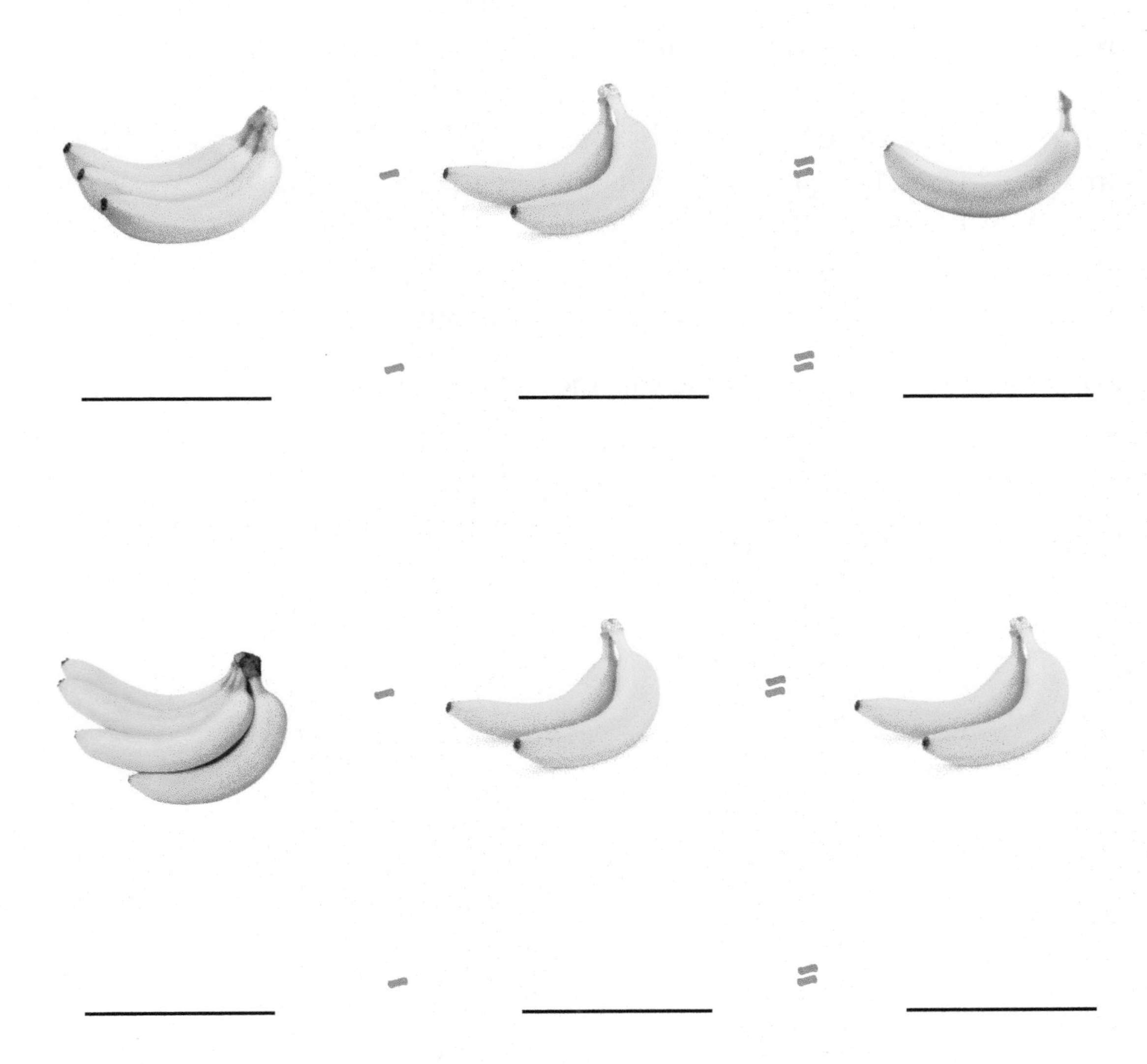

There are different ways to do subtraction. God made each of us differently. Each of us will subtract numbers in different ways. But the answer should always be the same.

Here are four ways to get the right answer when you subtract:

1. **Memory.** You can memorize 3 – 1 = 2. And 5 – 3 = 2. That's why you will do memory practice using flash cards. Practicing math with flash cards will help you memorize!
2. **Imagination.** At the beginning of the lesson, you imagined trees in the yard. Imagine 3 trees. Imagine cutting down 1 tree. Now, imagine how many trees are left. Only 2!
3. **Drawing.** You can make little drawings of things (like trees) on paper.
4. **Stones.** You can use your small stones from your math toolbox.

Use any of these four ways to do these subtraction exercises:

7 – 4 = 3   10 – 5 = ________

11 – 8 = ________   3 – 2 = ________

6 – 1 = ________   12 – 4= ________

Student Exercises

Let's practice subtraction! First, count the number of yummy things in each picture. Write the numbers under the pictures. Now read the equation out loud. For the first one, you would say, "3 cherries minus 2 cherries equals 1 cherry."

Imagine that you are taking away some of the cherries, bananas, or cookies because you have eaten them. How many are left? That's the number you see after the equal sign.

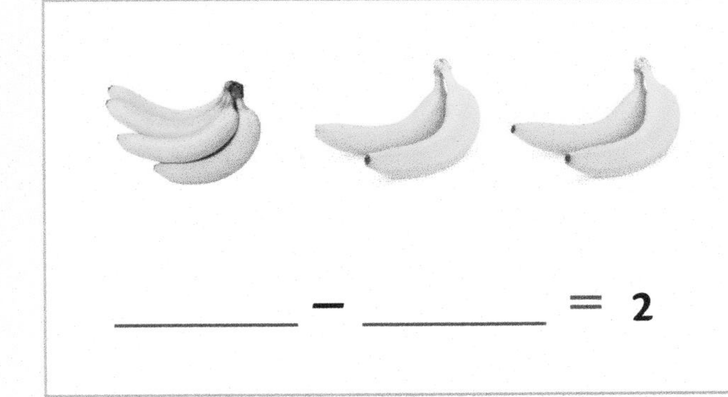

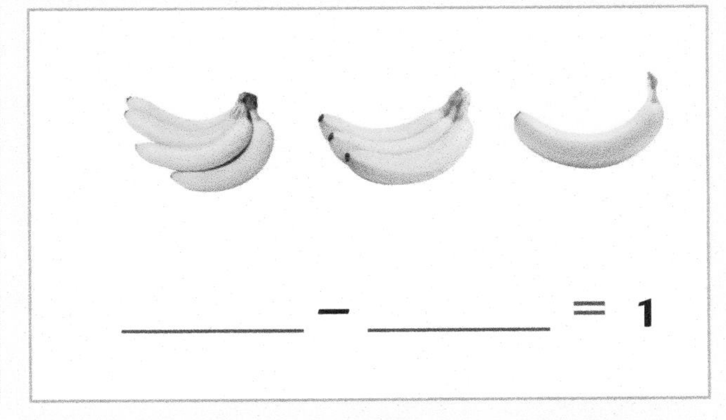

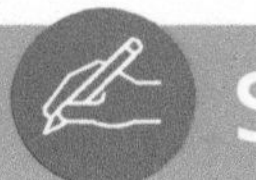

DAY 38 Practice

Student Exercises

Try these subtraction exercises. Can you describe the pattern of subtracting 0 to your parent or teacher? It's a nice one!

$1 - 0 =$ 1	$5 - 0 =$ ______
$12 - 0 =$ ______	$7 - 0 =$ ______
$2 - 0 =$ ______	$6 - 0 =$ ______
$0 - 0 =$ ______	$8 - 0 =$ ______
$3 - 0 =$ ______	$9 - 0 =$ ______

How can we split these big numbers into smaller numbers? For the first exercise, take 4 stones out of your tool box. Split the stones into two groups or sets. Count the members of each group. Write these numbers in the blanks. Finish by reading your addition exercise out loud. There will be several right answers for these exercises. Try to do the exercises without your stones, but you can use them if you need help.

4 = 1 + 3

4 = ______ + ______

7 = ______ + ______

7 = ______ + ______

10 = ______ + ______

10 = ______ + ______

11 = ______ + ______

11 = ______ + ______

You can count kittens in a litter. You can count family members sitting on a couch. You have counted the members of sets. God made kittens. He made your family. God made numbers too! Now you can find numbers living together in sets. How many numbers are living together in these sets? Count the number of numbers in each set.

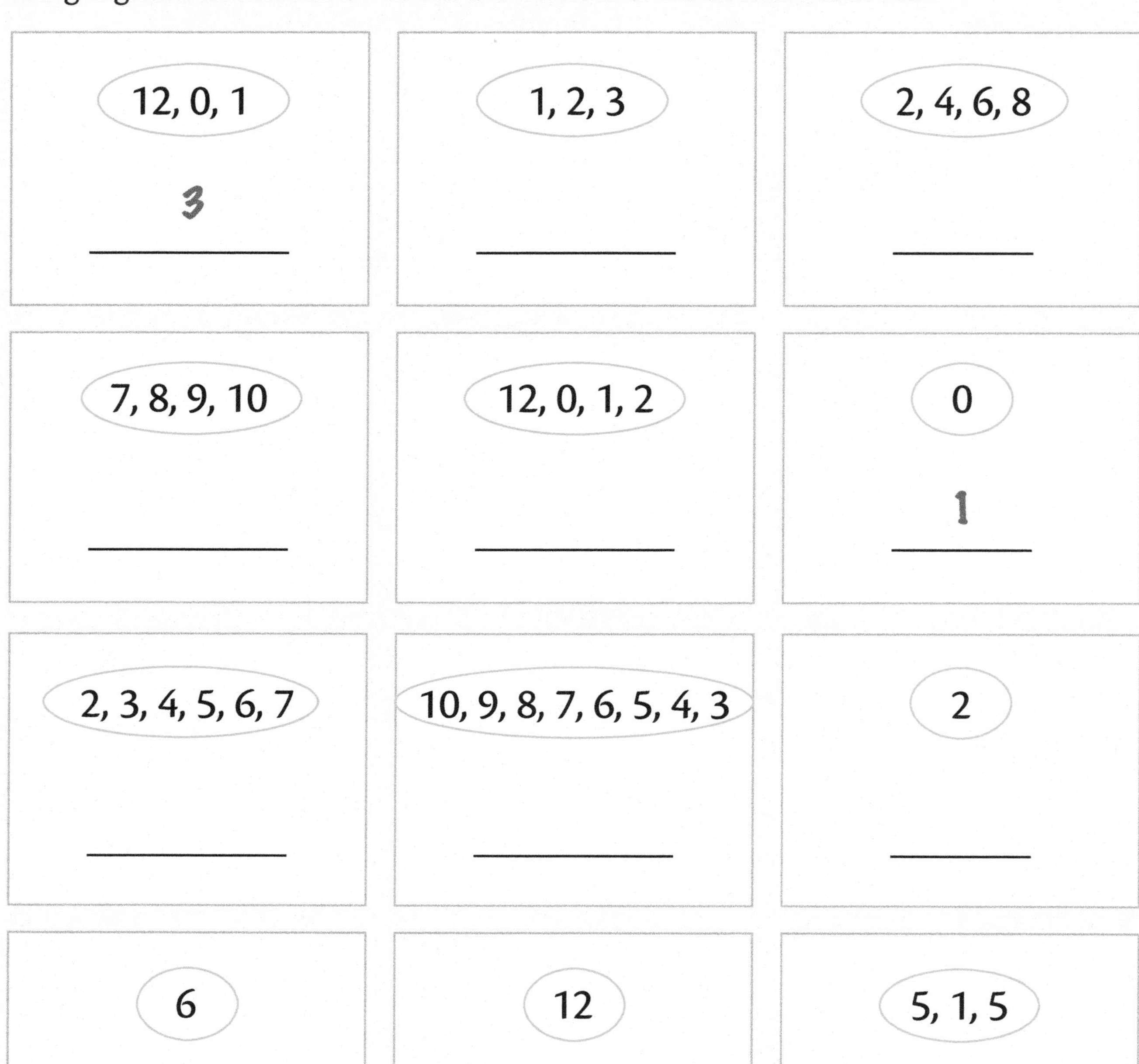

Comparing Sets DAY 39

This lesson compares sets in God's world by examining two pictures. It is followed by two pages of new and review exercises. This will require about 20 minutes of instruction from the parent/teacher.

Prayer

Thank God for something you have learned. Ask Him to help you as you do this lesson.

Memory

Spend a few minutes with addition flash cards (add only numbers 1 - 6 to each other).

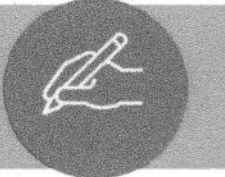

Lesson

Here are two sets of beautiful animals—all created by God! Compare these sets. These two sets share one member in common. Which animal shows up in both pictures? Draw a circle around this guy.

Each exercise below has two sets. Do you see anything that is shared by both sets? Draw what they share in the answer oval to show a new set. If they do not share anything, the new set is an empty set! You won't draw anything in the answer oval.

Answer:

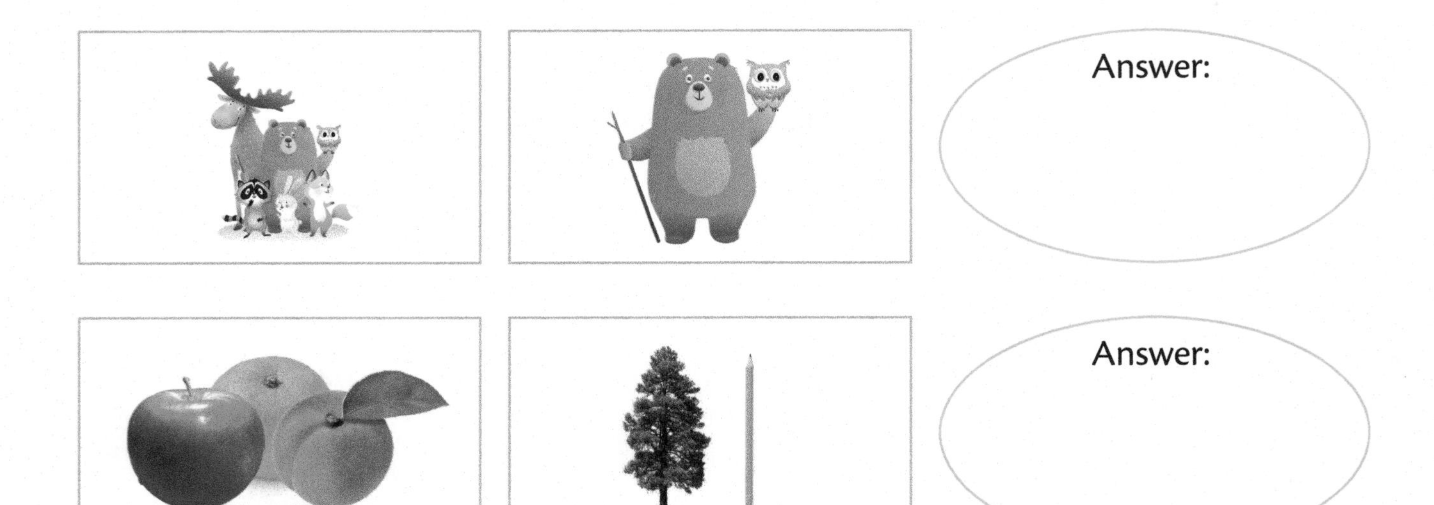

This is answer is an empty set. The sets do not share anything.

We can compare two sets of numbers too! In math, we put sets in **brackets** { }. Look at the two sets below. Each set is in brackets. Which number is shared by both sets?

Set 1: {1, 2} Set 2: {2, 3}

Can you find which number is shared by both sets? In math, this is called intersection. The number shared by both sets is 2! You can write the answer this way:

{2}

Let's try one more. Compare these sets. Which number is shared by both sets?

{0, 1} and {3, 4}

Do these sets have anything in common (shared)? There aren't any numbers shared in these two sets. The answer is none! You can write the answer this way:

{ }

This is called an **empty set**. There are no numbers in our new set.

Student Exercises

Try these subtraction exercises from memory. Pretty soon you'll get excited when you add and subtract with 0. "I can do those! No exercise. Thank You, God."

4 – 0 = _______	2 – 0 = _______
9 – 0 = _______	10 – 0 = _______
5 – 0 = _______	12 – 0 = _______
3 – 0 = _______	11 – 0 = _______
6 – 0 = _______	8 – 0 = _______

Your mom gave you 3 cookies. You shared 2 cookies with your friends. How many cookies do you have left?

☐ - ☐ = ☐

Jesus healed 10 lepers. Only 1 came back to thank Him. How many lepers forgot to thank Jesus?

☐ - ☐ = ☐

A very poor family comes to your church. Your family gave the poor family 2 loaves of bread. Others gave them 3 loaves of bread. How many loaves did they receive?

☐ + ☐ = ☐

Your friend received 6 little cars for his birthday. But now he only has 2 cars. How many has he lost? Can you share your toy cars with him now?

☐ - ☐ = ☐

Now compare these sets of numbers. What numbers do they have in common? Write these numbers on the blank lines. You are making new sets of numbers.

Sets {3, 4} and {4, 5} share = {_____4_____}

Sets {1, 2} and {1, 3} share = {__________}

Sets {4, 6} and {4, 6} share = {__________}

Sets {1, 5, 9} and {1, 6, 9} share = {__________}

Sets {4, 6, 10} and {4, 6, 10} share = {__________}

Sets {1, 2, 3} and {6, 7, 8} share = {__________}

Extra Challenge

We can show the sharing or intersection of sets using something called a Venn diagram. You can see that 3 and 4 belong to the first circle, and 4 and 5 belong to the second circle. Which number is shared?

Now, you try making a Venn diagram with this story exercise.

Abigail (A), Billy (B), and Christina (C) play the piano. Christina (C), David (D), and Edward (E) play the violin. Fill in the Venn diagram with the letters of the children that play the instruments.

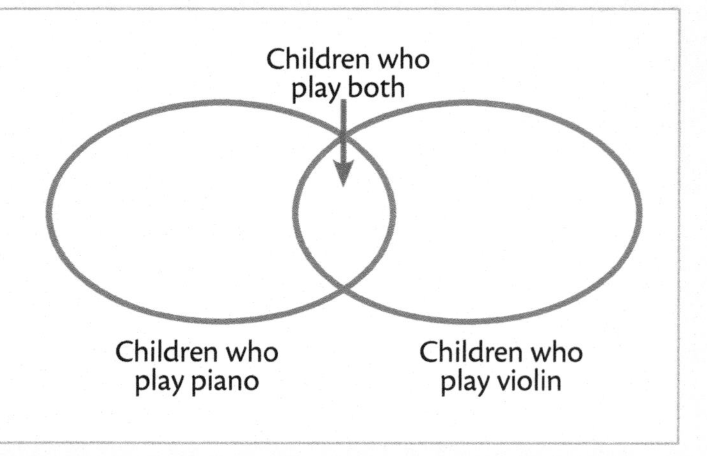

Anna (A), Benjamin (B), Catrina (C), and Daniel (D) like to play soccer. Anna (A), Eduardo (E), and Fredrick (F) like to swim. Fill in the Venn diagram with the letters of the children that enjoy these sports.

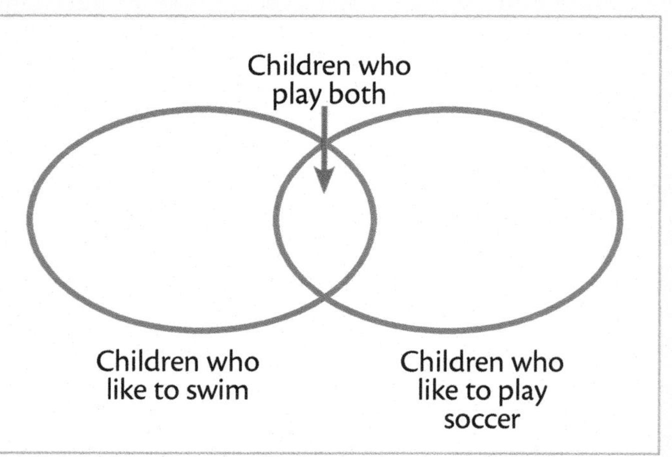

Student Exercises

These subtraction exercises are all the same kind. Take 1 away. How many are left? Can you tell what pattern they follow? This is like counting backwards! Can you do these in your head from memory?

$10 - 1 =$ ________	$6 - 1 =$ ________
$1 - 1 =$ ________	$9 - 1 =$ ________
$2 - 1 =$ ________	$8 - 1 =$ ________
$3 - 1 =$ ________	$5 - 1 =$ ________
$12 - 1 =$ ________	$9 - 1 =$ ________

You have been subtracting numbers that were written in a row. Big boys and big girls do their math in columns like the ones below. Now you try it! Subtract the small number from the big number. We want to find another small number. Try to answer these from memory, or by using your imagination, or by using your little stones.

8 – 5 ______	8 – 4 ______	8 – 3 ______
9 – 4 = ______	9 – 5 ______	9 – 7 ______
12 – 0 ______	12 – 1 ______	12 – 2 ______
2 – 2 ______	2 – 1 ______	2 – 0 ______

Dice Battle! DAY 41

This lesson explores addition and subtraction with a game. The parent/teacher may determine the length of this lesson by adjusting the length of the track and the number of times the game is played.

Prayer

Thank God for something you have learned. Ask Him to help you as you do this lesson.

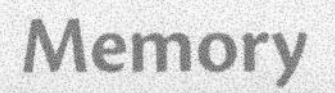

Memory

Spend a few minutes with addition flash cards (add only numbers 1 - 6 to each other).

Lesson

Let's play a game today! You will need these things from your math toolbox. Ask your parent/teacher if you need help counting the coins.

- 40 coins

- Two dice

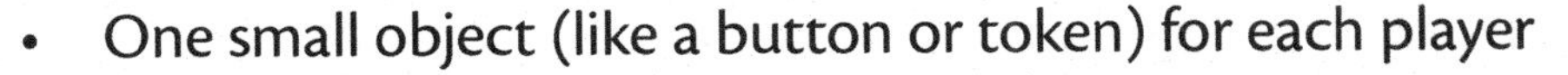

- One small object (like a button or token) for each player

Use your 40 coins to make a loop on the table or the floor. This is your racetrack. Each coin will make a space for the track. Next, mark the starting position on the racetrack. Set the buttons or tokens at the beginning of the track. These are your race cars.

Here are the rules:

1. Player 1 rolls the two dice. He or she adds the numbers and writes down the sum.
2. Player 2 rolls the two dice. He or she adds the numbers and writes down the sum.
3. The player with the highest sum gets to move his or her car forward. Let's say my roll was a sum of 7, and your roll was a sum of 9. Nine is higher than 7 so you are the one who gets to move your car. How far do you move your car? Find the difference of the players' sums to know how many places to move. $9 - 7 = 2$ You get to move forward 2 spaces.
4. The game continues until one player reaches the finish line.

Have fun!

DAY 42 Practice

Student Exercises

Try these subtraction exercises. Subtracting 1 from a number is just like counting backwards. 6 minus 1 is . . . think 8, 7, 6 . . . 5. 6 – 1 = 5! Answer these questions from memory. Count backwards in your head if you can't figure them out.

4 – 1 = ______	8 – 1 = ______
9 – 1 = ______	3 – 1 = ______
5 – 1 = ______	2 – 1 = ______
1 – 1 = ______	12 – 1 = ______
6 – 1 = ______	7 – 1 = ______

So far, you have learned addition and subtraction. Addition makes bigger numbers. Subtraction makes smaller numbers. God made numbers to work together. Sometimes numbers work together to make bigger numbers. Sometimes numbers work together to make smaller numbers.

For these exercises, you need to decide how these numbers work together. How do 2 and 8 work together to make 10? Of course, they must be added! Write "+" or "-" in each circle to show how they work together.

2 (+) 8 = 10	10 ◯ 8 = 2
5 ◯ 1 = 4	4 ◯ 1 = 5
10 ◯ 3 = 7	3 ◯ 7 = 10
8 ◯ 3 = 11	11 ◯ 3 = 8
3 ◯ 1 = 4	4 ◯ 3 = 1

Student Exercises

Write in the missing numbers in the order that God made for them. Sometimes you will have to count down (or backward) to find the right number. Sometimes you will have to count up (or forward) to find the number that comes next.

2, ____3____, ____4____

________, 5, ________

________, ________, 8

________, 9, ________, ________

________, ________, 11, ________

________, ________, 4

0, ________, ________, ________

________, ________, ________, ________, 6

Finding the Number That Fits DAY 43

This lesson introduces equation solving, and will challenge your child's thinking skills. It is followed by two pages of new and review exercises. This will require about 20 minutes of instruction from the parent/teacher.

Prayer

Thank God for something you have learned. Ask Him to help you as you do this lesson.

Memory

Spend a few minutes with addition flash cards (add any one-digit numbers).

Lesson

Do you like puzzles? Today you are going to do math puzzles. You will learn to find the missing piece in the math equation. This is like finding the missing puzzle piece.

You've already learned about addition equations. An equation is a math sentence written using numbers.

$$5 + 5 = 10$$

What if the equation has a missing piece? What is missing in this equation?

$$\_\_\_ + 6 = 10$$

We need to find the missing piece. Here is one way to read this:

What plus 6 is 10?

We need to add something to 6 to get 10, but how much do we add? You can guess. Try guessing some numbers. Do they fill in the missing piece?

$$5 + 6 = 10$$
$$3 + 6 = 10$$
$$4 + 6 = 10$$

Which of these is correct? That's right! The last one is the right answer. Go ahead and write 4 for the missing number.

You can also use your little stones from your math toolbox. Let's use them for another exercise: What plus 7 equals 10?

$$\_\_\_\_ + 7 = 10$$

Count out 7 stones. How many more stones do you need to add to make 10? That's right—3!

You can also draw pictures of little stones, to find the missing piece.

$$\_\_\_\_ + 8 = 10$$

What plus 8 equals 10? Draw 8 little stones or something else. How many more of these do you need to draw to make 10? Of course the answer is 2!

$$2 + 8 = 10$$

Student Exercises

Find the missing number. You can try guessing, using your stones, or drawing pictures if you need help.

What plus 7 equals 10? 3

What plus 3 equals 10? ______

What plus 9 equals 12? ______

What plus 3 equals 12? ______

What plus 3 equals 10? ______

What plus 1 equals 8? ______

What plus 2 equals 11? ______

Let's practice subtraction! First, count the numbers of cherries, bananas, and cookies in each picture. Write the numbers under them.

Now read the equation out loud. For the first one, you would say, "5 cherries minus 3 cherries equals 2 cherries."

Imagine that you are taking away the cherries, bananas, or cookies because you have eaten them. How many are left? That's the number you see after the equal sign.

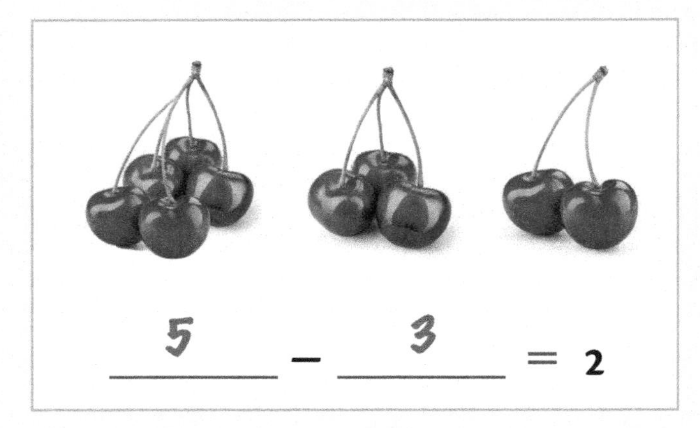

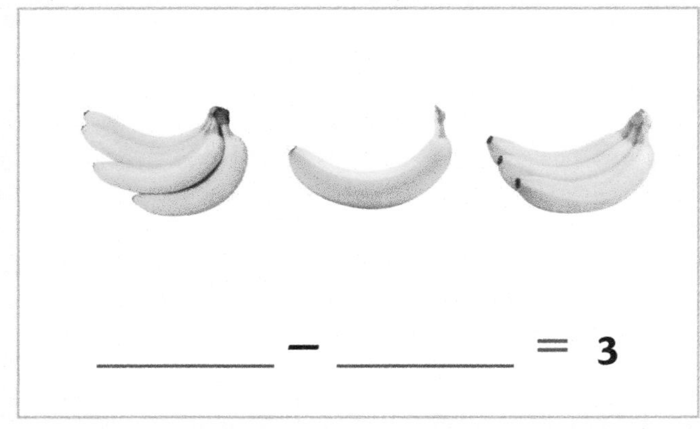

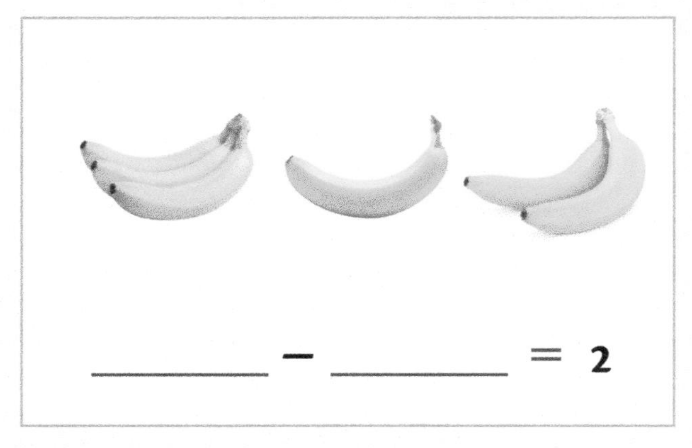

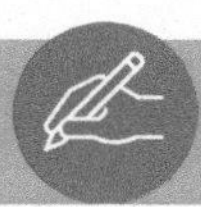 Student Exercises

Here are some subtraction exercises to practice today. Let's take 1 away and see how many are left! Are you getting faster at answering these? God helps you learn how to do this!

$10 - 1 =$ ______	$1 - 1 =$ ______
$7 - 1 =$ ______	$2 - 1 =$ ______
$8 - 1 =$ ______	$11 - 1 =$ ______
$9 - 1 =$ ______	$3 - 1 =$ ______
$5 - 1 =$ ______	$4 - 1 =$ ______

Let's practice subtraction! How does God make smaller numbers? Let's find out. Subtract the small number from the big number. Use your imagination first. Imagine 5 things. Take away 4 of them. How many are left? That's right! There is 1 left!

You can also use your little stones. Write the answers for the first six exercises on the line. Write the answers for the last six exercises under the line.

$5 - 4 =$ 1	$5 - 3 =$ ______
$5 - 1 =$ ______	$8 - 5 =$ ______
$8 - 3 =$ ______	$8 - 0 =$ ______

10 − 1 ______	10 − 4 ______	10 − 10 ______
4 − 1 ______	4 − 2 ______	4 − 4 ______

What are we doing with these equations? Are we adding or subtracting? Remember, addition makes bigger numbers. Subtraction makes smaller numbers. Decide how these numbers work together. How do 5 and 5 work together to make 10? Of course, they must be added! Write "+" or "-" in each circle to show how the numbers work together.

10 ○ 5 = 5

8 ○ 2 = 6

6 ○ 2 = 8

9 ○ 3 = 6

3 ○ 6 = 9

10 ○ 4 = 6

4 ○ 6 = 10

4 ○ 8 = 12

12 ○ 8 = 4

DAY 45 God Shows Us Opposites

This lesson explores opposites in Scripture and will require about 20 minutes of instruction from the parent/teacher.

Prayer

Thank God for something you have learned. Ask Him to help you as you do this lesson.

Memory

Spend a few minutes with addition flash cards (add any one-digit numbers).

Lesson

Read the following Bible verses or look them up in your Bible. Can you find the words that are opposite to each other?

> God called the light Day, and the darkness He called Night. So the evening and the morning were the first day. (Genesis 1:5)

> The Lord makes poor and makes rich;
> He brings low and lifts up. (1 Samuel 2:7)

> A wise son makes a father glad, but a foolish man despises his mother. (Proverbs 15:20)

> The wise woman builds her house, but the foolish pulls it down with her hands. (Proverbs 14:1)

1. Let's talk about these verses. What makes the opposite words different from each other?
2. Which of these words are like addition? Some words point to things that are getting bigger or better. Which of these words are like subtraction? These words talk about things that are worse. They talk about things that are less or negative (bad).

Student Exercises

Try these addition exercises from memory. When we do exercises like these, we say that we are making doubles. Can you figure out what we mean by that? Maybe your parent or teacher can help!

4 + 4 = ________	8 + 8 = ________
9 + 9 = ________	10 + 10 = ________
5 + 5 = ________	2 + 2 = ________
1 + 1 = ________	0 + 0 = ________
6 + 6 = ________	3 + 3 = ________

Make your own sets of numbers. If the number in the box says 2, make a set of 2 numbers. Use two different numbers in your sets. Get creative with your numbers. Include different patterns in your sets. You might use big and small numbers. Or, you might try even and odd numbers.

2 { 1, 4 }	2 { }
4 { }	4 { }
5 { }	5 { }
1 { }	1 { }
1 { }	1 { }

Each exercise starts with two sets of numbers. Compare these sets. Which numbers do they have in common? Write these numbers in the brackets. This makes a new set of numbers!

Sets {6, 5} and {5, 4} share = { 5 }
Sets {1, 2, 3} and {1, 2, 3} share = { }
Sets {7, 8} and {6 7} share = { }
Sets {4, 5, 6} and {8, 7, 6} share = { }
Sets {12, 11, 10} and {12, 0, 1} = { }
Sets {9} and {10} share = { }

DAY 47 A World of Opposites

This lesson explores opposites in God's natural creation, and is followed by one page of review exercises. This will require about 15 minutes of instruction from the parent/teacher.

Prayer

Thank God for something you have learned. Ask Him to help you as you do this lesson.

Memory

Spend a few minutes with addition flash cards (add any one-digit numbers).

Activity

You can find opposites everywhere in God's creation. You can also see opposites in a picture. Here is a picture by an artist named Pieter Bruegel the Elder. The picture is called "Hunters in the Snow." Do you see opposites? What is high? What is low? Do you see hot and cold? What are some more opposites? Hint: Look at the size of the dogs.

Read each sentence. Write or say another sentence using the opposite word. Does the first sentence use addition (+)? The second sentence will use subtraction (–). Does the first sentence use subtraction (–)? The second one will use addition (+). You might like to act out each sentence as you say it out loud.

1. The boy pours water **out of (–)** the jar onto his bean plant. He uses the hose to put more water **i**___ **(+)** his jar.
2. The rabbit digs **under (–)** the fence. But the squirrel climbs **o**__________ **(+)** the fence.
3. The girl goes **to (+)** the flower. But the butterfly flies away **f**___________**(–)** the flower.
4. **After (+)** we pray, we will sing a song. **B**_______________ **(–)** we sing a song, we will pray.
5. If we turn **left (–)**, we will be driving to the church. But if we turn **r**______________ **(+)** we will be going to Grandma's house.
6. At the zoo, I saw a **big (+)** elephant. I also saw a **l**________________ **(–)** poison dart frog.
7. The drive to the grocery store is **short (–)**. But the drive to the ocean is **l**_______________ **(+)**.
8. The library is a **quiet (–)** place. But a city street can be a **l**_________________ **(+)** place.

Now, take a look at the painting below. Can you find at least 3 opposites in this picture? The first is easy: There is a boy and a girl!

Student Exercises

Find the missing numbers! Try to find the missing pieces using your imagination. You can also use your little stones from the math toolbox.

What plus 6 equals 8?	2 + 6 = 8
What plus 2 equals 8?	______ + 2 = 8
What plus 2 equals 10?	______ + 2 = 10
What plus 8 equals 10?	______ + 8 = 10
What plus 0 equals 12?	______ + 0 = 12
What plus 2 equals 14?	______ + 2 = 14
What plus 4 equals 9?	______ + 4 = 9

Student Exercises

Add these numbers together to find a bigger number. Try to answer these from memory. You can use your little stones if you need help. Write the answers for the first six exercises **on** the line. Write the answers for the last six exercises **under** the line.

1 + 1 = 2	2 + 2 = ______
3+3= ______	4 + 4 = ______
5+5 = ______	0+0 = ______

1 + 3 ______	3 + 5 ______	5 + 7 ______
4 + 2 ______	4 + 6 ______	6 + 6 ______

Let's practice subtraction! Some of these exercises are a little harder than the ones you've done before. They are bigger numbers.

Subtract the small number from the big number. Try using your imagination first. Picture 11 of something in your mind. Take away 5 of them. How many are left? That's right! There are 6 left!

You can also use your little stones to do these exercises. Write the answer **on** the line for the first six exercises. Write the answers for the last six exercises **under** the line.

$11 - 5 =$ 6	$11 - 3 =$ ______
$11 - 1 =$ ______	$9 - 5 =$ ______
$9 - 3 =$ ______	$9 - 1 =$ ______

$7 - 5$ = ______	$7 - 3$ = ______	$7 - 1$ = ______
$5 - 5$ = ______	$5 - 3$ = ______	$5 - 1$ = ______

Go Serve with Subtraction DAY 49

This lesson integrates math into everyday life. This is an essential element to learning. The child is encouraged to apply God's patterns and wisdom to life in the home and community. Identify ways to make math part of your everyday life. The following are suggestions or examples, but other ideas may be added to the list.

Prayer

Thank God for something you have learned. Ask Him to help you as you do this lesson.

Memory

Spend a few minutes with addition flash cards (add any one-digit numbers).

Jesus wants us to love one another. Serving one another is a way to love. We can use math to serve God. We can use math to serve each other. How can you use subtraction in your home? Here are some ideas, but you could also add one of your own.

1. Keep track of the snacks . . . and fruits. How many packages of snacks, bananas, apples, or tomatoes did your parents buy at the store? Mark the day when your parents bought these foods. Wait one, two, or three days. Count how many snacks, bananas, apples, or tomatoes that are left. How many did you eat? When will your family need to buy more?

2. Use subtraction when you play games. Go bowling! When someone knocks down 3 pins, how many are left? Remember, you always start with 10 pins.

Play baseball or softball. Let's say one team has 5 runs. The other team has 2 runs. How many runs does the losing team need to get to catch up?

Play any kind of game that counts points. Which team is behind? Keep track of how many points the team needs to catch up.

3. Serve fewer people. Plan a meal when some members of your family are not going to be there. Count the members of your family. When 1 or 2 people will not be eating dinner with the rest of the family, how many are left? How many dishes will you need to use to set the table?

We need a big number
to count all of these elk!

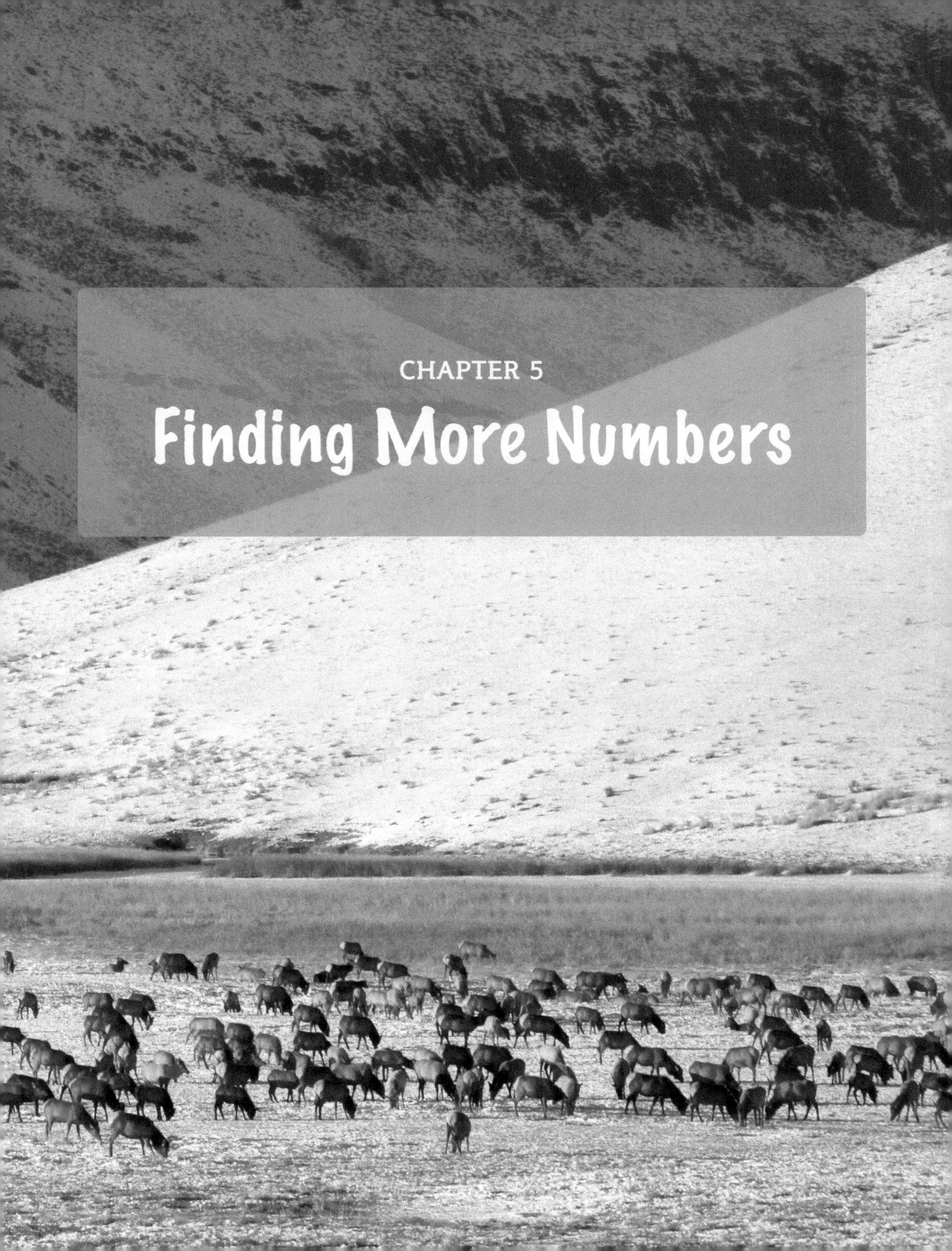

CHAPTER 5

Finding More Numbers

Introduction

So far, we have looked at God's numbers, big and small. Zero (0) is the smallest number. Is 12 the biggest number? No! Numbers can get bigger and bigger.

Count the geese in this picture. Are there more than 12 geese? Let's count past 12. Your teacher will help you.

1, 2, 3, 4, 5, 6, 7, 8, 9, 10, 11, 12, 13, 14, 15, 16. . .

In Psalm 40, David needed some big numbers. In fact, he couldn't find a number big enough for God's works. He needed a lot more than 12 or 16!

> Many, O LORD my God, are Your wonderful works . . .
> They are more than can be numbered. (Psalm 40:5)

Let's study more about these big numbers! With math, we can explore as far as we can see. We can explore as much as we can imagine. We are always asking God, "Is there more to find?" Have you looked up into the sky at night? How many stars can you count? There are too many to count, but this verse says that God counts them all. In fact, He has named every one of the stars!

> Lift up your eyes on high,
> And see who has created these things,
> Who brings out their host by number;
> He calls them all by name, by the greatness of His might
> and the strength of His power; not one is missing. (Isaiah 40:26)

Adding in Chunks DAY 50

This lesson introduces some of the bigger numbers as the child comes to better understand God's world. It is followed by an activity and one page of new exercises. This will require about 20 minutes of instruction from the parent/teacher.

Prayer

Our Father in Heaven, thank You for making our minds able to understand Your world. Please help us learn more about You and Your world today. Amen.

Memory

Spend a few minutes with subtraction flash cards (1 - 0, 2 - 0 . . . series).

Lesson

There are so many stars in the sky! There are so many fish in the seas! How do we count them all? We could make up numbers like this:

1, 2, 3, 4, 5, 6, 7, 8, 9, #, *, @, !, ?, $, %, &. . .

But how would you memorize all those numbers? Math doesn't make new numbers like this. Math puts numbers into sets. Let's call them **chunks**.

Look at the picture of the clover. You can see 1 little clover leaf by itself. It is not a full clover. You can also see two full clovers. Each full clover has 4 leaves.

We have three sets of leaves. The 1 clover leaf is in a set all by itself. This is the first set. Then we have a full clover with a set of 4 leaves. We could call this a chunk of 4 leaves. This is the second set. The other full clover also has 4 leaves. This is the third set. The full clovers each have a set or "chunk" of 4 leaves.

Great! However, there are A LOT more numbers than this. Do you want to memorize a name and a symbol for every single number? No, thank you! We need to do something different.

How many little leaves do you see altogether? Can you count them? Yes! You have nine little leaves.

You can also add the little leaves together. You can add the chunks of 4. What is 4 + 4? It's 8! But you still have one set left. It has one leaf in it. What is 8 + 1? It's 9! Altogether, there are nine leaves.

You can do the same thing with your little stones. Or you can use your imagination. Lay out your stones like the picture below. You will have one stone all by itself. You will also have two chunks of 4 stones each. That makes nine stones altogether. Can you count them?

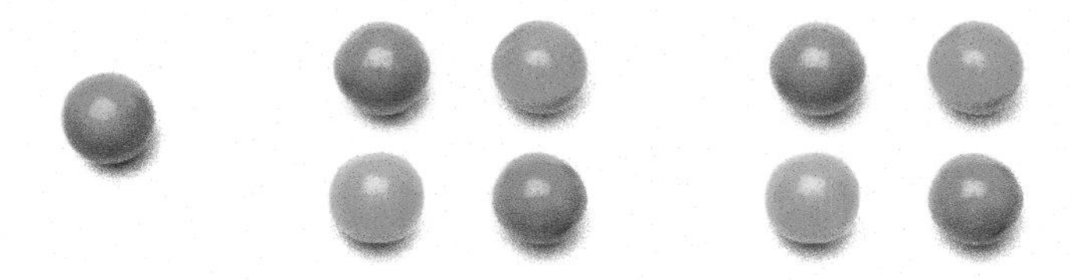

Look at all these big blocks and little blocks. We bring little blocks together to make chunks. These are the sizes of chunks we normally use for numbers. You can see the little 1-block. This is a block of 1. We call these single blocks "1s" ("ones").

You can see 10 little blocks in a line. These make up the 10-block. We will call this block a chunk of 10, or 10s ("tens").

You can see 100 little blocks (in a big flat chunk). These make up the 100-block. This is a chunk of 100.

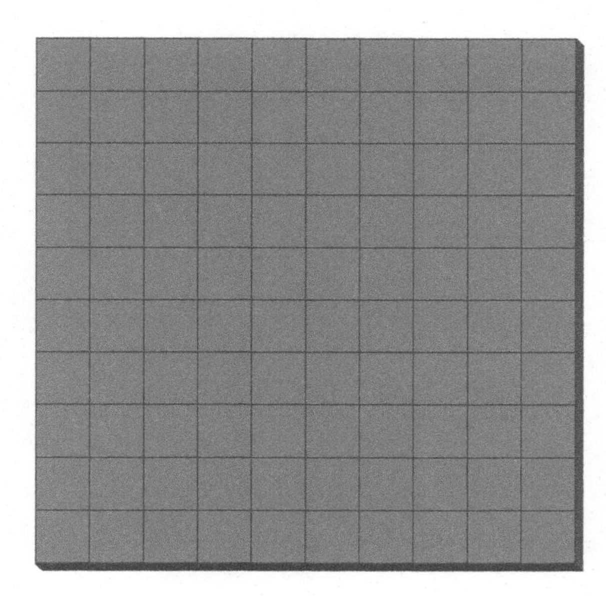

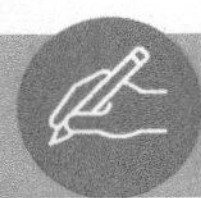

Student Exercises

How many chunks and blocks are in these sets? Count the chunks of 10. Write that number on the first line. Then count the number of singles. Write that number on the second line.

_______ 10s

_______ 1s

_______ 10s

_______ 1s

_______ 10s

_______ 1s

_______ 10s

_______ 1s

Find the sizes of these sets. First, count the single dots that are alone. Write the number in the blank called "singles." Then count the chunks (groups of dots that are sitting close to each other). If the exercise tells you to count chunks of 2, look for chunks of 2. If it says to look for chunks of 3, look for chunks of 3. Write the number of chunks in the blank called "chunks of . . ." Finish by counting all the dots. Write the number in the blank called "size."

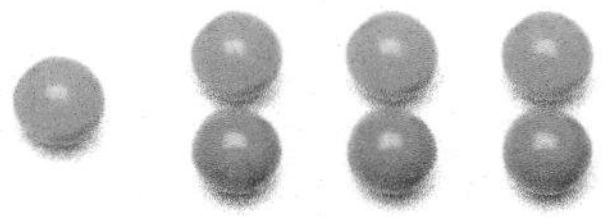

Singles: ________ Size:

Chunks of 2: ________ ________

Singles: ________ Size:

Chunks of 2: ________ ________

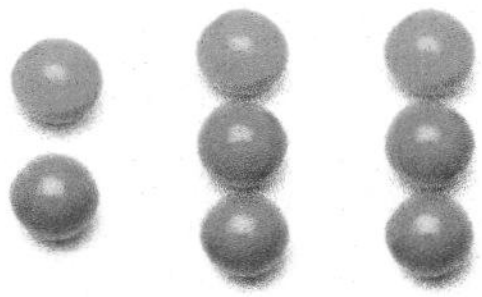

Singles: ________ Size:

Chunks of 3: ________ ________

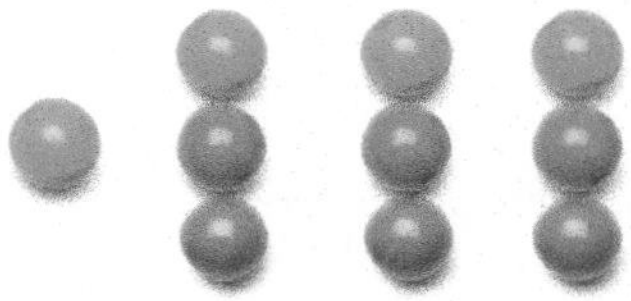

Singles: ________ Size:

Chunks of 3: ________ ________

Singles: ________ Size:

Chunks of 4: ________ ________

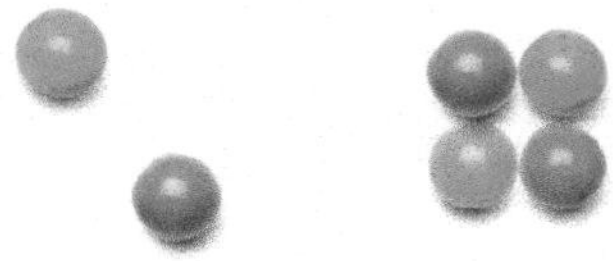

Singles: ________ Size:

Chunks of 4: ________ ________

DAY 51 Practice

Student Exercises

Make these sets! Draw dots or shapes to make each set. Then count the dots or shapes to figure out the size of the set. How many are there in the set?

1 single and 2 chunks of 2
1 single and 4 chunks of 2
2 singles and 1 chunk of 3
2 singles and 2 chunks of 3
3 singles and 1 chunk of 4
1 single and 2 chunks of 4
0 singles and 2 chunks of 5

Let's practice subtraction! Subtract the small number from the big number. Try using your imagination for the first one. Picture 4 of something in your mind. Take away 3 of them. How many are left? That's right! There is 1 left!

You can also use your little stones to do these exercises. Write the answer on the line for the first six exercises. Write the answers for the last six exercises under the line.

$4 - 3 =$ 1	$4 - 1 =$ ______
$4 - 0 =$ ______	$6 - 6 =$ ______
$6 - 4 =$ ______	$6 - 2 =$ ______

$\begin{array}{r} 5 \\ -\ 4 \\ \hline \end{array}$	$\begin{array}{r} 5 \\ -\ 1 \\ \hline \end{array}$	$\begin{array}{r} 5 \\ -\ 3 \\ \hline \end{array}$
$\begin{array}{r} 12 \\ -\ 2 \\ \hline \end{array}$	$\begin{array}{r} 12 \\ -\ 4 \\ \hline \end{array}$	$\begin{array}{r} 12 \\ -\ 7 \\ \hline \end{array}$

DAY 52 Counting with Blocks

This lesson introduces two-digit numbers in the decimal system using physical manipulatives. It is followed by two page of exercises, new and review. This will require about 30 minutes of instruction from the parent/teacher.

Highly Recommended: For this lesson and subsequent lessons, the teacher/parent is urged to consider purchasing a set of math manipulatives such as Decimal Blocks (available at generations.org) or something equivalent.

Prayer

Pray your own prayer of thanksgiving and praise to God. Pray for His help on this lesson.

Memory

Spend a few minutes with subtraction flash cards (1 - 0, 2 - 0 . . . series).

Lesson

Today we are going to use your colored blocks to learn more about numbers. We are going to use the little single 1-blocks and some 10-blocks. Remember, the 10-block is a chunk of 10 single blocks.

First, let's review the numbers we've learned so far:

1, 2, 3, 4, 5, 6, 7, 8, 9, 10, 11, 12 . . .

Now, let's use our blocks to make each of these numbers. Start with the number 1. Which block will you use for 1? Can you use chunks of 10 to make 1? No! Can we use a single block? Yes! The number 1 is made with 1 single block. Let's try 2. Use your blocks to make 2.

You must use singles to make numbers 1 through 9. You cannot use chunks of 10 (or the 10-block).

Now let's look at the number 10. Use your blocks to make 10. You can make 10 using single blocks. But you can also make 10 using a chunk of 10. Here's what 10 looks like:

Was it easier to use a chunk of 10 than to lay out 10 single blocks? Now let's make the number 11 using the blocks. What is 11 made of? Can you use any chunks of 10? Yes! You can use 1 chunk of 10. Then you must add something to number 10. What will you add? You must add another single block. 1 chunk of 10 and 1 single block makes 11.

Now, let's make 12 using the blocks. 12 is 1 chunk of 10 and 2 singles. Here's what it looks like:

Are you seeing a pattern now? We can find our next number by adding 1 more single. What do you think comes after 12? Hint: Start with 1 chunk of 10 and 3 singles. What number did you make? We call this number "13" ("thirteen").

After that comes 14, 15, 16, 17, 18, 19, and 20. Here's what 20 looks like:

We are now at 20. We use 2 chunks of 10 (2 of the 10-blocks). We can always add a single to find our next number. This is how you count up to really big numbers! You need bigger numbers to count the number of sheep in this herd.

Now, you are ready to learn about big numbers. Here is a big number:

24

What do we call this number? It's not "two-four." We call this number "twenty-four." Now, we will use blocks to show how it is done. Remember, each big block is a chunk of 10 little blocks. You can count them if you would like to. This is what 24 looks like:

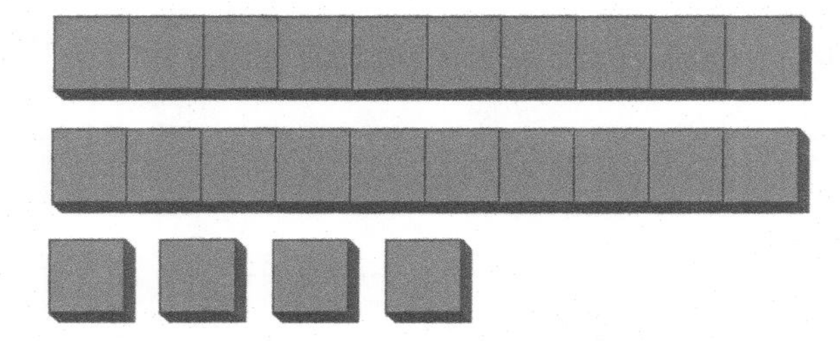

Here are 2 chunks of 10 and 4 singles who sit alone, all by themselves. When you see the number 24, you need to think of 2 chunks of 10, and 4 singles.

Here is how to say the number for 2 chunks of 10: Twenty! How would you say the number for 3 chunks of 10? Or 4 chunks of 10? That's thirty and forty. Then comes fifty, sixty, seventy, eighty, and ninety.

Let's try a few more numbers. When you see the number 35, you need to think of 3 chunks of 10, and 5 singles. What about 45, 62, and 70? How many chunks of 10 do you need to make up these numbers? How many singles?

45: ______ chunks of 10, and ______ singles

62: ______ chunks of 10, and ______ singles

70: ______ chunks of 10, and ______ singles

Do you notice another pattern? The first number tells us how many chunks of 10 we need. The second tells us how many singles we need.

You're doing great! Let's practice our numbers from 1 through 20 again.

See if you can count forwards from 1 to 20. Then count backwards from 20 down to 1!

Student Exercises

How many chunks of 10 do you need to make these numbers? How many singles (1s) do you need? Write the answers in the blanks.

43 → ________ 10s
→ ________ 1s

34 → ________ 10s
→ ________ 1s

25 → ________ 10s
→ ________ 1s

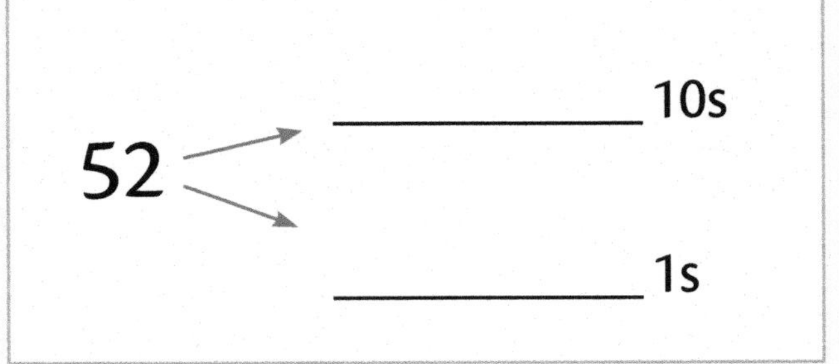

39 → ________ 10s
→ ________ 1s

93 → ________ 10s
→ ________ 1s

17 → ________ 10s
→ ________ 1s

71 → ________ 10s
→ ________ 1s

God made an order for bigger numbers too. Let's put these bigger numbers in order!

Sometimes you will have to count down (or backward) to find the right number. Sometimes you will have to count up (or forward) to find the number that comes next. Make chunks of 1s and 10s with your blocks if you need help.

24, __25__, __26__

______, 28, ______

______, ______, 32

______, 30, ______, ______

______, ______, 33, ______

______, ______, 37

37, ______, ______, ______

______, ______, ______, ______, 42

Practice DAY 53

Student Exercises

Let's start this practice lesson by counting to 20 together!
1 . . . 2 . . . 3 . . . 4 . . . 5 . . . 6 . . . 7 . . . 8 . . . 9 . . . 10 . . . 11 . . . 12 . . . 13 . . . 14 . . . 15 . . . 16 . . 17 . . . 18 . . . 19 . . . 20. And let's do it one more time!

How many chunks of 10 do you need to make these numbers? How many singles (1s) do you need? Write the answers in the blanks.

38 → ________ 10s → ________ 1s	44 → ________ 10s → ________ 1s
57 → ________ 10s → ________ 1s	80 → ________ 10s → ________ 1s
9 → ________ 10s → ________ 1s	99 → ________ 10s → ________ 1s
50 → ________ 10s → ________ 1s	2 → ________ 10s → ________ 1s

Sometimes you will have to count down (or backward) to find the right number.
Sometimes you will have to count up (or forward) to find the number that comes next.
Use your blocks if you need help.

67, 68, 69

______, 70, ______

______, ______, 74

______, 50, ______, ______

______, ______, 60, ______

______, ______, 72

60, ______, ______, ______

______, ______, ______, ______, 50

Now let's read Jesus' parable of the talents together:

> "For the kingdom of heaven is like a man traveling to a far country, who called his own servants and delivered his goods to them. And to one he gave five talents, to another two, and to another one, to each according to his own ability; and immediately he went on a journey. Then he who had received the five talents went and traded with them, and made another five talents. And likewise he who had received two gained two more also. But he who had received one went and dug in the ground, and hid his lord's money. After a long time the lord of those servants came and settled accounts with them. So he who had received five talents came and brought five other talents, saying, 'Lord, you delivered to me five talents; look, I have gained five more talents besides them.' His lord said to him, 'Well done, good and faithful servant; you were faithful over a few things, I will make you ruler over many things. Enter into the joy of your lord.' He also who had received two talents came and said, 'Lord, you delivered to me two talents; look, I have gained two more talents besides them.' His lord said to him, 'Well done, good and faithful servant; you have been faithful over a few things, I will make you ruler over many things. Enter into the joy of your lord.' Then he who had received the one talent came and said, 'Lord, I knew you to be a hard man, reaping where you have not sown, and gathering where you have not scattered seed. And I was afraid, and went and hid your talent in the ground. Look, there you have what is yours.' But his lord answered and said to him, 'You wicked and lazy servant, you knew that I reap where I have not sown, and gather where I have not scattered seed. So you ought to have deposited my money with the bankers, and at my coming I would have received back my own with interest. So take the talent from him, and give it to him who has ten talents." (Matthew 25:14-28)

Answer these questions. The first servant started with 5 talents. Then he added 5 more talents. What was the sum of his talents?

$$5 + 5 = \_\_\_\_\_\_$$

The second servant started with 2 talents. He added 2 more talents. What was the sum of his talents?

$$2 + 2 = \_\_\_\_\_\_$$

How many talents did the third servant have? That's right: 1. Do you remember how many talents the first servant had when his lord returned? That's right: 10! The master gave the third servant's talent to the first servant. How many talents did the first servant have at the end of the parable?

$$10 + 1 = \_\_\_\_\_\_$$

DAY 54 More Counting by 10s With Blocks

This lesson uses manipulatives to introduce numbers up to 100. The section includes two pages of exercises. This will require about 15 minutes of instruction from the parent/teacher.

Prayer

Pray your own prayer of thanksgiving and praise to God. Pray for His help on this lesson.

Memory

Spend a few minutes with subtraction flash cards (1 - 0, 2 - 0 . . . series).

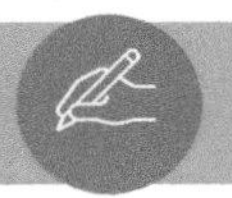

Lesson

Now, let's use our colored blocks to make the number 32. How many chunks of 10 do we need to make 32? How many 1-blocks (singles) do we need? We need 3 chunks of 10 and 2 singles.

Let's lay them out like this:

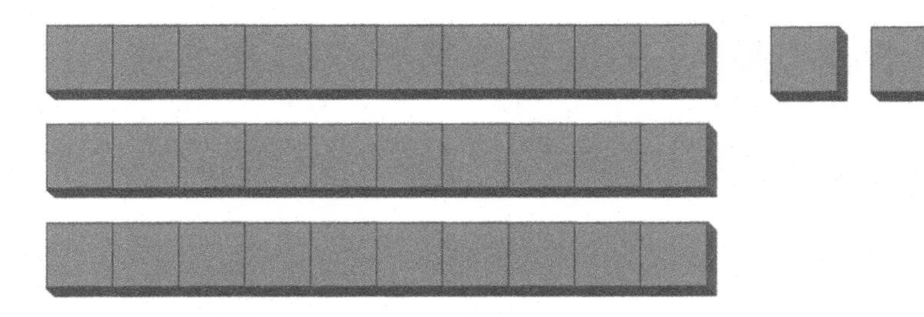

Now, take your colored blocks and lay out the following numbers:

34	21	83
75	30	36
66	48	50

Now let's use our blocks to make other big numbers. Remember, that 2 chunks of 10-blocks is "twenty," 3 chunks of 10 blocks is "thirty," 4 chunks of 10 blocks is "forty." Make these numbers and practice saying them three times: 20, 30, 40, 50, 60, 70, 80, 90, 100.

Try creating your own sets using the blocks, and have the child identify the number. For additional challenge, you may also use coins of 1 and 10 denominations for this exercise.

 Student Exercises

Try these addition exercises from memory. Are any giving you a hard time? Pray and ask God to teach you his number patterns. He is always listening to His children.

$4 + 4 =$ ________	$9 + 9 =$ ________
$8 + 8 =$ ________	$10 + 10 =$ ________
$5 + 5 =$ ________	$2 + 2 =$ ________
$1 + 1 =$ ________	$0 + 0 =$ ________

DAY 55 Practice

Student Exercises

Let's start this practice lesson by counting to 30 together!
1 . . . 2 . . . 3 . . . 4 . . . 5 . . . 6 . . . 7. . . 8 . . . 9 . . . 10 . . . 11 . . . 12 . . . 13 . . . 14 . . . 15 . . . 16 . . .
17 . . . 18 . . . 19 . . . 20 . . . 21 . . . 22 . . . 23 . . . 24 . . . 25 . . . 26 . . . 27 . . . 28 . . . 29 . . . 30.
And let's do it one more time!

Sometimes you will have to count down (or backward) to find the right number.
Sometimes you will have to count up (or forward) to find the number that comes next.
You may use your blocks if you need help.

90, ___91___, ___92___

_______, 80, _______

_______, _______, 70

_______, 60, _______, _______

_______, _______, 50, _______

_______, _______, 40

30, _______, _______, _______

_______, _______, _______, _______, 20

Try these addition exercises from memory. When we do exercises like these, we say that we are making doubles. Can you figure out what we mean by that? Maybe your parent or teacher can help!

$$\begin{array}{r} 10 \\ +\ 10 \\ \hline \end{array}$$

$$\begin{array}{r} 1 \\ +\ 1 \\ \hline \end{array}$$

$$\begin{array}{r} 7 \\ +\ 7 \\ \hline \end{array}$$

$$\begin{array}{r} 2 \\ +\ 2 \\ \hline \end{array}$$

$$\begin{array}{r} 8 \\ +\ 8 \\ \hline \end{array}$$

$$\begin{array}{r} 0 \\ +\ 0 \\ \hline \end{array}$$

$$\begin{array}{r} 9 \\ +\ 9 \\ \hline \end{array}$$

$$\begin{array}{r} 3 \\ +\ 3 \\ \hline \end{array}$$

$$\begin{array}{r} 5 \\ +\ 5 \\ \hline \end{array}$$

DAY 56 Bigger Numbers in the Bible

This lesson explores how Jesus used numbers to quantify the greatness of His miracles. It is followed by one page of review exercises. This will require about 15 minutes of instruction from the parent/teacher.

Prayer

Pray your own prayer of thanksgiving and praise to God. Pray for His help on this lesson.

Memory

Spend a few minutes with subtraction flash cards (1 - 0, 2 - 0 . . . series).

Let's read these Bible verses together. They talk about two of Jesus' miracles:

> And He left them, and getting into the boat again, departed to the other side. Now the disciples had forgotten to take bread, and they did not have more than one loaf with them in the boat. Then He charged them, saying, "Take heed, beware of the leaven of the Pharisees and the leaven of Herod." And they reasoned among themselves, saying, "It is because we have no bread." But Jesus, being aware of it, said to them, "Why do you reason because you have no bread? Do you not yet perceive nor understand? Is your heart still hardened? Having eyes, do you not see? And having ears, do you not hear? And do you not remember? When I broke the five loaves for the five thousand, how many baskets full of fragments did you take up?" They said to Him, "Twelve." "Also, when I broke the seven for the four thousand, how many large baskets full of fragments did you take up?" And they said, "Seven." So He said to them, "How is it you do not understand?" (Mark 8:13-21)

Jesus fed a lot of people with just a little bit of bread and fish. We know He did this at least two times. He fed 5,000 people in the first miracle. Of the bread, He used only 5 loaves. Then, He fed 4,000 in a second miracle. He used just 7 loaves of bread. They even had extra food left over after everybody had finished eating!

How many baskets of leftovers did they have the first time? __________

How many baskets of leftovers did they have the second time? __________

How many baskets of leftovers did they take up? _______ + _______ = _______

Do you know why the disciples were so upset? They forgot to bring food with them on their boating trip.

Student Exercises

Let's start this practice lesson by counting to 30 together!
1 . . . 2 . . . 3 . . . 4 . . . 5 . . . 6 . . . 7. . . 8 . . . 9 . . . 10 . . . 11 . . . 12 . . . 13 . . . 14 . . . 15 . . . 16 . . .
17 . . . 18 . . . 19 . . . 20 . . . 21 . . . 22 . . . 23 . . . 24 . . . 25 . . . 26 . . . 27 . . . 28 . . . 29 . . . 30.
And let's do it one more time!

How many chunks of 10 do you need to make these numbers? How many singles (1s) do you need? Write the answers in the blanks.

43 → ________ 10s
→ ________ 1s

68 → ________ 10s
→ ________ 1s

25 → ________ 10s
→ ________ 1s

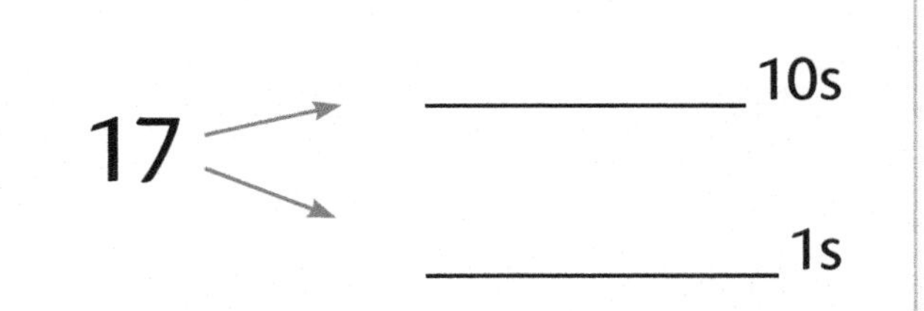

39 → ________ 10s
→ ________ 1s

95 → ________ 10s
→ ________ 1s

52 → ________ 10s
→ ________ 1s

75 → ________ 10s
→ ________ 1s

DAY 57 Adding & Subtracting Bigger Numbers

This lesson will explore addition and subtraction in double digits. The child may use manipulatives, coins, or other objects to work these exercises. The section includes two pages of exercises. This will require about 30 minutes of instruction from the parent/teacher.

Prayer

Pray your own prayer of thanksgiving and praise to God. Pray for His help on this lesson.

Memory

Spend a few minutes with subtraction flash cards (1 - 1, 2 - 1 . . . series).

Lesson

Now that you have learned about big numbers, it's time to use them for adding and subtracting!

Let's add 9 and 6, using our blocks.

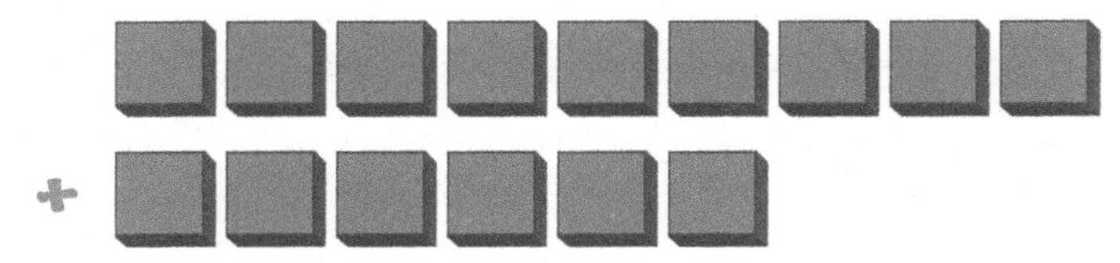

This answer is going to be a big number—bigger than 10. How many blocks do we have to add to the set of 9 single blocks to make a chunk of 10? Can we move one of the singles from the set of 6 over to the set of 9 single blocks? Now we have one chunk of 10, and 5 singles. That's 15!

How many chunks of 10 (10-blocks) are there in this set? Just 1. How many single blocks are there in the set? There are 5! We still have the same number of blocks that we started with. But now we have one chunk of 10 and 5 singles. That's 15 altogether! This means 9 + 6 = 15.

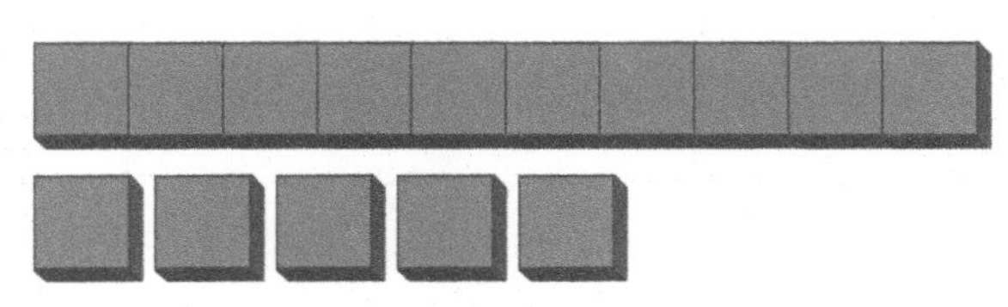

Let's try another addition exercise: 11 + 3. First we will lay out the set for the number 11 as 1 chunk of 10 (a 10-block) and 1 single. That makes 11. Then, we will add 3 single 1s, for a set of 3.

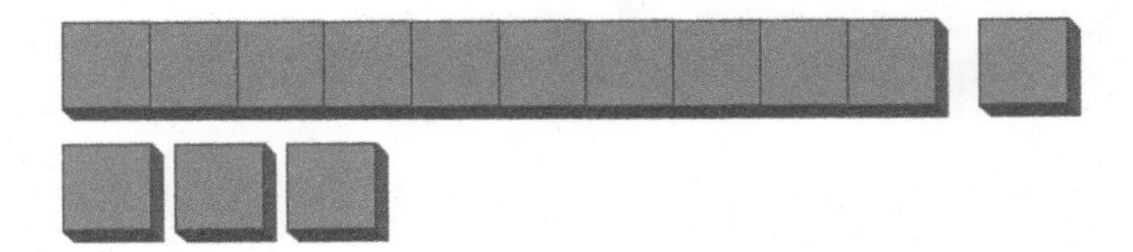

Now let's rearrange the blocks. Let's put all the single 1s together. Now, we have 1 chunk of 10, and 4 single 1s. That's 14 altogether! This means that 11 + 3 = 14. Look carefully at the blocks above. We can make another equation from this: 10 + 4 = 14.

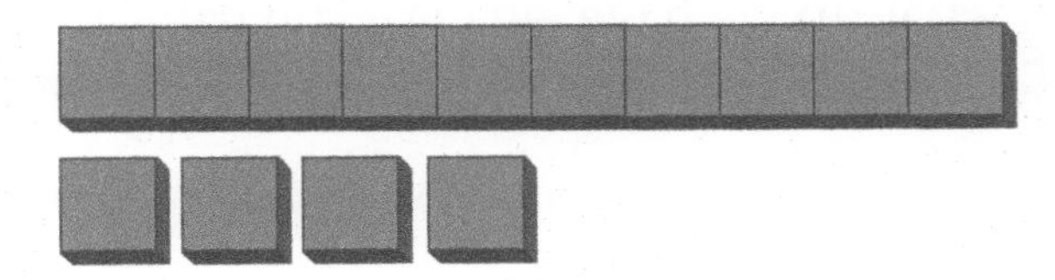

Now let's try a subtraction exercise, using your blocks: What is 12 - 5?

This time, we start out with 12 blocks. We need to take 5 blocks away. How many blocks will be left? Let's find out.

Look at your 12 blocks. How many single 1-blocks can you take away from the set of 12? That's right: 2! Go ahead and remove the 2 single blocks. Now how many more do you need to take away? That's right. You need to take away 3 more singles from the chunk of 10. (You can cover up 3 of them using your hand.) Now you have taken away 5 blocks. How many are left? That's right: 7! The answer is: 12 – 5 = 7.

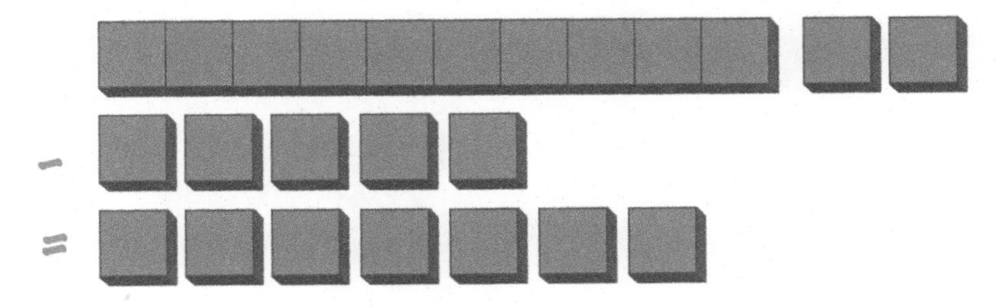

Here is one more way to do this subtraction exercise. Break off 5 blocks from the set of 12 in this picture, and how many are left? That's right! 7!

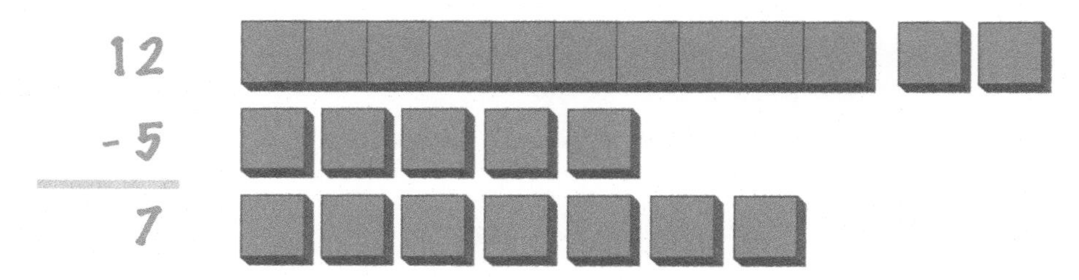

Sometimes, when we subtract from bigger numbers, we don't have to take away any singles from the chunk of 10. Use your blocks for this exercise: 16 - 5. Start out with 1 chunk of 10 and 6 singles. Take away 5 singles, and what do we have left? We still have 1 chunk of 10 left, and 1 single. This means that 16 - 5 = 11.

Try a few more on your own using your blocks, coins, or stones.

12 + 6	13 − 5
11 + 3	9 + 6
16 − 9	14 − 8
11 − 6	10 + 5

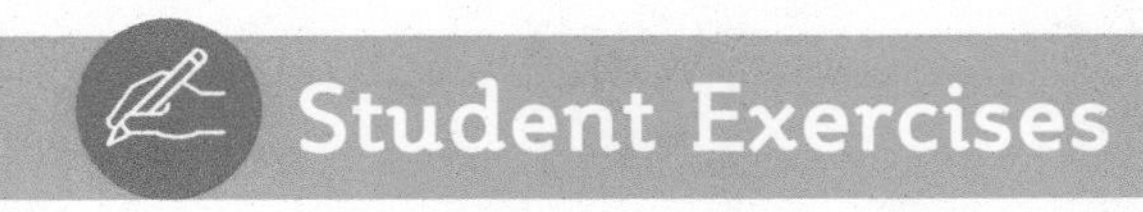

Use your blocks or coins to add or subtract these numbers.

8 + 5 ___	6 + 9 ___	7 + 7 ___
17 − 8 ___	15 − 8 ___	16 − 8 ___
12 − 4 ___	14 − 7 ___	14 − 5 ___

Each exercise starts with two sets of numbers. Compare these sets. Which numbers do they have in common? Write these numbers in the space provided. This makes a new set of numbers!

Sets {3, 4} and {4, 5} share = {__4__}

Sets {1, 2, 3} and {2, 3, 4} share = {______}

Sets {4, 6, 8} and {8, 10, 12} share = {______}

Sets {19, 20, 21} and {17, 18, 19} share = {______}

Sets {38, 39, 40} and {40, 41, 42} share = {______}

Sets {59, 60, 61} and {89, 90, 91} share = {______}

Practice DAY 58

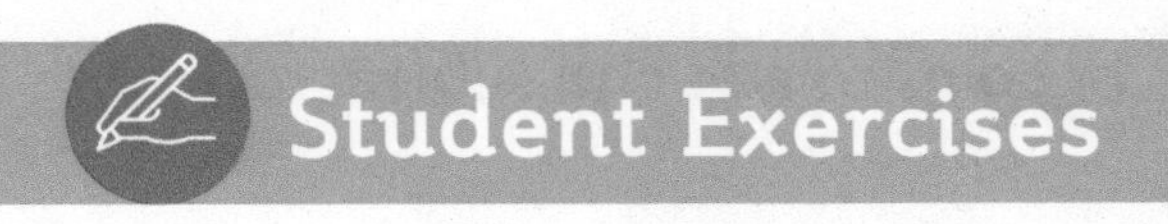

Let's add and subtract blocks. Use your own blocks if you would like. Write the numbers in the boxes.

____

+ ____ ?

= 13

____

+ ____

= ____ ?

____

+ ____

= ____ ?

____

+ ____

= ____ ?

____

+ ____

= ____ ?

Let's practice counting to 40! Then use your blocks or coins to add and subtract. Work on memorizing each of the exercises as you go.

Let's count some big numbers today! Let's start our exercises by counting to 40 together! 1 . . . 2 . . . 3 . . . 4 . . . 5 . . . 6 . . . 7 . . . 8 . . . 9 . . . 10 . . . 11 . . . 12 . . . 13 . . . 14 . . . 15 . . . 16 . . . 17 . . . 18 . . . 19 . . . 20 . . . 21 . . . 22 . . . 23 . . . 24 . . . 25 . . . 26 . . . 27 . . . 28 . . . 29 . . . 30 . . . 31 . . . 32 . . . 33 . . . 34 . . . 35 . . . 36 . . . 37 . . . 38 . . . 39 . . . 40! And let's do it one more time!

$14 + 2 = 16$	$14 + 4 =$ ___	$14 + 6 =$ ___
$6 + 5 =$ ___	$6 + 7 =$ ___	$6 + 9 =$ ___
$15 - 8 =$ ___	$15 - 9 =$ ___	$15 - 10 =$ ___
$13 - 5 =$ ___	$13 - 6 =$ ___	$13 - 7 =$ ___

Numbers in Really Big Groups DAY 59

This lesson explores addition and subtraction with higher two-digit numbers, and is followed by three pages of new and review exercises. This will require about 20 minutes of instruction from the parent/teacher.

Prayer

Pray your own prayer of thanksgiving and praise to God. Pray for His help on this lesson.

Memory

Spend a few minutes with subtraction flash cards (1 -1, 2 - 1 . . . series).

Lesson

Here is a number line. Can you fill in the rest of the numbers on this line? You can add and subtract numbers on a number line.

45 46 47 48 49

What is 45 + 4? Put your pencil on the number 45. Count up 4 places. What number did you stop at?

What is 49 - 5? Put your pencil on 49. Count down 5 places. What number did you stop at?

What is 42 + 9? Put your pencil on 42. Count up 9 places. What number did you stop at?

Student Exercises

First, draw a line to connect the dots below. Then fill in the rest of the numbers on the number line. Start with 66.

Use this number line to do the exercises. Remember, you will count up (to the right) for addition. You will count backwards (to the left) for subtraction.

66 ____ ____ ____ ____ ____ ____ ____ ____ ____

66 + 4 = 70	66 + 6 = ____	66 + 8 = ____
74 – 5 = ____	74 – 6 = ____	74 – 7 = ____
68 + 0 = ____	70 + 0 = ____	72 + 0 = ____

Let's practice counting to 40! Then use your blocks or coins to add and subtract. Work on memorizing each of the exercises as you go.

Let's count some big numbers today! Let's start our exercises by counting to 40 together! 1 . . . 2 . . . 3 . . . 4 . . . 5 . . . 6 . . . 7 . . . 8 . . . 9 . . . 10 . . . 11 . . . 12 . . . 13 . . . 14 . . . 15 . . . 16 . . . 17 . . . 18 . . . 19 . . . 20 . . . 21 . . . 22 . . . 23 . . . 24 . . . 25 . . . 26 . . . 27 . . . 28 . . . 29 . . . 30 . . . 31 . . . 32 . . . 33 . . . 34 . . . 35 . . . 36 . . . 37 . . . 38 . . . 39 . . . 40! And let's do it one more time!

$4 + 7 = 11$	$5 + 7 = $ ___	$9 + 5 = $ ___
$4 + 8 = $ ___	$5 + 9 = $ ___	$9 + 7 = $ ___
$14 - 5 = $ ___	$13 - 6 = $ ___	$16 - 7 = $ ___
$14 - 7 = $ ___	$13 - 6 = $ ___	$16 - 8 = $ ___

DAY 60 Practice

Student Exercises

Fill in the rest of the numbers on the number line. Start with 36. You will need to count backwards. Then use the number line to do the exercises.

Remember, you will count up (to the right) for addition. You will count backwards (to the left) for subtraction.

27 ____ ____ ____ ____ ____ ____ ____ ____ 36

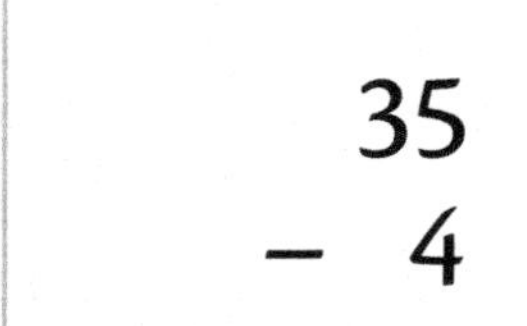

$35 - 4 = 31$	$35 - 5 =$ ____	$35 - 6 =$ ____
$28 + 3 =$ ____	$28 + 4 =$ ____	$30 - 3 =$ ____
$33 - 5 =$ ____	$33 - 6 =$ ____	$36 - 7 =$ ____

Use the pictures to make your own subtraction equations. How many members do you see in each set? There are 5 ducks in the first set. How many would you like to take away (or subtract)? Write your equation. Then make up your own story!

5 – 4 = 1

"When the ducklings grow up, momma duck will be all alone!"

_____ – _____ = _____

_____ – _____ = _____

_____ – _____ = _____

_____ – _____ = _____

DAY 61 Dividing Things Up into Parts

This lesson introduces fractions. It is followed by 2 pages of exercises. This will require about 20 minutes of instruction from the parent/teacher.

Prayer

Pray your own prayer of thanksgiving and praise to God. Pray for His help on this lesson.

Memory

Spend a few minutes with subtraction flash cards (1 - 1, 2 - 1 . . . series).

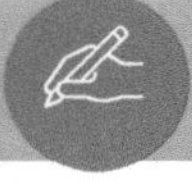

Lesson

Have you ever looked up into the sky at night and seen half a moon shining? We call this a "half moon." God made the moon look like a whole pie sometimes, and a half pie at other times. The moon looks this way because the sun is shining on it sideways, and the shady half is too dark for us to see.

Look at this glass. It is half full of water.

Here is half an apple and half a watermelon!

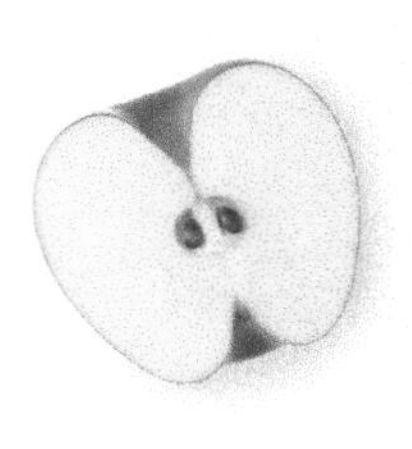

Do you see two **halves** here? We call these same-sized parts. Put the 2 halves together and you get one whole! One half and one half make one whole thing!

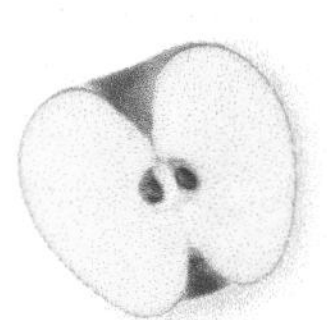

Circle all the 2 halves that make a whole in these pictures:

  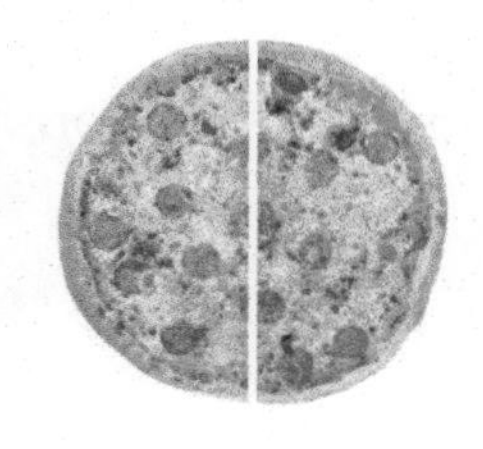  

Let's divide these shapes into 2 same-sized parts. These same-sized parts are 2 halves. Can you draw a line and pretend to cut these shapes in 2 same-sized pieces?

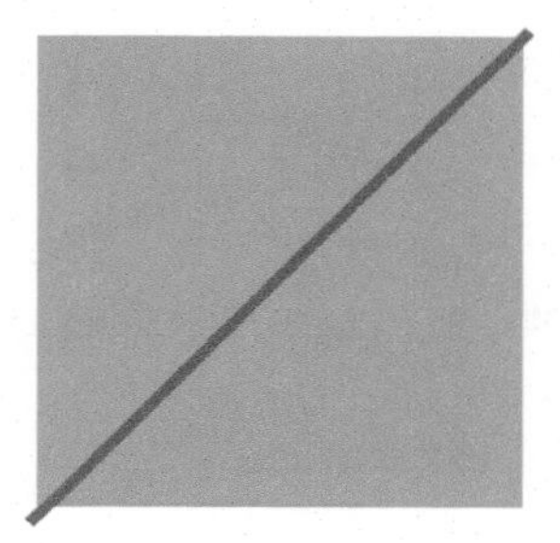  

Now let's divide each of these pies into 3 pieces! Which of these two pies were cut so that everybody gets the same-sized piece of pie?

If the pie is cut into 3 same-sized pieces, we call the pieces **thirds**. If a pie is cut into 4 same-sized pieces, we call the pieces **fourths**.

Student Exercises

Color 1/2 (half) of each of these shapes with a different color: yellow, green, red, blue, or orange. Leave the other half of each shape blank.

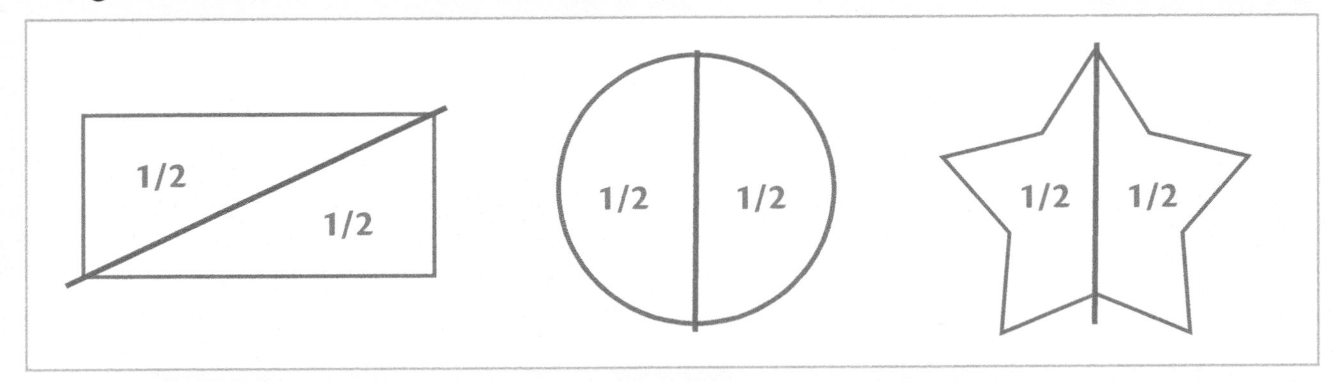

Color 1/3 of each of these shapes with a color. Choose from yellow, green, red, blue, or purple. Leave the other two spaces of each shape blank.

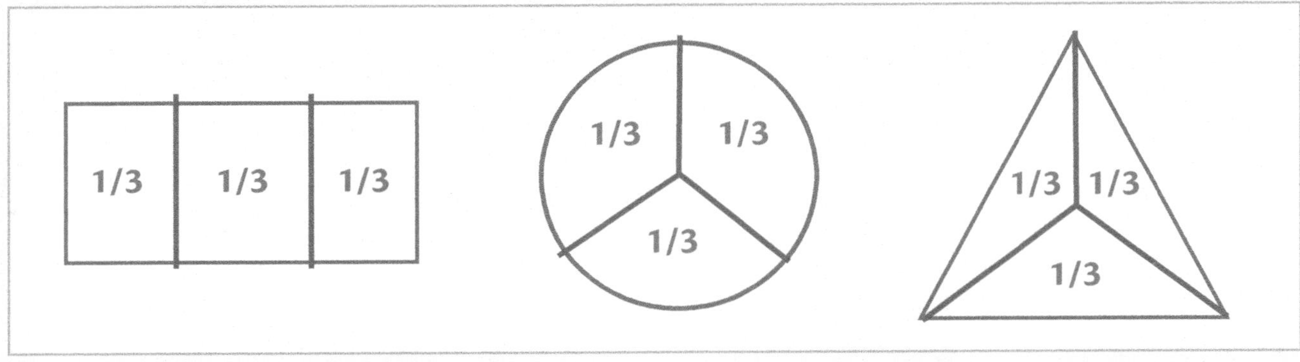

Now, color 1 of the halves in the first shape below. Color 2 of the fourths in the next two shapes. Be sure to color two of the fourths that sit right next to each other. We call this 2/4. Can you say "two fourths?" Does 2/4 look like 1/2?

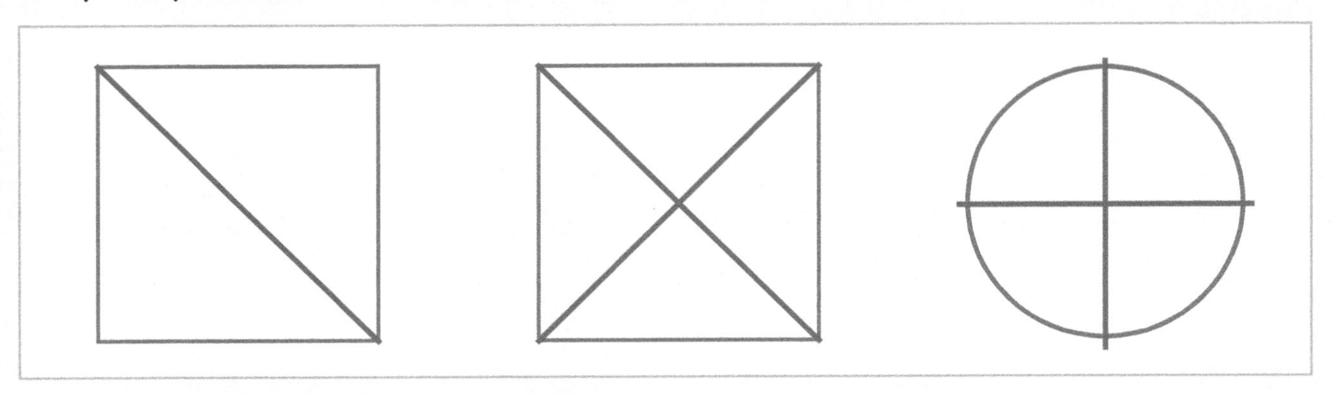

Addition makes bigger numbers. Subtraction makes smaller numbers. God made numbers to work together. Sometimes numbers work together to make bigger numbers. Sometimes numbers work together to make smaller numbers. For these exercises, you need to decide how these numbers work together. How do 1 and 2 work together to make 3? Of course, they must be added! Write "+" or "-" in each circle to show how they work together.

1 (+) 2 = 3	3 ◯ 2 = 5
4 ◯ 2 = 2	2 ◯ 2 = 0
5 ◯ 3 = 2	3 ◯ 2 = 5
7 ◯ 4 = 11	11 ◯ 4 = 7
2 ◯ 5 = 7	7 ◯ 2 = 9

DAY 62 Practice

Student Exercises

Use your blocks or coins to practice this addition and subtraction. Work on memorizing each of the exercises as you go.

$$\begin{array}{r} 10 \\ +\ 3 \\ \hline 13 \end{array} \qquad \begin{array}{r} 8 \\ +\ 7 \\ \hline \end{array} \qquad \begin{array}{r} 8 \\ +\ 4 \\ \hline \end{array}$$

$$\begin{array}{r} 12 \\ +\ 3 \\ \hline \end{array} \qquad \begin{array}{r} 7 \\ +\ 6 \\ \hline \end{array} \qquad \begin{array}{r} 8 \\ +\ 6 \\ \hline \end{array}$$

$$\begin{array}{r} 18 \\ -\ 6 \\ \hline \end{array} \qquad \begin{array}{r} 15 \\ -\ 7 \\ \hline \end{array} \qquad \begin{array}{r} 13 \\ -\ 5 \\ \hline \end{array}$$

$$\begin{array}{r} 18 \\ -\ 8 \\ \hline \end{array} \qquad \begin{array}{r} 15 \\ -\ 4 \\ \hline \end{array} \qquad \begin{array}{r} 13 \\ -\ 8 \\ \hline \end{array}$$

Pretend that your parents have given you the things in the picture. They ask you to share with a new friend from church. Below, make two equal parts — one for you and one for the other child. Draw a circle around your share and another circle around the other child's share. We call each share 1/2 or one half of the full amount. Can you say "one half?"

How many candies do you see all together? ________

Write the number of candies each person gets below.

Your name ______________________________ ________

Friend's name ___________________________ ________

How many cookies do you see all together? ________

Write the number of cookies each person gets below.

Your name ______________________________ ________

Friend's name ___________________________ ________

How many pencils do you see all together? ________

Write the number of pencils each person gets below.

Your name ______________________________ ________

Friend's name ___________________________ ________

Cut this pie into 4 equal pieces for you and three other children. Everybody gets the same-sized piece. Everybody gets 1/4 or one quarter of the pie. Can you say "one-fourth?" That's a big piece of pie!

DAY 63 Go Help with Bigger Numbers!

This lesson integrates math into everyday life. This is an essential element to learning. The child is encouraged to apply God's patterns and wisdom to life in the home and community. Take a break from memory work and academic exercises, and identify ways to make math part of your everyday life. The following are suggestions or examples, but other ideas may be added to the list.

> If the ax is dull . . .
> Then [one] must use more strength;
> But wisdom brings success. (Ecclesiastes 10:10)

The Bible talks a lot about wisdom. Here it tells us that wisdom helps us to work. It helps us to work well. It helps us to get work done. What happens if you try to cut wood with a dull ax? You have to work very hard! It's easier to cut wood with a sharp ax. It is wiser too! You are learning a lot about math. Now you have learned about bigger numbers. You have more wisdom. You have new math skills to help you serve God better. You can serve others more and do it well. Here are some more ways you can use math in everyday life! Choose one of the following activities and use your skills to serve others and glorify God.

1. **Counting eggs.** Eggs usually come in cartons of 12 (or 10). How many eggs does your family eat in a week? Does your family usually buy sets of 10 eggs or 12? Use that number to figure out how many cartons of eggs you need each week. You can use your blocks or coins to help you.
2. **Packages of drinks.** Other foods and drinks (like bottled water) come in chunks too. Sometimes they come in sets of 12. Sometimes they come in sets of 24. Sometimes they come in sets of 36! How many bottles or cans of a certain drink does your family need in a week? Can you figure out how many packages of drinks you need to buy to give you the right number? Add when you can. Count if you need to.
3. **How much do you weigh?** How much does your brother or sister weigh? How much do you weigh? Weigh each other once a month for three months. Use subtraction to find out how much you have grown in three months. Use subtraction to find out how much your brother or sister has grown. Your father and mother want to make sure that you are healthy. This is one way to watch your health.

4. Bathroom supplies. Usually, Mom wants to keep extra supplies in the bathroom. You can be in charge of keeping everything supplied. Mom might want five extra rolls of toilet paper in each bathroom. How many rolls will you need in the house? How many rolls of toilet paper do you use in a month? Can you keep track of this too?

5. Too many clothes. Does your family have extra clothes they do not use? Ask your family if there are extra clothes they do not wear any more. You could collect these. You could share them with the poor.

6. List five reasons why you are thankful for math. Or list five ways you can use math. Write them out here (or draw pictures)!

Our first adventure
in measuring God's
world.

CHAPTER 6

Long, Longer, Longest

Introduction

You are halfway through your first math book! Praise God! You are learning so many good things. Hopefully, you are learning to like math. It is a gift of God.

We use math to help us get things done. You've already learned that numbers tell us the sizes of sets. Numbers can also be used for measuring things!

God likes to measure things. That's the way He has made this world. God measures the oceans in His hands! He knows how much water is in the ocean!

> [The Lord] has measured the waters in the hollow of his hand
> and marked off the heavens with a span,
> enclosed the dust of the earth in a measure
> and weighed the mountains in scales
> and the hills in a balance. (Isaiah 40:12, ESV)

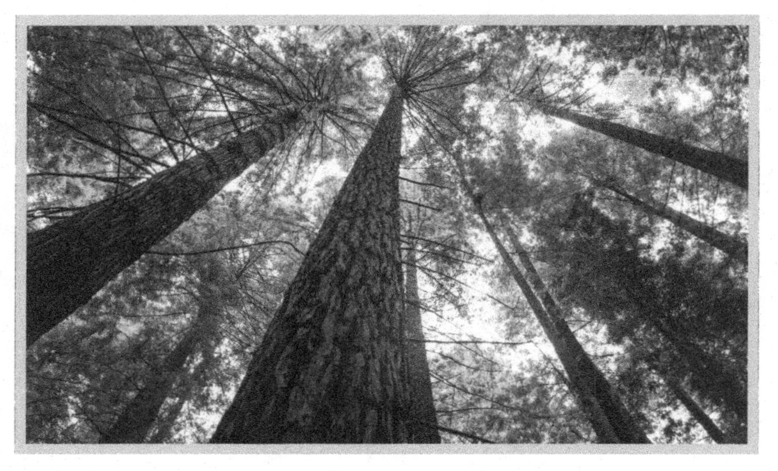

God wants *us* to measure things too. We measure tall things and short things. We measure heavy things and light things. We measure when we count apples at the store. We can weigh the apples on scales too. God doesn't want us to make mistakes when we measure things. He doesn't want us to cheat. He wants us to be especially honest when we weigh out apples at the store.

> "You shall have a perfect and just weight, a perfect and just measure." (Deuteronomy 25:15)

Measuring is serious work. But measuring is fun too! Let's start measuring things!

God's Straight World DAY 64

This lesson introduces measurements in the real world, and is followed by one page of review exercises. This will require about 20 minutes of instruction from the parent/teacher.

Prayer

Our Father in Heaven, thank You for helping me to learn. You have given us good tools for learning. They help us everyday. Thank You for our math tools. Show us how we can take good care of the things You give us. Amen.

Memory

Spend a few minutes with subtraction flash cards (only exercises whose answer is 6 or less).

Lesson

How long is your pencil? How long is that hotdog? How far is it from the chair to the wall? These are called **lengths**. Let's learn how to measure lengths first!

Hold your hand up in front of you. Which finger is the **longest**? Which is the **shortest**? Is your pinky the same length as your thumb?

People around the world use two different ways to measure lengths. Their rulers (measuring sticks) have two ways to measure. Look at the ruler in the picture. Do you see the large numbers? The big 1, 2, and 3 show us **inches**. The little 1, 2, and 3 show us **centimeters**. One inch is longer than one centimeter. People in China usually use centimeters to measure things. People in America usually use inches to measure things.

How many centimeters do you see on the ruler? How many inches do you see? Are there more centimeters than inches? Yes! That's because a centimeter is smaller than an inch and more of them fit on a ruler.

Activity

Start measuring your world! Find a ruler. Measure the length of things in the room. Your parent/teacher can help you with this. You might find things that don't line up exactly at a number on your ruler. Just use the closest number if this happens.

1. Let's start with inches. Find things in your room or in the house that are about 1 inch long. Then find things that are about 2 inches long. Then find things that are about 3 inches long. Keep measuring until you've found some things that are 12 inches long.
2. Next, measure things using centimeters. Can you find things that are 5 centimeters long? What about 10 centimeters? Can you find things that are 15 cm? Keep measuring until you've found some things that are 20 centimeters long.
3. Find a set of three things that are all the same length. Find another set of five things that are the same length.
4. Pick one thing like a box or a book. Find three edges where you can measure different lengths.

You have just used numbers to measure things! This is how we compare the lengths of things.

We cannot measure everything God has made. This world is just too big. That's what God says in Job 38:4-5:

"Where were you when I laid the foundations of the earth?
Tell Me, if you have understanding.
Who determined its measurements?
Surely you know!
Or who stretched the [ruler] upon it?"

The answer is clear: God!

Student Exercises

Let's measure these to the nearest inch! Write your answer on the blank line.

Parent/Teacher Note: You may use centimeters if so desired.

Length ___________

Length ___________

Length ___________

Length ___________

Length ___________

DAY 65 Practice

Student Exercises

Try to do these addition and subtraction exercises from memory. You may also use your blocks, stones, or coins if you need help. This will help you get better at adding and subtracting!

$\begin{array}{r} 5 \\ +\ 4 \\ \hline \end{array}$	$\begin{array}{r} 3 \\ +\ 0 \\ \hline \end{array}$	$\begin{array}{r} 7 \\ +\ 4 \\ \hline \end{array}$
$\begin{array}{r} 7 \\ -\ 2 \\ \hline \end{array}$	$\begin{array}{r} 12 \\ -\ \ 5 \\ \hline \end{array}$	$\begin{array}{r} 8 \\ -\ 6 \\ \hline \end{array}$
$\begin{array}{r} 8 \\ +\ 5 \\ \hline \end{array}$	$\begin{array}{r} 10 \\ +\ \ 6 \\ \hline \end{array}$	$\begin{array}{r} 6 \\ +\ 9 \\ \hline \end{array}$
$\begin{array}{r} 14 \\ -\ \ 7 \\ \hline \end{array}$	$\begin{array}{r} 15 \\ -\ \ 5 \\ \hline \end{array}$	$\begin{array}{r} 16 \\ -\ \ 8 \\ \hline \end{array}$

God made many numbers. The numbers in this exercise are just a few of God's numbers! Sometimes you will have to count down (or backward) to find the missing number. Sometimes you will have to count up (or forward) to find the number that comes next. Use singles and chunks of 10 from your colored blocks if you need help.

25, __26__, __27__

______, 35, ______

______, ______, 40

______, 50, ______, ______

______, ______, 55, ______

______, ______, 60

70, ______, ______, ______

______, ______, ______, ______, 80

DAY 66 God's Curvy World

This lesson measures rounded objects in God's world, and is followed by an activity and one page of review exercises. This will require about 20 minutes of instruction from the parent/teacher.

Prayer

Pray your own prayer of thanksgiving and praise to God. Pray for His help on this lesson.

Memory

Spend a few minutes with subtraction flash cards (only exercises whose answer is 6 or less).

Lesson

Let's keep measuring things in God's great big world! In our last lesson, we measured **straight** things. This time we will measure **curvy** things.

You can use a string or a soft sewing tape measure to measure curvy things. If you use a string, your parent/teacher will show you how to do it.

God made the earth round. Have you ever wondered how far it is around the whole earth? You would need a really big tape measure to figure it out. The earth is 25,000 miles (40,000 km) around at the equator (its middle)! If you were flying in a jet, it would take you 50 hours to travel around the world.

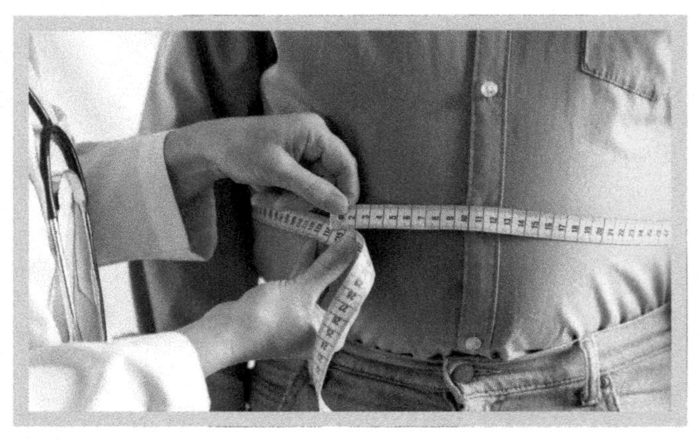

Activity

Let's have some fun measuring God's curvy world!

Measure your wrist. Wrap the string or soft tape measure around your wrist. Pinch and hold the place where the string or tape measure meets its beginning. If you use a tape measure, find the closest number to your pinched place. If you use a string, you will need to place it over a ruler to see what number your pinched place is closest to. This number is about how big your wrist is.

Measure your chest. Measure your waist. Measure your neck. This is how your parents can figure out what size you need when they buy shirts and pants for you.

Measure your head. This will tell you your hat size.

You might also want to measure other curved things like:

- Cups, jars, and bottles
- Musical instruments such as a violin
- Other things around the house such as a candle holder
- Toys such as round blocks

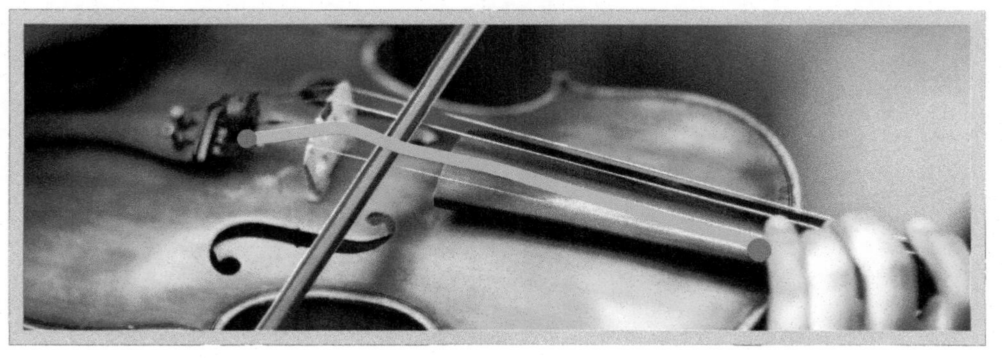

Student Exercises

Look at these sets. Circle the one shape that is different from the others in each set. Can you explain why it is different? One of these exercises has two answers. God makes things different in different ways.

Discuss the group with the red circle, green square, and red square. What is it that makes each of these look different from the others in the group?

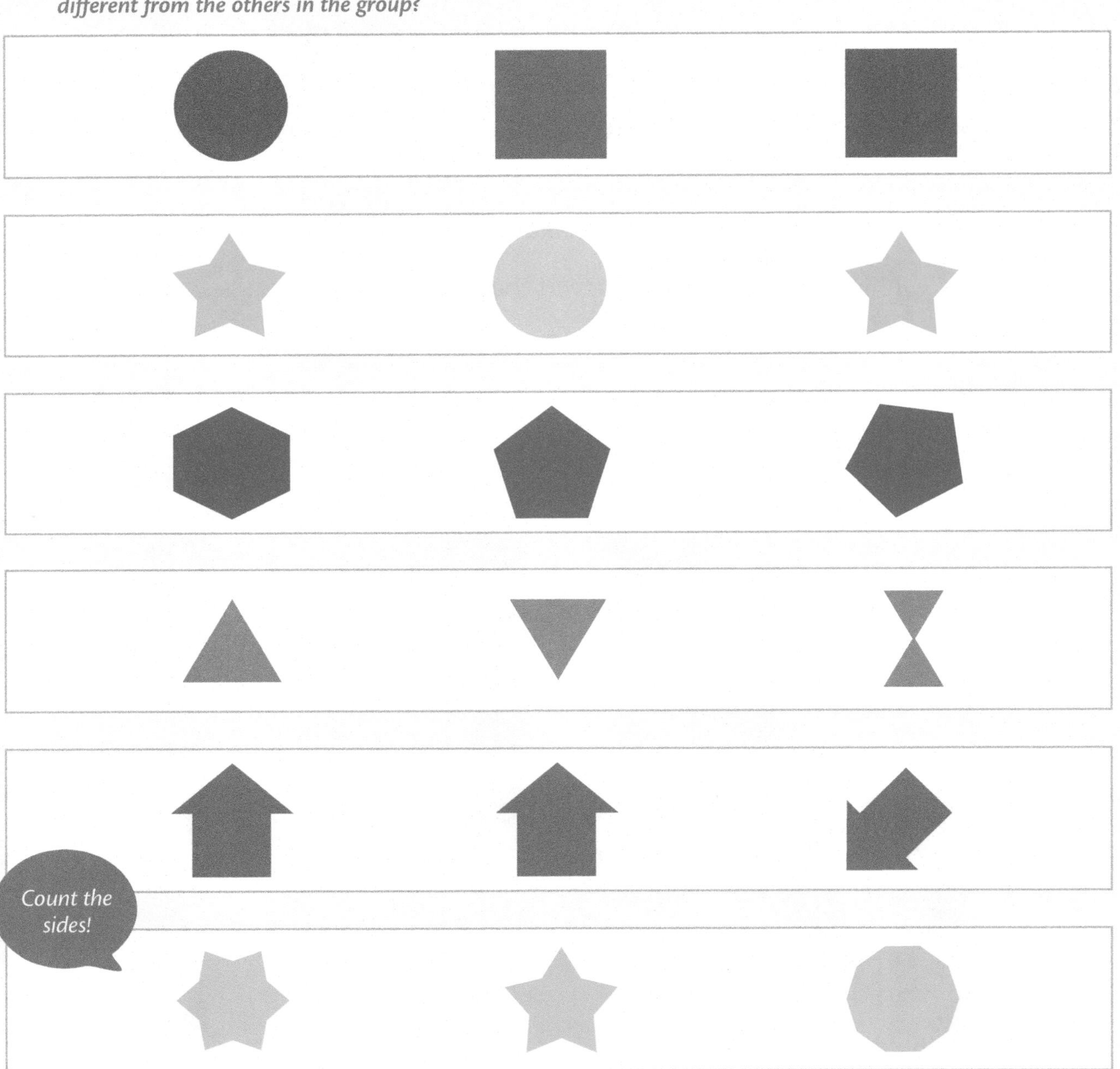

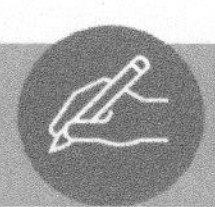

Student Exercises

How are these numbers working together? Are they using addition or subtraction? Remember, addition makes bigger numbers. Subtraction makes smaller numbers.

Decide how these numbers are working together. How do 6 and 6 work together to make 12? Of course, they must be added! Write "+" or "-" in each circle to show how they work together.

6 (+) 6 = 12	12 ○ 1 = 11
7 ○ 3 = 10	11 ○ 2 = 9
10 ○ 2 = 8	2 ○ 5 = 7
9 ○ 3 = 6	4 ○ 1 = 5
2 ○ 2 = 4	7 ○ 4 = 3

It's fun making doubles. It's a great pattern! There is just something nice about it! Can you figure out these addition exercises from memory?

$\begin{array}{r} 10 \\ +\ 10 \\ \hline \end{array}$	$\begin{array}{r} 1 \\ +\ 1 \\ \hline \end{array}$	$\begin{array}{r} 6 \\ +\ 6 \\ \hline \end{array}$
$\begin{array}{r} 2 \\ +\ 2 \\ \hline \end{array}$	$\begin{array}{r} 8 \\ +\ 8 \\ \hline \end{array}$	$\begin{array}{r} 4 \\ +\ 4 \\ \hline \end{array}$
$\begin{array}{r} 9 \\ +\ 9 \\ \hline \end{array}$	$\begin{array}{r} 3 \\ +\ 3 \\ \hline \end{array}$	$\begin{array}{r} 5 \\ +\ 5 \\ \hline \end{array}$

Practice DAY 68

Student Exercises

Fill in the rest of the numbers on the number line. Start with 25. Then use the number line to do the exercises.

Remember, you will count up (to the right) for addition. You will count backwards (to the left) for subtraction.

25 ____ ____ ____ ____ ____ ____ ____ ____ ____ 35

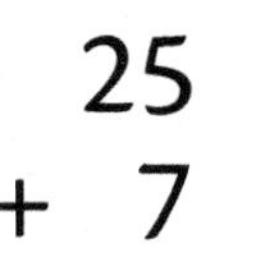

25 + 7 ____	26 + 4 ____	29 + 2 ____
33 − 4 ____	32 − 4 ____	31 − 6 ____
27 + 0 ____	32 − 0 ____	31 − 1 ____

These are addition and subtraction exercises. You can use your stones if you need help finding the missing number. For the addition exercises, ask, "How many stones do you have to add to get the bigger number?" Hint: What number is smaller? Start with that number. How many stones do you need to add to the small number to get the big number?

Let's do the first subtraction exercise together: What minus 3 equals 2? Use your imagination for this one. Pretend your dad gave you some pickles. But you can't remember how many he gave you because you already ate some. How many pickles did your dad give you? Suddenly you remember that you ate 3 pickles. Then you see there are only 2 pickles left on your plate. The 3 pickles in your stomach plus the 2 on your plate means your dad must have given you 5 pickles! 5 is the missing number. Now, you can say "5 pickles (from Dad) minus 3 pickles (eaten) equals 2 pickles (on the plate)."

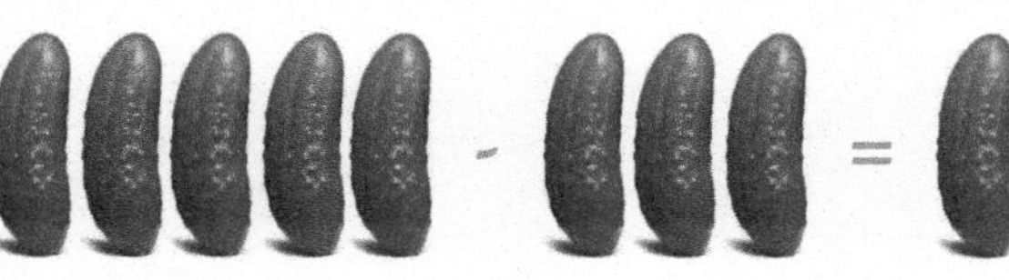

What minus 3 equals 2?	______ − 3 = 2
What plus 0 equals 0?	______ + 0 = 0
What plus 0 equals 1?	______ + 0 = 1
What minus 4 equals 3?	______ − 4 = 3
What plus 2 equals 4?	______ + 2 = 4
What plus 5 equals 5?	______ + 5 = 5
What plus 1 equals 6?	______ + 1 = 6
What minus 2 equals 7?	______ − 2 = 7

Lengths and Shapes DAY 69

This lesson explores more measurements in the real world and introduces multi-digit addition. It is followed by one page of new exercises. This will require about 20 minutes of instruction from the parent/teacher.

Prayer

Pray your own prayer of thanksgiving and praise to God. Pray for His help on this lesson.

Memory

Spend a few minutes with subtraction flash cards (subtracting from numbers 1 - 6).

Lesson

> "Have the people make an ark of acacia wood—a sacred chest 45 inches long, 27 inches wide . . ." (Exodus 25:10, paraphrase)

God told the children of Israel to build the Ark of the Covenant—a beautiful golden box to hold the Ten Commandments God gave them. He told them to make it out of acacia wood. He told them exactly what its measurements should be.

You have already learned how to measure the length of lines. This means we can measure all the way around any shape. First, we measure each side. Then we add these lengths together.

2 inches

3 inches 3 inches

2 inches

This rectangle has four sides. One side is 2 inches long. Another side is 3 inches long. The third side is 2 inches long. The last side is 3 inches long. How big is the whole shape? To find the length all the way around the shape, we need to add all the sides together. Our addition equation is: 2 + 3 + 2 + 3.

Let's start by adding the first two numbers together. What are the first two numbers? 2 and 3! 2 + 3 = 5. Now we cross out the 2 + 3. We write 5 where the first 2 and 3 used to be in the equation.

$$2 + 3 + 2 + 3 =$$
$$\cancel{2 + 3} + 2 + 3 =$$
$$5 + 2 + 3 =$$

Now add 5 and the next number: 5 + 2. What is 5 + 2? It's 7! Now we can cross out the 5 + 2. We write 7 where 5 and 2 used to be in the equation.

$$5 + 2 + 3 =$$
$$\cancel{5 + 2} + 3 =$$
$$7 + 3 = 10!$$

Now we finish by adding 7 + 3 to get 10. The length around the rectangle is 10 inches!

Now try these three addition exercises yourself. Find the sum of the first two numbers and cross them out. Write their sum in the first blank below the exercise. Then add that sum to the last number. What is your answer?

3+2+4 = ______	4+2+2 = ______	5+3+1 = ______
______+4 = ______	______+2 = ______	______+1 = ______

Student Exercises

Measure the sides of each shape in centimeters. How long is each side? Add the sides together to find the length around the whole shape. Write your answer inside each shape.

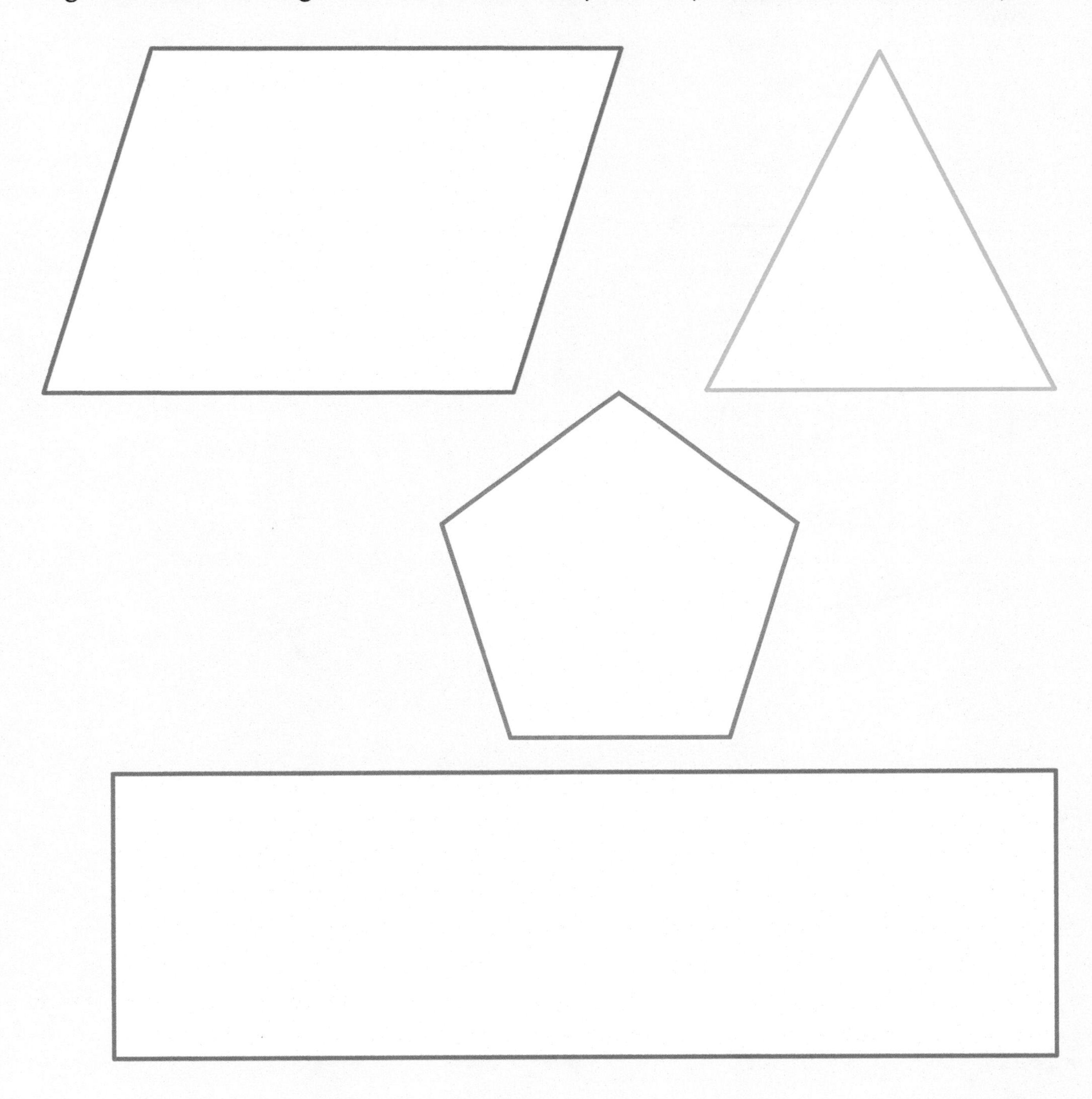

DAY 70 Practice

Student Exercises

Use a string to measure these lines. Lay the string over the curvy line as best as you can. Mark the length on the string. Then measure the length in inches using a ruler. Write the answer in the box.

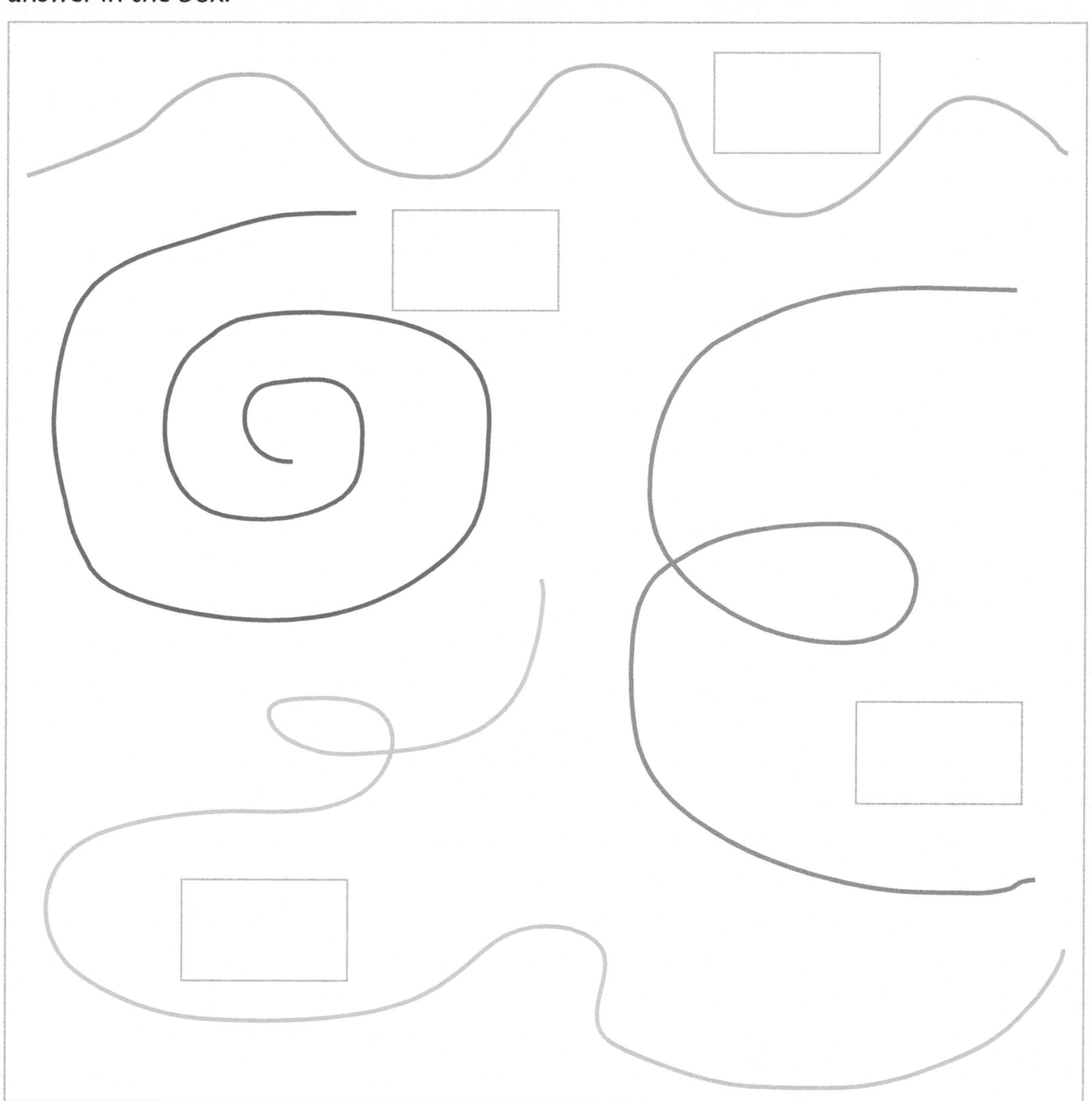

Let's add three numbers! Find the sum of the first two numbers and cross them out. Write their sum in the first blank below the exercise. Then add that sum to the last number. What is your answer?

2 + 4 + 1 = ______ _____ + 1 = ______	4 + 4 + 3 = ______ _____ + 3 = ______
1 + 2 + 3 = ______ _____ + 3 = ______	3 + 3 + 3 = ______ _____ + 3 = ______
6 + 2 + 1 = ______ _____ + 1 = ______	7 + 0 + 2 = ______ _____ + 2 = ______
5 + 7 + 1 = ______ _____ + 1 = ______	2 + 4 + 6 = ______ _____ + 6 = ______

Pretend that your parent has given you these little gifts. They ask you share with two friends. Draw a circle around your share. Draw another circle around one friend's share. Draw one more circle around the other friend's share. We call each share 1/3 of the full amount. Can you say "one third?"

How many cupcakes do you see altogether? ________
Write the number of cupcakes each person gets below.

Your name ____________________ ________

Friend's name ____________________ ________

Friend's name ____________________ ________

How many candied apples do you see altogether? Write the number of candied apples each person gets below.

You ________

Your first friend ________

Your second friend ________

How many chocolates do you see altogether? Write the number of chocolates each person gets below.

You ________

Your first friend ________

Your second friend ________

Draw lines to cut this sandwich into 3 equal parts for you and your 2 friends.

Cut this apple pie into 3 equal pieces for you and your 2 friends. Everybody gets the same-sized piece. Everybody gets 1/3 (one third) of the pie. Can you say one-third? That's a big piece of pie!

DAY 71 Huge Creations!

This lesson compares measurements to God's nature, and is followed by one page of review exercises. This will require about 15 minutes of instruction from the parent/teacher.

Prayer

Pray your own prayer of thanksgiving and praise to God. Pray for His help on this lesson.

Memory

Spend a few minutes with subtraction flash cards (subtracting from numbers 1 - 6)

Activity

So far, we've measured small things. We've used small inches and centimeters to measure small things. But God made big things that need to be measured too. God gave us big measurements (like feet and meters) to measure big things. One foot is 12 inches. One meter is 100 centimeters. That's a lot of centimeters!

Let's measure some big things now. How tall are you in inches? About how tall are you in feet? How tall is your mother or your brother or your sister?

How tall is the mother elephant in the picture? You can make a good guess by looking at the children. It looks like the mother elephant is about as tall as two children, one on top of the other. Let's say the two children are about the same height as you. How tall is that elephant? It's about as tall as your height plus your height again. Add: Your height in feet + your height in feet = Mama Elephant's height.

This is a whale shark. How long is that whale shark? Can you see the scuba divers in the picture? Let's say the scuba divers are 6 feet tall. How many scuba divers could lie down, lining up head to toe, along the whale shark's back?

Whale sharks can grow up to 60 feet in length! Ten scuba divers could line up head to toe on a whale shark that big.

How big is this castle? Can you see the little people walking on the beach? Yet the people would look even smaller to us if they climbed up to the castle walls far behind them. The castle is a fortress that protects people from the enemy. God is our fortress.

> God is our refuge and strength,
> A very present help in trouble.
> Therefore we will not fear,
> Even though the earth be removed,
> And though the mountains be carried into the midst of the sea . . . (Psalm 46:1-3)

> The LORD is my rock and my fortress
> and my deliverer . . . (Psalm 18:2)

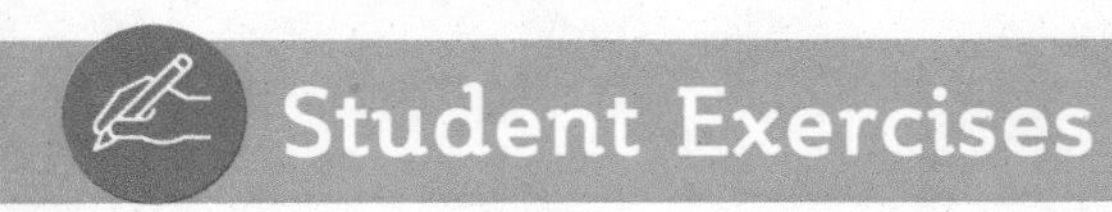

Student Exercises

Try to do these addition and subtraction exercises from memory. You may also use your blocks, stones, or coins if you need help. This will help you get better at adding and subtracting. Pray that God will help you with the harder exercises.

$\begin{array}{r} 6 \\ +\ 6 \\ \hline \end{array}$	$\begin{array}{r} 5 \\ +\ 1 \\ \hline \end{array}$	$\begin{array}{r} 2 \\ +\ 8 \\ \hline \end{array}$
$\begin{array}{r} 7 \\ -\ 3 \\ \hline \end{array}$	$\begin{array}{r} 12 \\ -\ 10 \\ \hline \end{array}$	$\begin{array}{r} 10 \\ -\ 4 \\ \hline \end{array}$
$\begin{array}{r} 8 \\ +\ 6 \\ \hline \end{array}$	$\begin{array}{r} 11 \\ +\ 4 \\ \hline \end{array}$	$\begin{array}{r} 9 \\ +\ 5 \\ \hline \end{array}$
$\begin{array}{r} 16 \\ -\ 5 \\ \hline \end{array}$	$\begin{array}{r} 13 \\ -\ 5 \\ \hline \end{array}$	$\begin{array}{r} 14 \\ -\ 8 \\ \hline \end{array}$

Student Exercises

Measure the sides of each shape in centimeters. How long is each side? Add the sides together to find the length around the whole shape. Write your answer inside each shape.

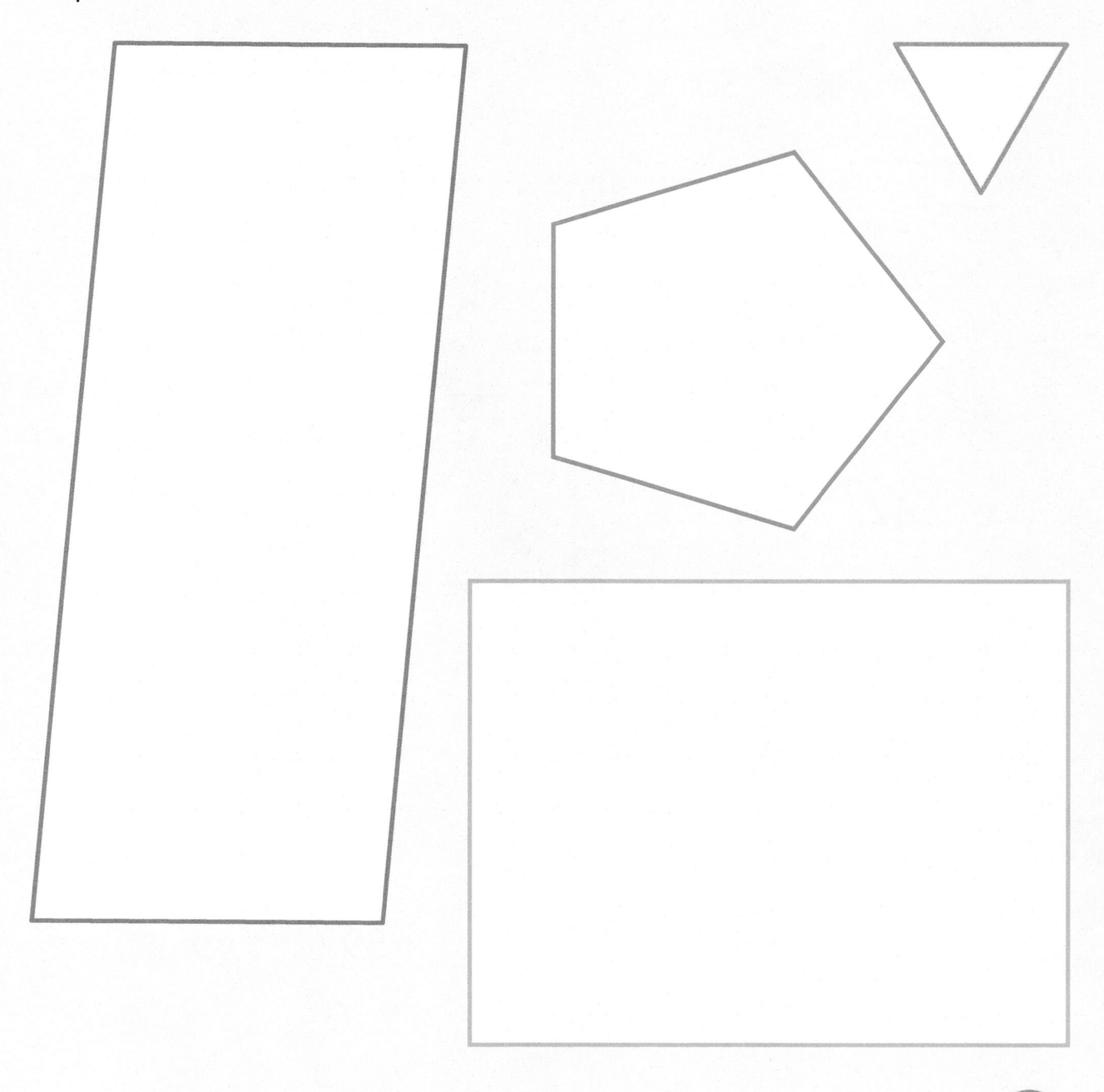

Compare these numbers! God made some numbers bigger than other numbers. Which number is bigger? Which number is smaller? Fill in the blank with the correct symbol: < (smaller than) or > (bigger than). Then read the math sentence. Remember, the baby shark eats the bigger number!

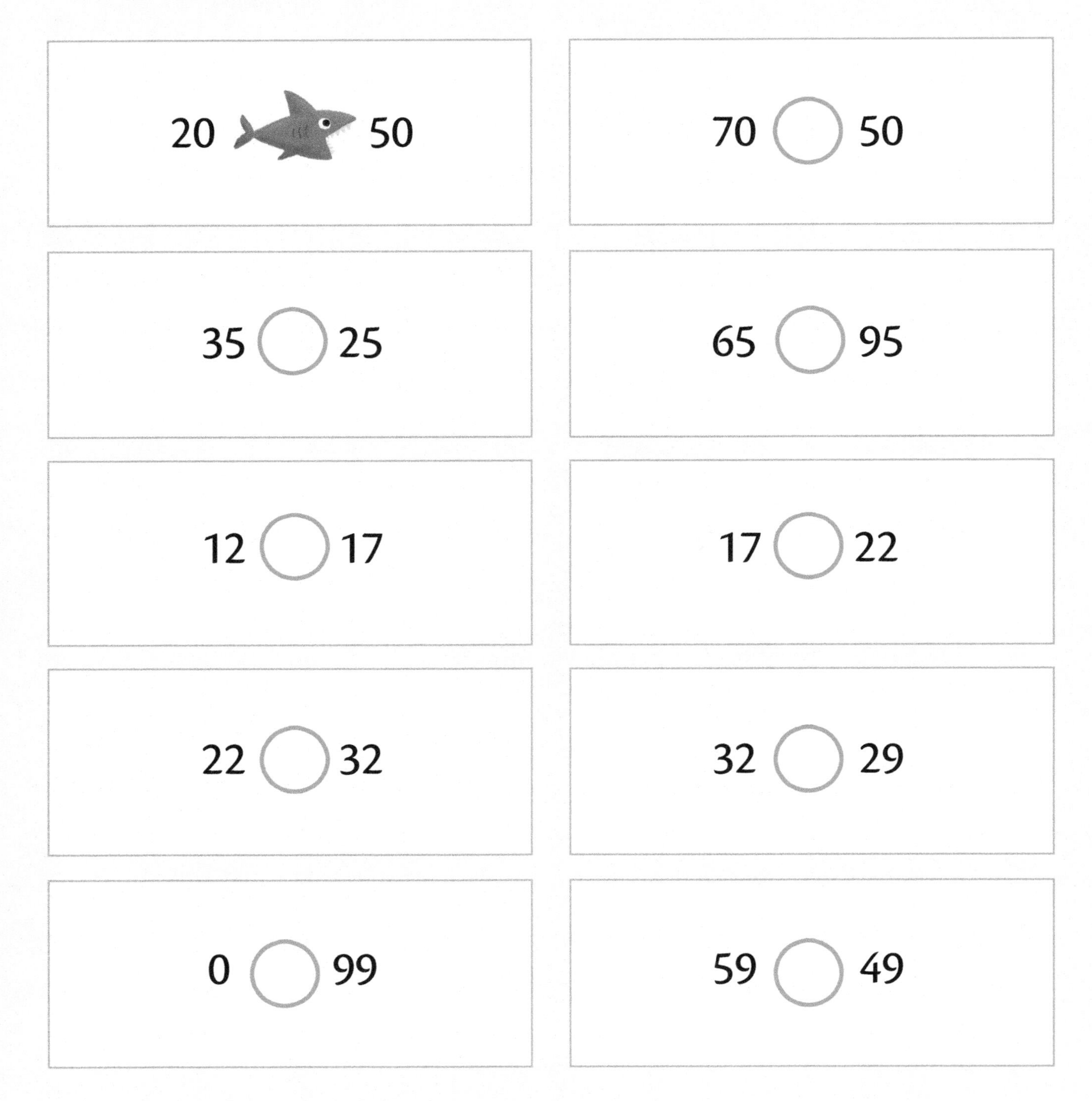

Practice DAY 73

Student Exercises

How many chunks of 10 do you need to make these numbers? How many singles (1s) do you need? Write the answers in the blanks.

35 → ________ 10s
35 → ________ 1s

42 → ________ 10s
42 → ________ 1s

58 → ________ 10s
58 → ________ 1s

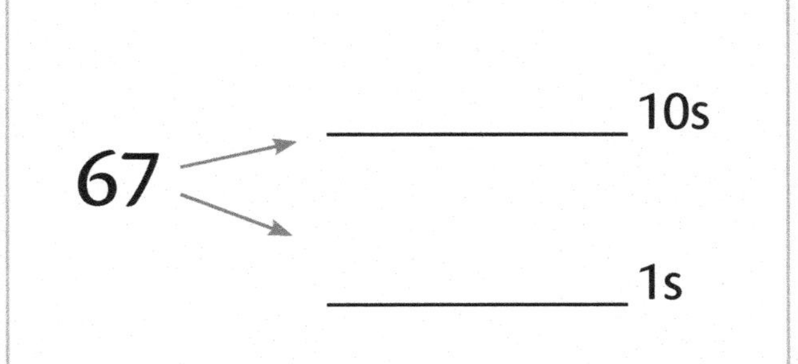

67 → ________ 10s
67 → ________ 1s

70 → ________ 10s
70 → ________ 1s

83 → ________ 10s
83 → ________ 1s

90 → ________ 10s
90 → ________ 1s

9 → ________ 10s
9 → ________ 1s

Let's add three numbers! Find the sum of the first two numbers and cross them out. Write their sum in the first blank below the exercise. Then add that sum to the last number. What is your answer?

$5 + 1 + 3 =$ 9 6 $+ 3 =$ 9	$4 + 3 + 2 =$ _____ _____ $+ 2 =$ _____
$3 + 0 + 2 =$ _____ _____ $+ 2 =$ _____	$5 + 3 + 0 =$ _____ _____ $+ 0 =$ _____
$3 + 3 + 3 =$ _____ _____ $+ 3 =$ _____	$4 + 4 + 4 =$ _____ _____ $+ 4 =$ _____
$2 + 2 + 6 =$ _____ _____ $+ 6 =$ _____	$6 + 4 + 1 =$ _____ _____ $+ 1 =$ _____

Go Measure God's World! DAY 74

This lesson integrates math into everyday life. This is an essential element to learning. The child is encouraged to apply God's patterns and wisdom to life in the home and community. Take a break from memory work and academic exercises, and identify ways to make math part of your everyday life. The following are suggestions or examples, but other ideas may be added to the list.

Activity

Let's use our new math skills to serve God! God tells us:

> Let all things be done decently and in order. (1 Corinthians 14:40)

God loves order! Do you remember the verse about the Ark of the Covenant?

> "Have the people make an ark of acacia wood—a sacred chest 45 inches long, 27 inches wide, and 27 inches tall." (Exodus 25:10, paraphrase)

God wanted the people to make the ark a special size. He told them how long the ark should be. He told them how wide the ark should be. He also told them how tall it should be. God wanted the people to make the ark in a special way.

Measuring is one way we can bring order. The people obeyed God by measuring. They followed His order.

You know how to measure things now. There are new ways to use this in your home. There are new ways to use this in your life. You can measure to bring order and beauty to your home and life!

1. Part of taking care of God's world is filling it with beauty.

Sometimes we make our homes beautiful by hanging pictures on the walls. Make sure that the pictures are centered. How far is it from one side of the picture to the edge of the wall? How far is it from the other side of the picture to the other edge of the wall? Are they the same? If so, the picture is centered. If not, do you know how to fix it?

Sometimes we make our homes beautiful by decorating. When we celebrate seasons and holidays, we are celebrating God's goodness. What holiday season is coming up? How can you help to decorate the home? Here are some ideas:

- When hanging decorations on the wall, measure the length between each decoration. Try to make a nice pattern.
- When making decorations, cut out patterns that are the same size. Measure and mark your cutouts to make sure they are all the same.

2. Make a jump rope for your friend, brother, or sister. How tall is that person? The jump rope should be 3 feet longer than the height of the person using it. This means if your friend is 4 feet tall, your rope needs to be 7 feet long. You can buy rope at the hardware store. Most stores sell rope by length. Finish by making the handles out of plastic pipe.

3. Help your father or mother fix or build something for the home. You might build a fence or table out of wood. Or you could put a fence post in the ground. Maybe the bed slats under the bed need to be fixed. You will need to measure the pieces of wood before you cut them.

You could also help replace the window blinds in your home. You will need to measure the height and width of the window so you can buy the right blinds.

4. Before buying clothes or shoes, you need to know your size.

Measure your chest, your waist, your legs' inseam, and your neck. Your parent/teacher can help you figure out your shirt and pants size using conversion charts online.

Find your shoe size by tracing your foot on paper. Your parent/teacher can help you find your shoe size by referring to a conversion chart online.

5. Keep track of your height. Measure your height for a few months or a year. Mark your height on a wall (or on paper attached to the wall). Praise God for your growth! This is another sign that you are healthy and strong.

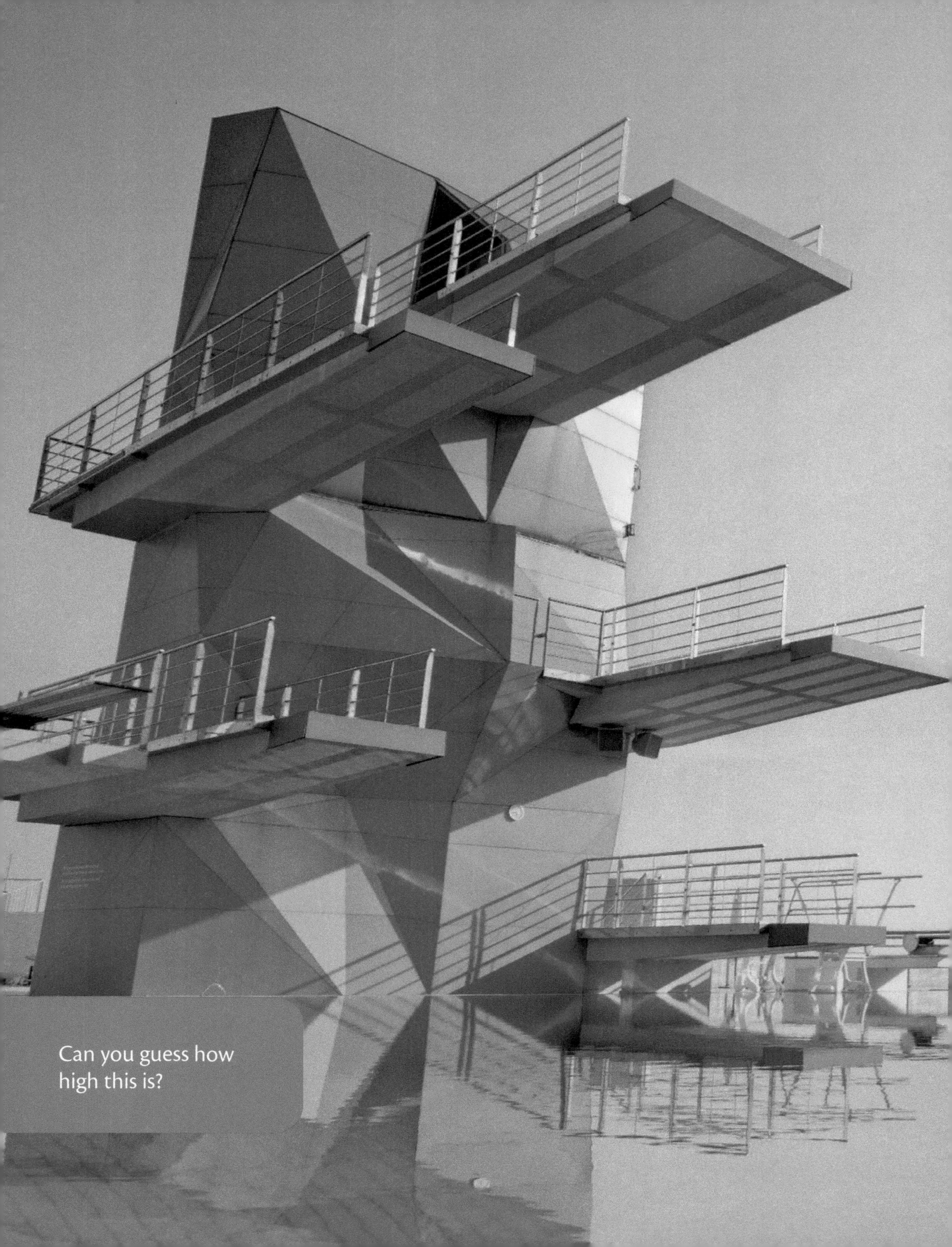

Can you guess how high this is?

CHAPTER 7

Good Guessing

Introduction

This lesson introduces approximations, and is followed by one page of review exercises. This will require about 20 minutes of instruction from the parent/teacher.

Have you noticed that most things don't measure exactly to whole inches or whole centimeters? Things are often a little bit longer or a little bit shorter than an exact inch. They might be a little bit longer or a little bit shorter than a full 2 inches. They might be a little bit longer or a little bit shorter than a full 5 inches. This means that sometimes you have to make a good guess about the inch it's closest to.

How tall is the boy in the picture? It looks like he is between 130 and 140 centimeters tall. How tall is he exactly? Can you tell? When you measure things, sometimes you will have to make a good guess. You want to come as close as possible. It looks like this boy is about 133 centimeters tall. That's a good guess! In this lesson, we will learn how to make good guesses.

Distance Between Numbers DAY 75

Prayer

Our Father in Heaven, Your world is so big that we cannot understand everything. Thank You for helping us to understand some things about it. We love You, and we need Your help today. Amen.

Memory

Spend a few minutes with addition flash cards (adding numbers 1 - 12 together).

Lesson

Today we will learn about **distance**. How far is it to church? How far is Grandfather and Grandmother's house from here? How many steps must you take to get to the bathroom?

Distance is the length between two things. We measure a lot of distances in inches, centimeters, **miles**, and **kilometers.**

> So when Jesus came, He found that [Lazarus] had already been in the tomb four days. Now Bethany was near Jerusalem, about two miles away. (John 11:17-18)

Jesus went to visit Mary and Martha. They lived in a town called Bethany. It was about two miles away from Jerusalem. Their brother, Lazarus, had died, but Jesus raised Lazarus from the dead.

Activity

Now let's find the distance between you and the things around you.

1. How many steps do you think it will it take you to get to the bathroom door? Take a guess.
2. Now, count your steps as you walk to the bathroom.
3. Find the difference between your guess and the actual steps you took. You will need to use subtraction.
4. Can you say this? "The distance between here and there was ________ steps."

Repeat this activity in two or three other places in your house. Guess first. Then count your steps.

Sometimes we use subtraction to figure out distances.

1. Lay out 8 stones in a line. Make them the same distance apart. Pretend that these are mile markers on a road. This means that the distance between each stone is a mile. These stones mark 7 miles.
2. Pretend that you have traveled 4 miles from your house. Lay 5 stones next to the 8 stones like this:

3. How many more stones do you need to have 7 stones? How many more miles do you have to travel to make it the whole 7 miles?
4. Use subtraction to figure out the distance. 8 stones minus 5 stones equals 3 stones.
5. Repeat the same exercise with a set of 9 stones and 3 stones.

Now let's measure distance using a 12-inch ruler.

1. Put your finger on the 10-inch mark. Do you see the 4-inch mark?
2. Ask yourself this question: What is the distance from 10 to 4?
3. Count the spaces between the marks as you move from 10 to 4. How many are there? Write the answer to this subtraction exercise:

   10 - 4 =________

   The distance from 10 inches to 4 inches is 6 inches.
4. What is the distance between the 9-inch mark and the 5-inch mark? Do you see the hiker standing on the rock in this picture? Think about the distance between the hiker and the mountains far away. That's a long way!

Length = __________

Student Exercises

What is the distance between these two points? Use subtraction to figure it out.

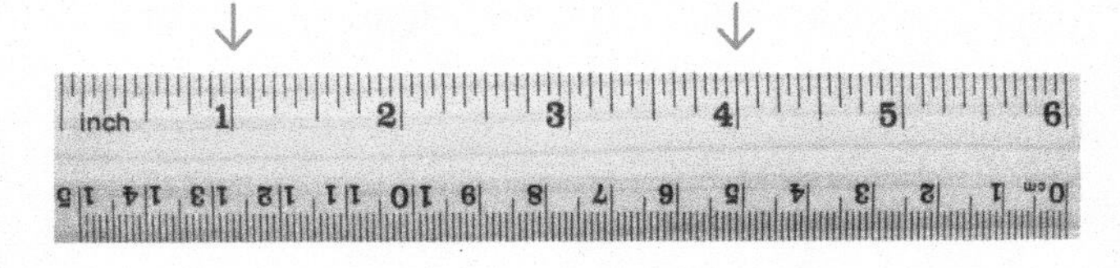

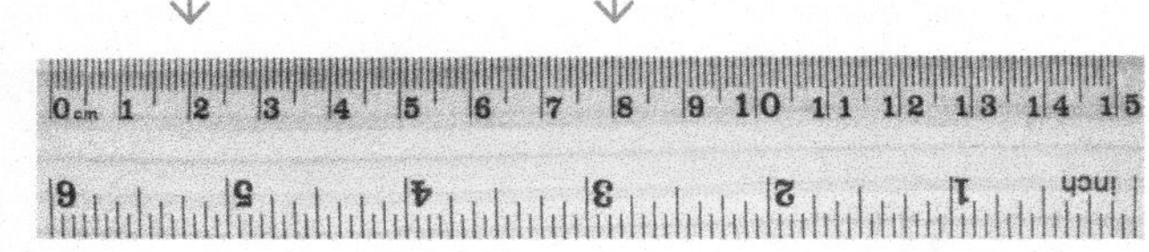

4 – 1 = _____ Distance = _____ 8 – 2 = _____ Distance = _____

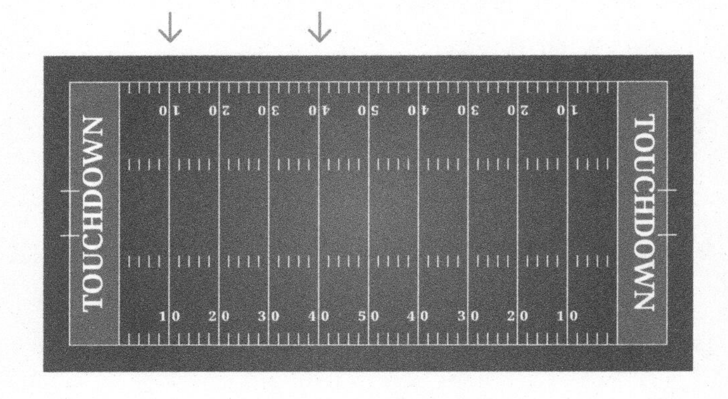

40 – 10 = _____ Distance = _____

Extra Challenge

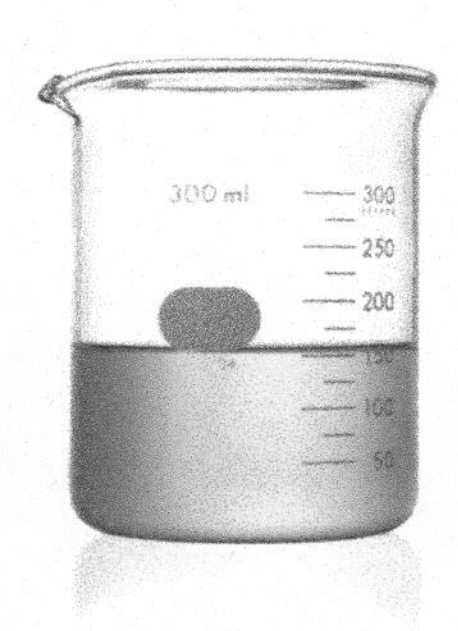

How much blue liquid is in this cup? _________ ml

Let's find the difference between the morning temperature and the afternoon temperature on these thermometers!

Cold in the morning Hot in the afternoon

What is the difference in temperature?

_________ – _________ = _________
afternoon morning

Practice DAY 76

Student Exercises

Compare these numbers! God made some numbers bigger than other numbers. Which number is bigger? Which number is smaller? Fill in the blank with the correct symbol: < (smaller than) or > (bigger than). Then read the math sentence. Remember, the baby shark eats the bigger number!

32 43	60 ◯ 6
12 ◯ 21	65 ◯ 60
25 ◯ 17	10 ◯ 30
98 ◯ 89	45 ◯ 44
0 ◯ 7	55 ◯ 66

What small numbers can be used to make a bigger number? For the first exercise, take 12 stones out of your toolbox. Split the stones into two groups or sets. Count the members of each group. Write these numbers in the blanks. Finish by reading your addition equation out loud.

There will be several right answers for these exercises. Try to do the rest of the exercises without your stones, but you can use them if you need help. How many ways can you make the numbers 12 and 15?

12 = _____ + _____

12 = _____ + _____

12 = _____ + _____

12 = _____ + _____

15 = _____ + _____

15 = _____ + _____

15 = _____ + _____

15 = _____ + _____

Finding Distances & Differences in Real Life DAY 77

This lesson explores distances and differences with real-life applications, and is followed by two pages of review exercises. This will require about 20 minutes of instruction from the parent/teacher.

Prayer

Pray your own prayer of thanksgiving and praise to God. Pray for His help on this lesson.

Memory

Spend a few minutes with addition flash cards (adding numbers 1 - 12 together).

Lesson

We use subtraction to find the distance or the difference between two measurements. You can use subtraction for distances and differences in real life too.

What is the distance between 5 inches and 8 inches? You can find the difference on the ruler by counting. Or you can find the distance by using subtraction:

$$8 - 5 = 3$$

Practice your subtraction! Figure out the differences between these distances on your ruler.

5 and 12

1 and 4

4 and 6

7 and 2

11 and 7

3 and 9

Now compare your height with your brother, sister, father, or mother.

1. Measure your height in inches.
2. Measure your brother, sister, father, or mother's height in inches. Pick someone who is taller than you!
3. How many inches do you have to grow to become the height they are now?

The next time you go to the store, help your parents weigh the fruit or vegetables.

1. First, ask your parents how many pounds (or kilograms) of the fruit or the vegetable they want to buy. Let's say they want to buy 5 pounds of apples.
2. At the store, find the apples and place some on the scale. How many pounds (or kilograms) do you have? If you don't have enough, how many more pounds do you need to make 5 pounds? You can use subtraction to help you.

Maybe you put too many apples on the scale! How many pounds do you need to take away to only buy 5 pounds? You can use subtraction for this too!

The man on the bicycle needs to get back down the mountain. How long would it take him to ride down the mountain? Perhaps your parent or teacher can guess how far he will need to ride to get to the valley below.

Student Exercises

God used chunks of 10 and singles (1s) to make these numbers! How many chunks of 10 did He use? How many singles (1s) did He use? Write the answers in the blanks.

5 → ________ 10s → ________ 1s

13 → ________ 10s → ________ 1s

24 → ________ 10s → ________ 1s

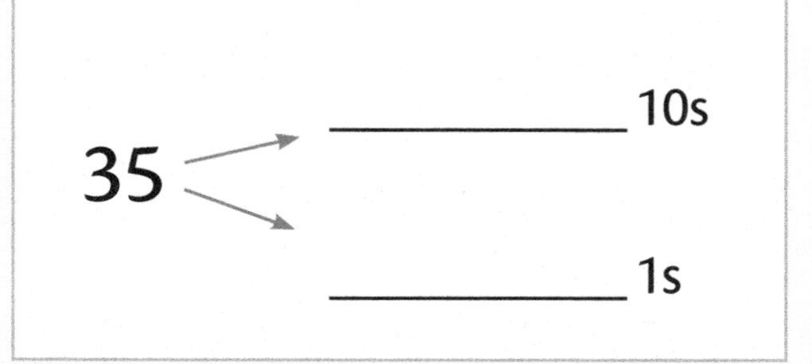

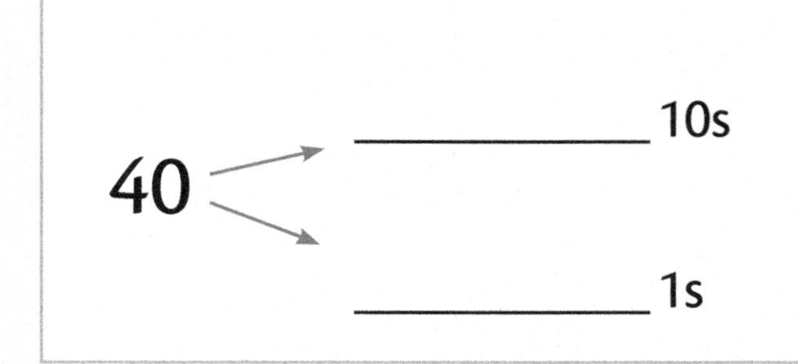

50 → ________ 10s → ________ 1s

68 → ________ 10s → ________ 1s

71 → ________ 10s → ________ 1s

Each exercise has two sets of numbers. Compare these sets. Which numbers do they have in common? Write these numbers in the space provided. This makes a new set of numbers! If they don't have any numbers in common, you will make an empty set.

Sets {2, 4, 6} and {4, 6, 8} share = {______}

Sets {1, 2, 3} and {1, 2, 4} share = {______}

Sets {5, 10, 15} and {10, 20, 30} share = {______}

Sets {0, 1, 2, 3} and {90, 91, 92, 93} share = {______}

Sets {2, 4, 6, 8} and {0, 4, 8, 12} share = {______}

Sets {12, 15, 18, 21} and {6, 12, 18, 24} share = {______}

Student Exercises

Find the distance between the two numbers for each exercise. Subtract the smaller number from the bigger number. You can also use a ruler if you need to find your answer by counting.

3 and 4	8 and 9
3 and 2	6 and 5
2 and 4	5 and 3
10 and 12	9 and 11
7 and 11	8 and 4

Try to do these addition and subtraction exercises from memory. You may also use your blocks, stones, or coins. This practice will help you get better at adding and subtracting! If you're having trouble, don't forget to ask God for help.

$5 + 3 =$ ____	$6 + 4 =$ ____	$7 + 5 =$ ____
$7 - 0 =$ ____	$6 - 4 =$ ____	$5 - 3 =$ ____
$7 + 6 =$ ____	$7 + 8 =$ ____	$9 + 7 =$ ____
$15 - 4 =$ ____	$15 - 6 =$ ____	$15 - 8 =$ ____

Rounding DAY 79

This lesson introduces the mathematical concept of rounding to the student, and is followed by two pages of review exercises. This will require about 15 minutes of instruction from the parent/teacher.

Prayer

Pray your own prayer of thanksgiving and praise to God. Pray for His help on this lesson.

Memory

Spend a few minutes with addition flash cards (adding numbers 1 - 12 together).

Lesson

Some people make wild guesses. They turn out to be wrong. But you can learn to make good guesses in math. **Rounding** is one way to make a good guess. Rounding is what we did when we measured things and made a good guess about the nearest inch.

There are two ways to do rounding. You can round up (make it more). You can also round down (make it less).

10 11 12 13 14 15 16 17 18 19 20

This is part of a number line. Can you find the number 17? Let's learn how to round 17 to the nearest 10. Think of chunks of 10.

How do we do that? Start by asking, Which two 10s is the number 17 between?" The answer is: The number 17 is between 10 and 20. Then we ask, "Is 17 closer to 10 or to 20 on the number line?" Of course, it is closer to 20. The distance between 10 and 17 is much bigger than the distance between 17 and 20. So we will round **up** to 20. We say: "17, rounded to the nearest 10, is 20."

Let's round the number 13 to the nearest 10. 13 is between 10 and 20. What is the distance between 13 and 10? What is the distance between 13 and 20? Can you count it out on the number line? 13 is closer to 10 than it is to 20. This time, we round **down** to

10. So 13, rounded to the nearest 10, is 10.

Now, try some rounding yourself. Round these numbers to the nearest 10.

79 3 64

Is 79 closer to 70 or to 80? Is 3 closer to 0 or to 10? Is 64 closer to 60 or 70? Will you round up or down? (We will use an arrow next to a number that needs to be rounded to the nearest 10.)

There is a special rule for rounding numbers that are halfway between two 10s. Rule: If the number is halfway between (like the number 15), we will always round up. So we will always round 15 up to 20. Let's practice with a few numbers that end with 5. What 10 do we round to for each of these numbers?

25 65 45

Student Exercises

Let's do these fun story exercises!

Pretend you have 6 strawberries. You share 4 of these strawberries with 2 friends. Write some friends' names on the blank lines. How many strawberries will you give each friend?

Name of two friends ______________________ How many for each? __________

How many strawberries will you have left?

__6__ – ________ = ________

You have 8 apples left after your friends ate 3 of your apples. How many apples did you have at the beginning?

________ + ________ = ________

It was time for Cami's birthday party. Cami's mom was setting up the chairs—one for Cami and one for each friend. But then Cami heard that 2 of her friends were sick and wouldn't be able to come! Soon her other friends arrived and Cami sat down with them at the table. Cami and her friends filled 5 of the seats. There were 2 empty seats that reminded them to pray for their sick friends. How many children would have been sitting at the table if everyone was well?

________ + ________ = ________

Mom bought 12 cheeseburgers. The family ate 8. How many are left?

________ – ________ = ________

Here are some sets of numbers. Put the numbers in each set in the order God made for them. Put them in order from smallest to biggest.

{45, 72, 27, 63, 36}

________, ________, ________, ________, ________

{17, 34, 72, 48, 95}

________, ________, ________, ________, ________

{25, 45, 15, 65, 55}

________, ________, ________, ________, ________

Student Exercises

Round each number to the nearest 10. Sometimes you will round up. Sometimes you will round down. Which numbers do not need to be rounded up or down?

16 → ______	23 → ______
30 → ______	47 → ______
65 → ______	72 → ______
71 → ______	15 → ______
85 → ______	80 → ______

Find the distance between the two numbers for each exercise. Subtract the smaller number from the bigger number. You can also use a ruler if you need to find your answer by counting.

7 and 4 _______	10 and 7 _______
3 and 5 _______	3 and 1 _______
8 and 6 _______	4 and 6 _______
2 and 7 _______	7 and 12 _______
5 and 5 _______	6 and 6 _______

Rounding Lengths DAY 81

This lesson explores estimations using measurements of length, and is followed by one page of review exercises. This will require about 20 minutes of instruction from the parent/teacher.

Prayer

Pray your own prayer of thanksgiving and praise to God. Pray for His help on this lesson.

Memory

Spend a few minutes with subtraction flash cards (subtracting from numbers 1 - 12).

Lesson

We learned about rounding to the nearest 10 in the last lesson. Rounding is a way of guessing. Now we will make good guesses for lengths. We will round using a ruler.

Here is a picture of a pencil. Let's measure the pencil using the ruler in the picture.

Is the pencil 4 inches long? No! Is the pencil 5 inches long? No! Do you see the long mark halfway between 4 and 5? We call this a "half" mark. The pencil stops halfway between 4 and 5. You can round up the pencil's length to 5 inches. You can call it 5 inches long.

If the length is longer than the halfway mark between two numbers (right in the middle), you will round up.

If the length is shorter than the halfway mark between two numbers, you will round down.

If the length is right in the middle–at the halfway mark–you will round up as we did with the pencil.

Activity

Now that you have learned about rounding, try measuring things around the house (or the area where you are studying). Use a string, a ruler, or a tape measure.

1. Measure three straight things. Round up or down to the nearest inch (or centimeter).
2. Measure three curvy things. Round up or down to the nearest inch (or centimeter).
3. Find an object (like a book). Take three measurements. Hint: Can you measure one long edge of the book? Can you measure a shorter edge? Can you measure an edge to show how thick the book is?

Round up or down to the nearest inch (or centimeter).

Student Exercises

Round each number to the nearest 10. Sometimes you will round up. Sometimes you will round down. Which numbers do not need to be rounded up or down? In this case, just write the same number over again.

65 → ______	22 → ______
76 → ______	60 → ______
55 → ______	57 → ______
31 → ______	36 → ______
25 → ______	20 → ______

DAY 82 **Practice**

Student Exercises

Each exercise has two sets of numbers. Compare these sets. Which numbers do they have in common? Write these numbers in the space provided. This makes a new set of numbers! If they don't have any numbers in common, you will make an empty set.

Sets {12, 15, 18} and {12, 20, 28} share = {________}

Sets {19, 20, 21} and {22, 21, 20} share = {________}

Sets {24, 25, 26} and {27, 28, 29} share = {________}

Sets {0, 1, 2, 3} and {1, 3, 5, 7} share = {________}

Sets {2, 6, 8, 10} and {2, 8, 14, 20} share = {________}

Sets {13, 19, 29, 31} and {19, 31, 13, 29} share = {________}

In some of these exercises, you'll see numbers that split up into two equal parts. For others, you'll take everything away and get 0. Can you do all these exercises from memory?

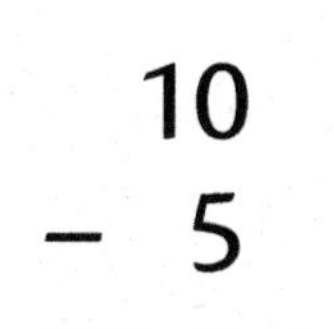

$$\begin{array}{r} 10 \\ -\ 5 \\ \hline \end{array}$$

$$\begin{array}{r} 6 \\ -\ 3 \\ \hline \end{array}$$

$$\begin{array}{r} 1 \\ -\ 1 \\ \hline \end{array}$$

$$\begin{array}{r} 9 \\ -\ 9 \\ \hline \end{array}$$

$$\begin{array}{r} 2 \\ -\ 1 \\ \hline \end{array}$$

$$\begin{array}{r} 8 \\ -\ 4 \\ \hline \end{array}$$

$$\begin{array}{r} 3 \\ -\ 3 \\ \hline \end{array}$$

$$\begin{array}{r} 5 \\ -\ 5 \\ \hline \end{array}$$

$$\begin{array}{r} 4 \\ -\ 2 \\ \hline \end{array}$$

DAY 83 Guessing Short Lengths

This lesson explores observing lengths and making educated guesses in the real world. It is followed by one page of review exercises. This will require about 10 minutes of instruction from the parent/teacher.

Prayer

Pray your own prayer of thanksgiving and praise to God. Pray for His help on this lesson.

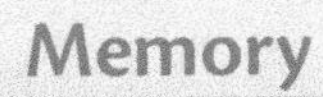

Spend a few minutes with subtraction flash cards (subtracting from numbers 1 - 12).

You've now learned about two ways of guessing. You've learned about rounding numbers and you've learned about rounding lengths. Today you will try another kind of guessing. You will guess the lengths of small things in the room. Then, you will measure each thing with a ruler. Will your guess be close to the real length?

Start with the picture of this mouse. How tall do you think the mouse is on the page? Take a guess and use inches. Now measure the height of the mouse with your ruler. You will have to round up or round down to the nearest inch. Write down your guess and your measurement.

Now repeat this with things around the room. Write down what you're looking at, what your guess is, and your measurement.

Object	Your Guess	Your Measurement
1. ____________	____________	____________
2. ____________	____________	____________
3. ____________	____________	____________
4. ____________	____________	____________

Student Exercises

Here are sets of numbers. Put the numbers in each set in the order God made for them. Put them in order from smallest to biggest.

{1, 31, 11, 21, 51}

________, ________, ________, ________, ________

{23, 31, 32, 33, 13}

________, ________, ________, ________, ________

{64, 56, 46, 54, 65}

________, ________, ________, ________, ________

DAY 84 Practice

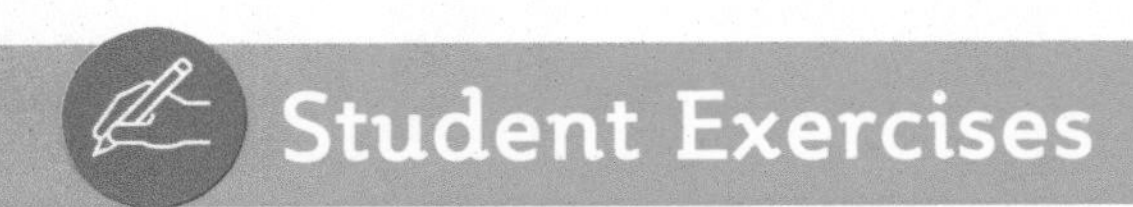

Student Exercises

Addition makes bigger numbers. Subtraction makes smaller numbers. God made numbers to work together. Sometimes numbers work together to make bigger numbers. Sometimes numbers work together to make smaller numbers.

For these exercises, you need to decide how these numbers work together. How do 11 and 1 work together to make 10? Of course, the smaller number must be subtracted! If you added 1 and 11, you would get . . . that's right! 12! But 12 is not the number we want.

Write "+" or "-" in each circle to show how the numbers work together.

$11 \bigcirc 1 = 10$	$2 \bigcirc 8 = 10$
$12 \bigcirc 4 = 8$	$6 \bigcirc 2 = 8$
$8 \bigcirc 2 = 6$	$9 \bigcirc 3 = 6$
$2 \bigcirc 2 = 4$	$5 \bigcirc 1 = 4$
$6 \bigcirc 0 = 6$	$8 \bigcirc 0 = 8$
$0 \bigcirc 6 = 6$	$0 \bigcirc 3 = 3$

How long is each side? You may have to round up or round down to get the number of centimeters for each side. Add the sides together to find the length around the whole shape. Write your answer inside each shape.

DAY 85 Guessing Long Lengths

This lesson explores longer lengths measured in feet and meters. This will require about 15 minutes of instruction from the parent/teacher.

Prayer

Pray your own prayer of thanksgiving and praise to God. Pray for His help on this lesson.

Memory

Spend a few minutes with subtraction flash cards (subtracting from numbers 1 - 12).

Activity

Today we will measure big things. Long lengths are measured in feet or meters. A meter is about as long as a tall man's legs. There are 100 centimeters in a meter. A foot is about as long as a big man's foot. There are 12 inches in a foot.

Before we go on in this lesson, answer these three questions. You will need to use a long measuring tape.

How many inches tall are you? __________

How tall is your mother or father in inches? __________

How many feet long is your house? __________

What measurement did you use to measure each of these things? Did you use inches? Did you use centimeters? No. It was easier to use feet or meters!

Now let's look at some really big things from the Bible.

Do you remember who the giant Goliath was? Goliath was 10 feet tall! Compare yourself and your parent to Goliath. Draw yourself, your parent, and Goliath on another piece of paper. How much taller was he than you? How much taller was he than your parent?

Og was another giant in the Bible. He did not like Israel. He fought a war with the people of God. Og was 13 feet tall. Compare Og to Goliath. Was Og taller than Goliath? Praise the Lord! God's people won the fights with these giants.

Do you remember the story about Noah's ark? Noah's ark was 450 feet long. Compare your house to Noah's ark. Draw your house next to Noah's ark on another piece of paper. Hint: If your house is 50 feet long, then Noah's ark would be as long as 9 of your houses put together.

Jesus walked on water for 30 furlongs. That's about 2,000 feet. That's a long way to walk on water!

You've measured a few things in feet! Now try to guess how big things are without measuring. Find something on your body (like your forearm) that is about 1 foot long. Guess the length of your leg, then measure it using your forearm. How many forearms long is your leg?

Explore your room, your house, or the yard. Guess the height, the length, or the width of big things all around you. How tall are these trees? Could they be 99 feet tall?

> "Can you search out the deep things of God?
> Can you find out the limits of the Almighty?
> They are higher than heaven—what can you do?
> Deeper than Sheol—what can you know?
> Their measure is longer than the earth
> And broader than the sea." (Job 11:7-9)

DAY 86 Practice

Student Exercises

In the exercises below, guess which is taller, bigger, or heavier. Put them in order of biggest to smallest, or most to least by writing 1st, 2nd, 3rd, and 4th.

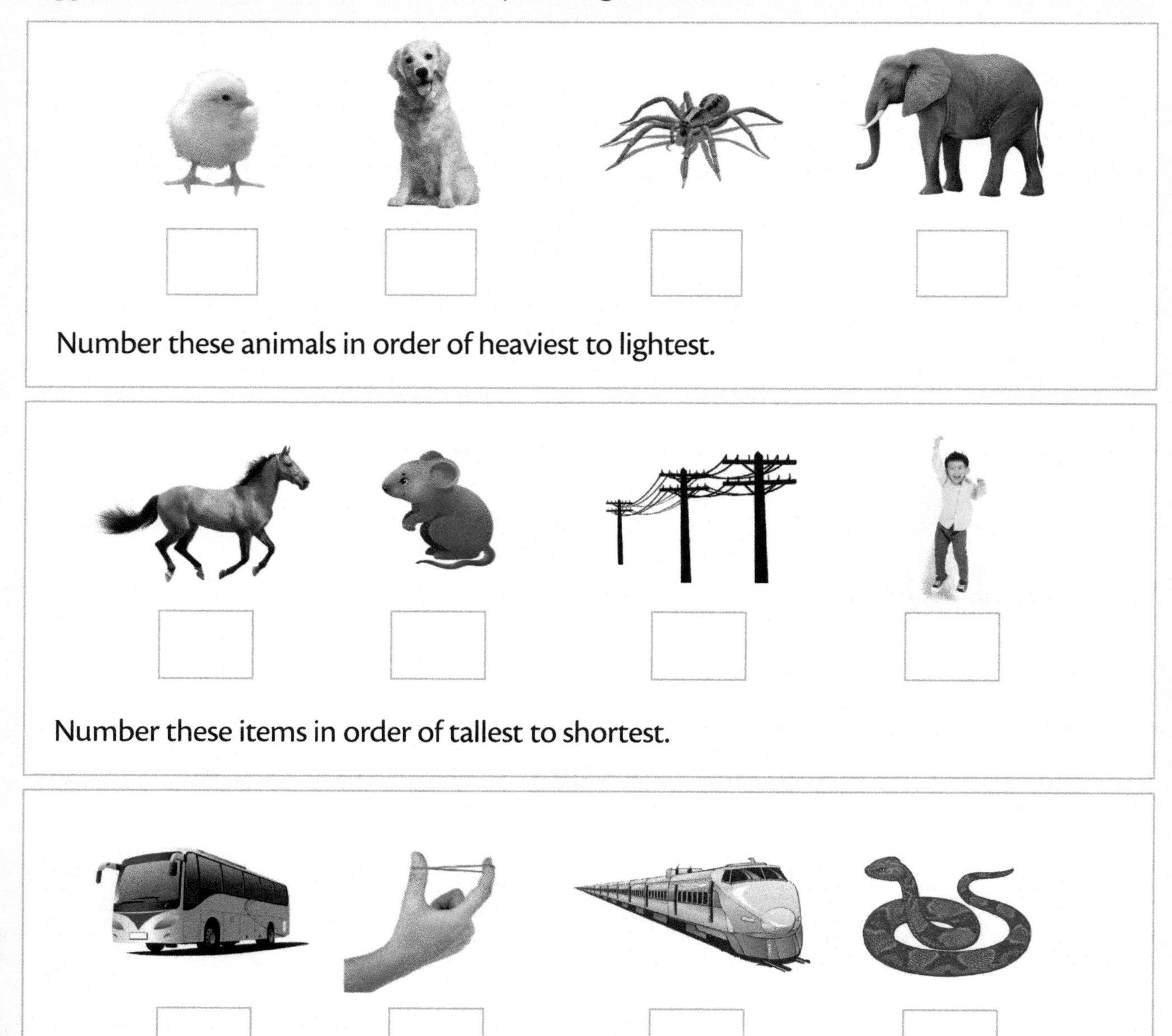

Let's add three numbers! Find the sum of the first two numbers and cross them out. Write their sum in the first blank below the exercise. Then add that sum to the last number. What is your answer?

3 + 4 + 1 7 + 1 = 8	1 + 4 + 4 ____ + ____ = ____
2 + 2 + 3 ____ + ____ = ____	1 + 4 + 1 ____ + ____ = ____
3 + 4 + 2 ____ + ____ = ____	0 + 1 + 1 ____ + ____ = ____
2 + 4 + 5 ____ + ____ = ____	6 + 0 + 1 ____ + ____ = ____

DAY 87 Go Guess in Your World!

This lesson integrates math into everyday life. This is an essential element to learning. The child is encouraged to apply God's patterns and wisdom to life in the home and community. So we will take a break from memory work and academic exercises, and identify ways to make math part of your everyday life. The following are suggestions or examples, but other ideas may be added to the list.

Activity

We guess measurements every day in real life. In fact, you will learn to make good guesses when you try doing it more often. God has made your mind able to learn to make better guesses as you go along. Let's try some things today that will help you with guessing.

1. **Proper portions.** Guess how much food you will eat for lunch or dinner. How much food do you put on your plate? How much food would you put on a little child's plate? God doesn't want you to waste your food, so don't serve yourself too much! Or maybe your family is having another family over for dinner. How much will they eat? You will need to fix more food for more people.
2. **Playing it safe.** Can you jump off that tall fence? Or can you jump out of that tree? Can you jump down 10 feet from the swing on the playground? If you jump from too high, you might break your leg. God wants you to take good care of your body, so you should be wise about what you can and cannot do.
3. **Working faster.** Guessing can be helpful when working around the house or doing chores. Which is faster: guessing a length or measuring it? Guessing! Let's say you are vacuuming for your mother. You will need to plug the cord into the wall. If you want the vacuum to reach every corner, which plug should you use?
4. **Quick cooking.** Good cooks must be good guessers! With help from your parents, you can learn to cook too. How much milk will you pour into your breakfast cereal? When you fry eggs, how much salt do you think you will need to add? Will you shake the salt shaker one time, two times, or three times? You need to add sugar when you make lemonade. How much sugar will you need to add? Mix in a little sugar first. Then taste it. Is the lemonade still too sour? Do you need to add a little more sugar? Guess how much more.

Did you know that some things can't be measured? Can you measure the sand on the beach? No! There is too much sand for us to measure. God tells us that His family will be too big for us to measure or count too!

> "Yet the number of the children of Israel shall be as the sand of the sea,
> Which cannot be measured or numbered." (Hosea 1:10)

Here's a fun game. Look for three numbers in a row going up and down or sideways that make an addition sentence, or subtraction sentence. Find at least 10 number sentences. Circle them when you see them. Two of them have already been found for you: 7 - 2 = 5 and 5 - 2 = 3!

3	4	7	2	5	2	1	3
3	2	5	3	2	6	7	2
6	6	12	5	3	8	8	5
10	5	5	7	6	3	9	6
7	5	2	3	5	2	3	11
4	5	9	4	4	5	9	5
5	3	4	1	3	8	12	4
9	8	5	3	7	5	4	1

What would you need to help you cross this desert?

CHAPTER 8

Making Bigger Numbers

Introduction

Some things are hard to do. It's hard to pull your tooth. It's hard to change a big tractor tire. It's hard to climb Mount Everest. Mount Everest is the tallest mountain in the world!

It would also be hard to walk across the Sahara Desert. That's the biggest desert in the world! But you could fly in an airplane across the desert instead. Airplanes are useful tools.

God gives us tools. Tools help us do hard things. God also gives us wisdom. Wisdom helps us to figure out how to use the tools.

In this chapter, we will learn how to add big numbers. God has given us a powerful tool to add big numbers.

God tells us about a man named Bezalel in the Bible. He designed the tabernacle for the Israelites in the Old Testament. God gave Bezalel the wisdom he would need for this big job.

> Then Moses called Bezalel and Aholiab . . . in whose heart the LORD had put wisdom . . . to come and do the work. (Exodus 36:2)

Adding Bigger Numbers DAY 88

This lesson introduces double-digit addition, and is followed by one page of review exercises. This will require about 20 minutes of instruction from the parent/teacher.

Prayer

Our Father in Heaven, thank You for Your love. Thank You for giving us tools to do things in Your world. Please help us to love You. Help us to love others as You have loved us. Amen.

Memory

Spend a few minutes with both addition and subtraction flash cards (mix and match).

Lesson

Now you know how to add 4 and 3 to get 7. How would you add 45 and 32? That is much harder, isn't it? How long would it take you to count 45 stones and add another 32 stones? It's much faster to add in chunks of 10 and 1s! Let's add 45 and 32 using blocks. Do you see the 45 blocks? They are made up of 4 chunks of 10 and 5 single 1s.

Do you see the 32 blocks? They are made up of 3 chunks of 10 and 2 single 1s.

45

+ 32

Here is the fast way to combine the chunks of 10 and the single 1s. Add the single 1s. Then add the chunks of 10. How many 1s do you have? How many chunks of 10? What's the answer to the addition exercise? That's right! The answer is 77!

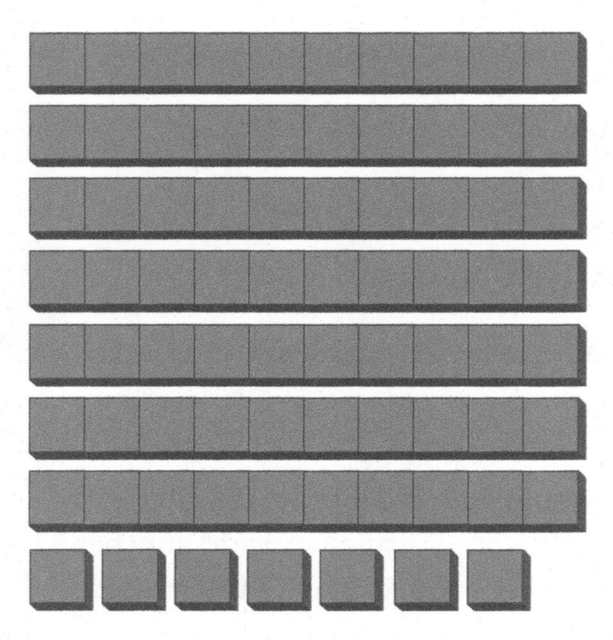

Now let's try another example. Add 35 and 17 using your blocks. How do you make 35 with your blocks? How do you make 17? This is what it should look like:

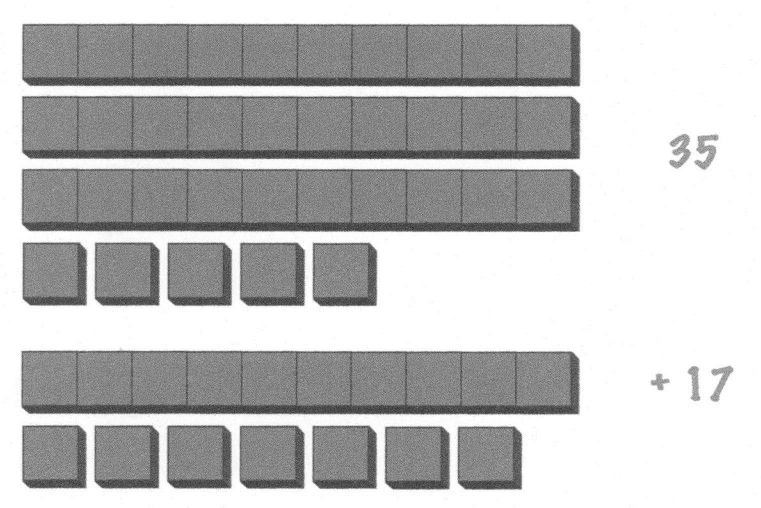

Add all of these blocks together. You get 12 single 1s. And you get 4 chunks of 10. Now, can you make another chunk of 10? Look at the 12 single blocks. 12 is bigger than 10! You can make another chunk of 10. Join 10 of your single blocks to make a chunk of 10. How many 1s do you have left? You have 2 single 1s left. Altogether, you have 5 chunks of 10 and 2 single 1s. This is 52.

Now, this is what it looks like when we have finished the exercise!

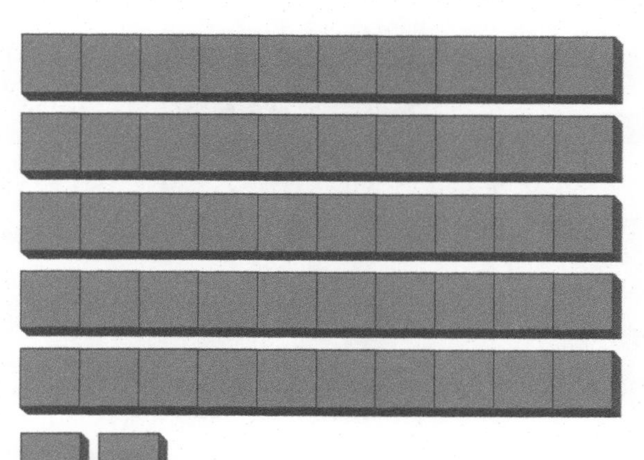

Use your blocks of 10 and your 1s to add the following big numbers. Remember, if you have more than 10 singles, you can turn 10 singles into a chunk of 10.

12 + 16	48 + 21	35 + 62

Student Exercises

Here are some sets of different things. Circle the one that is different from the others in the picture. Can you explain why it is different from the others?

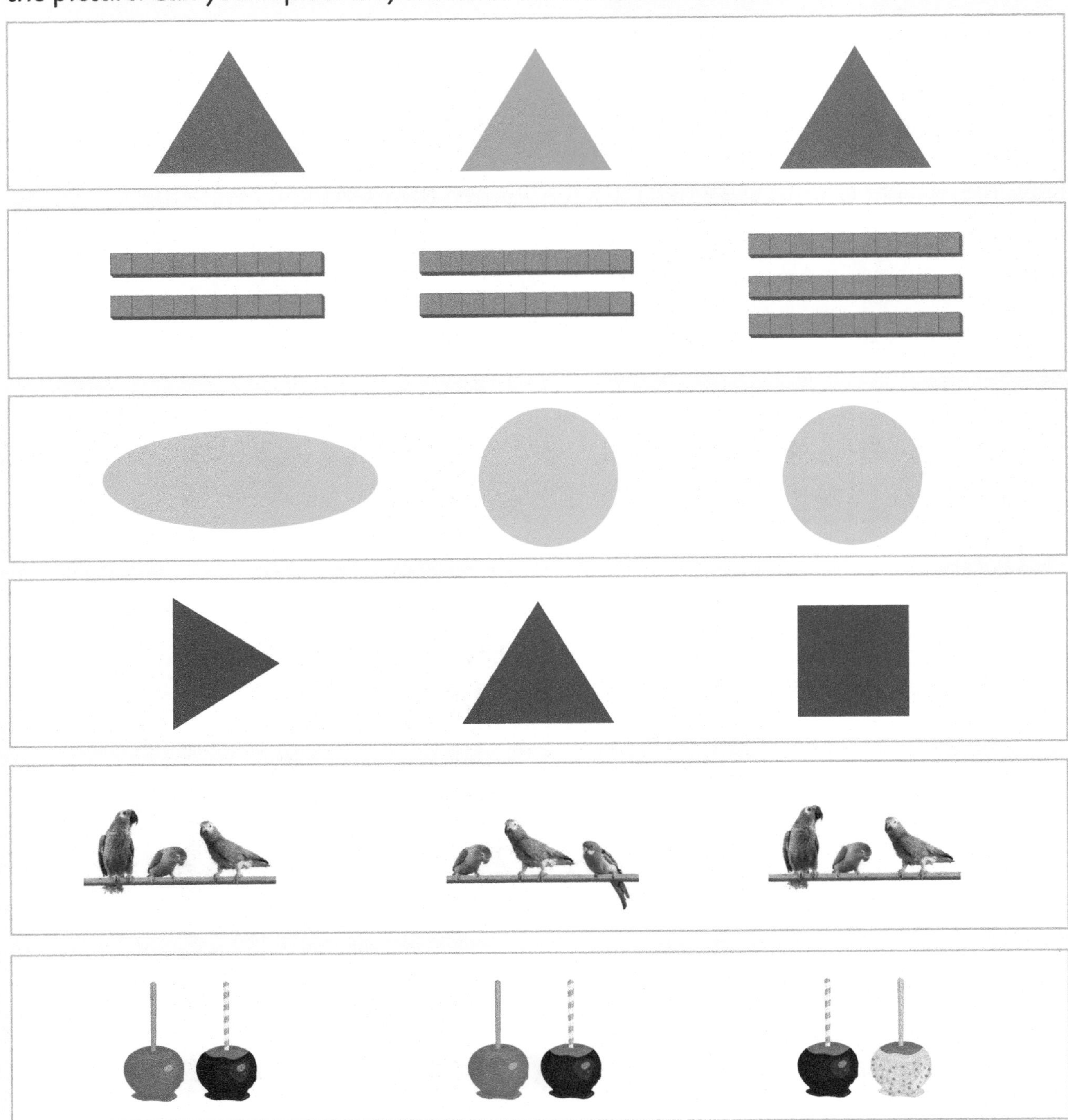

Let's review some easier addition problems. Try to do all of them from memory!

$$\begin{array}{r} 10 \\ +\ 4 \\ \hline \end{array}$$

$$\begin{array}{r} 11 \\ +\ 4 \\ \hline \end{array}$$

$$\begin{array}{r} 12 \\ +\ 4 \\ \hline \end{array}$$

$$\begin{array}{r} 10 \\ +\ 5 \\ \hline \end{array}$$

$$\begin{array}{r} 11 \\ +\ 5 \\ \hline \end{array}$$

$$\begin{array}{r} 12 \\ +\ 5 \\ \hline \end{array}$$

$$\begin{array}{r} 10 \\ +\ 6 \\ \hline \end{array}$$

$$\begin{array}{r} 11 \\ +\ 6 \\ \hline \end{array}$$

DAY 89 Practice

Student Exercises

Put these numbers in the right order. Put them in the order God made for them!

Find the smallest number. Write it in the first blank. What comes next? Write it in the next blank. What might the numbers stand for? Each set of numbers gives you something to imagine as you put them in order. God's world gives us lots of things to count and to measure!

You are...

63, 72, 62, 36, or 27 feet deep in the ocean!

________, ________, ________, ________, ________

There are...

73, 61, 97, 44, or 11 people at the carnival!

________, ________, ________, ________, ________

You are climbing a rock that is...

35, 47, 62, 25, or 63 feet high!

________, ________, ________, ________, ________

Round each of these numbers to the nearest 10. Sometimes you might have to round up. Sometimes you might have to round down. Remember, 5 is always rounded up.

73 → ________	65 → ________
7 → ________	49 → ________
94 → ________	16 → ________
55 → ________	2 → ________
20 → ________	30 → ________

DAY 90 A New Way to Add

This lesson explores more long addition using two-digit numbers, and is followed by two pages of review exercises. This will require about 30 minutes of instruction from the parent/teacher.

Prayer

Pray your own prayer of thanksgiving and praise to God. Pray for His help on this lesson.

Memory

Spend a few minutes with both addition and subtraction flash cards (mix and match).

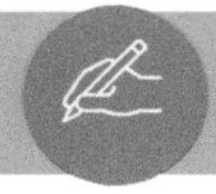

Lesson

Let's add some big numbers! This gets harder, but God gives us good ways to do this. He will help you! Using blocks or coins can take a long time. This tool gives us a faster way to add.

Let's stack the numbers on top of each other. You can pretend you are piling up blocks or coins. It looks like this:

10s	1s
6	3
+1	4

Be sure to write the plus sign. This reminds us that we are adding numbers together. It's also a good idea to always put the bigger number on top.

When you read a word, you go from left to right. When you say a big number (like 49) you also go from left to right ("forty-nine"). But when you add big numbers, you go from right to left! You add the 1s first. Then you add the 10s. Let's add the 1s for 63 + 14. 3 + 4 = 7. Write the 7 under the 3 and the 4.

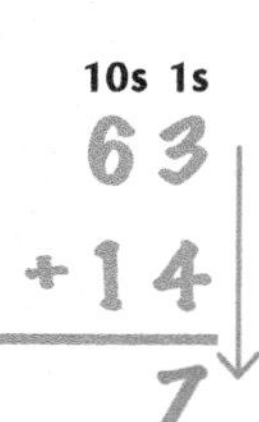

Then, add the numbers on the left. These are the chunks of 10. 6 + 1 = 7. Write this 7 under the 6 and the 1 to show that it's 7 chunks of 10. Now we can say 63 + 14 = 77!

$$\begin{array}{r} 63 \\ +\,14 \\ \hline 77 \end{array}$$

Let's try adding a small number and a big number. How do you think you would add this?

Let's start by adding the 1s. The first number has 2 single 1s, and the second number has 6 single 1s. Now we add them together: 2 + 6 = 8.

$$\begin{array}{r} 22 \\ +\,6 \\ \hline 8 \end{array}$$

Then we need to add the chunks of 10. The first number has 2 chunks of 10. How many chunks of 10 do you see in the second number? The number 6 doesn't have any chunks of 10. Here is one way we could write out the exercise:

We imagine that a 0 is in front of the 6. What is 2 plus 0? 2! We have now added our 1s and our chunks of 10. We found our answer: 28!

$$\begin{array}{r} 22 \\ +\,06 \\ \hline 28 \end{array}$$

Adding big numbers this way is called long addition. Add the following big numbers. You can check the answer using your blocks or coins.

$$\begin{array}{r} 34 \\ +\,13 \\ \hline \end{array} \qquad \begin{array}{r} 53 \\ +\,25 \\ \hline \end{array} \qquad \begin{array}{r} 78 \\ +\,21 \\ \hline \end{array}$$

Student Exercises

How can we split these big numbers into smaller numbers? You can use your blocks or stones to make two sets if you need help. How many blocks or stones are in each set?

5 = _____ + _____	6 = _____ + _____
7 = _____ + _____	8 = _____ + _____
9 = _____ + _____	10 = _____ + _____
11 = _____ + _____	12 = _____ + _____
13 = _____ + _____	14 = _____ + _____

Measure the sides of each shape in inches. How long is each side? Add the sides together to find the length around the whole shape. Write your answer inside each shape. Be careful: you may have to round your answer up or down!

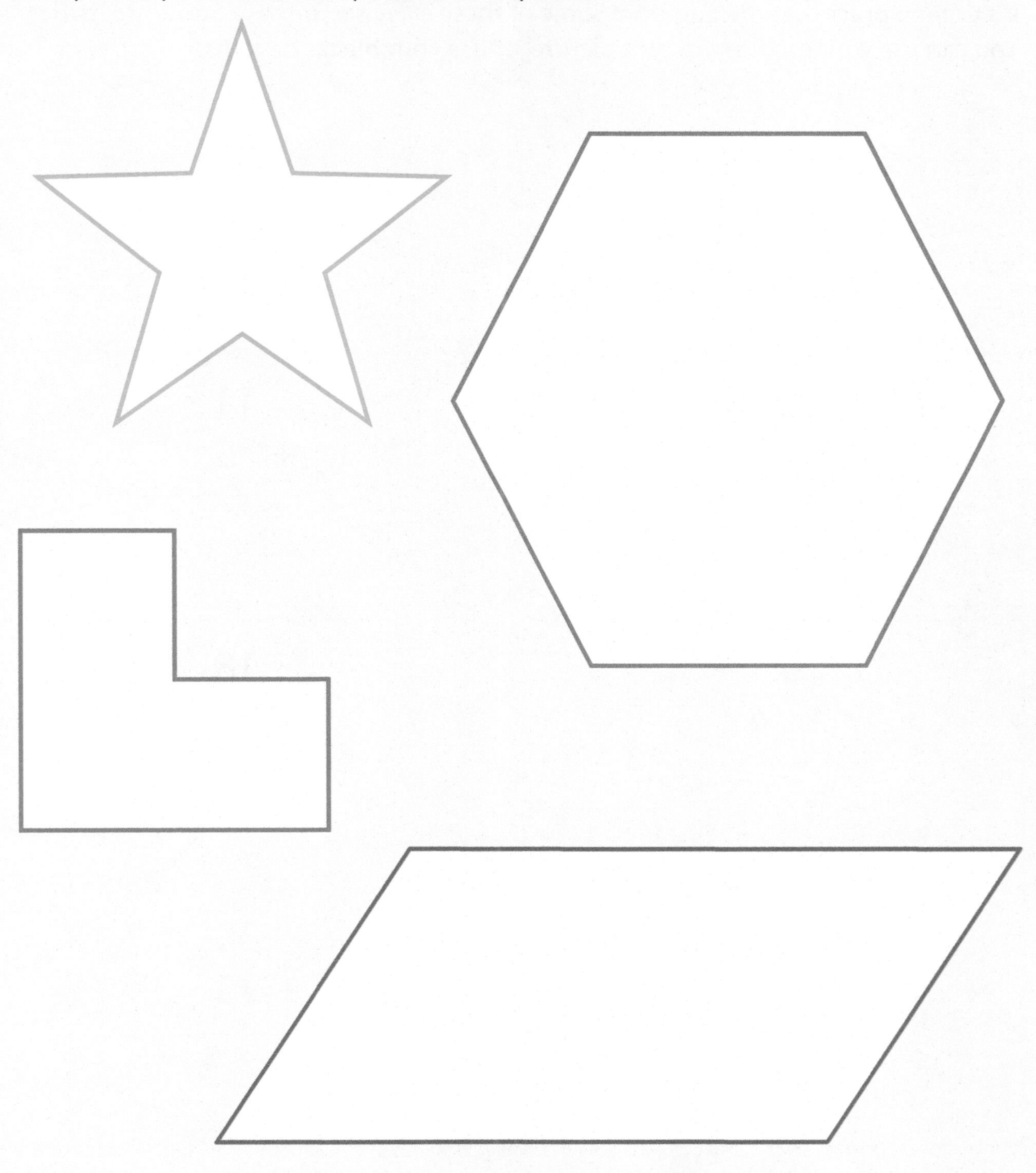

DAY 91

Practice

Student Exercises

It's time to practice more addition! Some of these exercises are easy. Some are harder. You can use your memory, draw a picture, or use your blocks or coins.

$$\begin{array}{r} 2 \\ +\ 5 \\ \hline \end{array}$$

$$\begin{array}{r} 6 \\ +\ 6 \\ \hline \end{array}$$

$$\begin{array}{r} 14 \\ +\ \ 0 \\ \hline \end{array}$$

$$\begin{array}{r} 11 \\ +\ \ 5 \\ \hline \end{array}$$

$$\begin{array}{r} 21 \\ +\ \ 4 \\ \hline \end{array}$$

$$\begin{array}{r} 18 \\ +\ \ 6 \\ \hline \end{array}$$

$$\begin{array}{r} 16 \\ +\ 71 \\ \hline \end{array}$$

$$\begin{array}{r} 35 \\ +\ 23 \\ \hline \end{array}$$

Find the missing numbers! Try to find the missing pieces in these equations. Use your imagination. You can also use your little stones from the math toolbox.

For the addition exercises, ask: "How many stones do you have to add to get the bigger number?" Hint: What number is smaller? Start with that number. How many stones do you need to add to the small number to get the big number?

Now let's look at the first subtraction exercise, "What minus 2 equals 7?" Ask, "How many stones do I need to start with? I need to take away two stones. And I need to have seven stones left. So how many stones should I start with?"

What plus 4 equals 5? __________

What plus 3 equals 6? __________

What minus 2 equals 7? __________

What minus 1 equals 6? __________

What plus 1 equals 10? __________

What plus 2 equals 8? __________

What minus 3 equals 9? __________

DAY 92 Patterns in Chess

This lesson explores mathematical patterns and distances in the game of chess. It is followed by one page of review exercises. This will require about 20 minutes of instruction from the parent/teacher.

Prayer

Pray your own prayer of thanksgiving and praise to God. Pray for His help on this lesson.

Memory

Spend a few minutes with both addition and subtraction flash cards (mix and match).

Activity

All day long you are moving. How do you get from one place to another? You have to move across a space. There is a special distance between each place. You use math to figure out the distance.

Have you ever played the game of chess? When you play chess, you move each of the pieces across the board. But each piece moves using a different pattern.

Pawn: The pawn can move one square at a time. It can only move forward.

Knight: The knight can move three squares at a time. The three squares have to be in an L shape. But the L shape can go any direction.

Rook: The rook can move all the way across the board, many squares at a time. It can move sideways, forward, or backward.

Bishop: The bishop can move all the way across the board diagonally.

Queen: The queen can move any direction. It can move all the way across the board.

King: The king can move one square at a time in any direction.

How Each Piece Can Move on the Chess Board

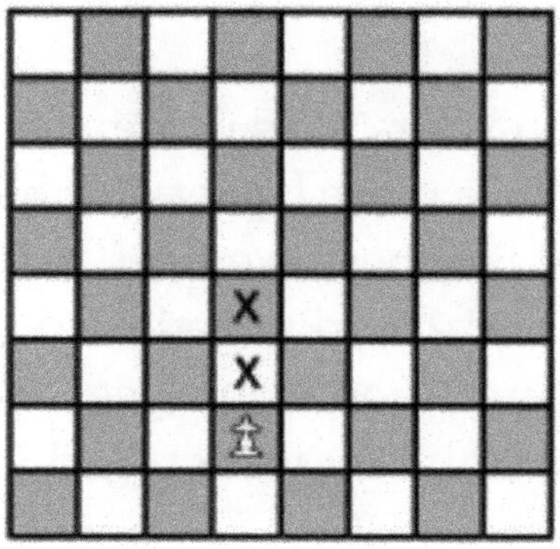

Pawn

Bishop

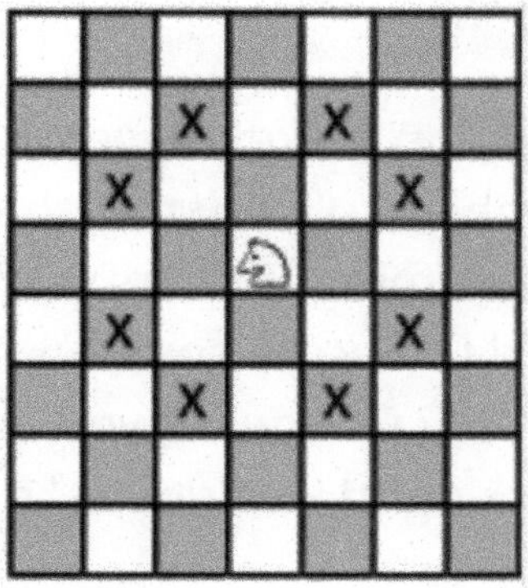

Knight

Rook

Queen

King

You can try this exercise using your own chess board. Or just imagine moving the chess pieces. How do you get the chess piece from one square to another?

1. Pick two squares. Can the pawn get from one square to the other? If so, how many moves will it take?
2. Pick two more squares. Can the knight get from one square to the other? If so, how many moves will it take?
3. Pick two squares. Can the rook get from one square to the other? If so, how many moves will it take?
4. Pick two squares. Can the king get from one square to the other? If so, how many moves will it take?

You can learn math by playing chess. See if you can learn to play the game. You will find that learning math can be fun. God made the world to work together in an orderly way, kind of like chess pieces on the board.

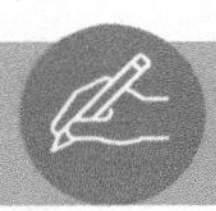

Student Exercises

Addition makes bigger numbers. Subtraction makes smaller numbers. God made numbers to work together. Sometimes numbers work together to make bigger numbers. Sometimes numbers work together to make smaller numbers. For these exercises, you need to decide how these numbers work together. How do 3 and 2 work together to make 1? Of course, you must use subtraction! If you added 3 and 2, you would get . . . that's right! 5! Write "+" or "-" in each circle to show how these numbers work together.

3 ◯ 2 = 1	6 ◯ 4 = 2
12 ◯ 9 = 3	10 ◯ 6 = 4
3 ◯ 2 = 5	3 ◯ 3 = 6
4 ◯ 3 = 7	10 ◯ 2 = 8
10 ◯ 1 = 9	5 ◯ 5 = 10

 Student Exercises

Let's add with small numbers and big numbers! Some of these will be easy. Some will be harder. Ask God to help you!

Do any of these exercises have numbers with chunks of 10? Remember, you need to add the numbers on the right first (the 1s). Then you will add the numbers on the left (the 10s).

$4 + 4 = $ ___	$3 + 3 = $ ___	$7 + 7 = $ ___
$8 + 8 = $ ___	$10 + 10 = $ ___	$30 + 30 = $ ___
$12 + 13 = $ ___	$24 + 22 = $ ___	$43 + 51 = $ ___

God made some numbers bigger than others. Put these numbers in the order God made for them! Use the symbols bigger than (>) and smaller than (<). Remember, the baby shark eats the bigger number!

2 ○ 4 ○ 6

5 ○ 4 ○ 3

15 ○ 10 ○ 5

20 ○ 50 ○ 60

12 ○ 13 ○ 32

51 ○ 50 ○ 49

20 ○ 25 ○ 43

78 ○ 80 ○ 82

A New Way to Subtract

DAY 94

This lesson explores subtraction with two-digit numbers, and is followed by one page of review exercises. This will require about 35 minutes of instruction from the parent/teacher.

Prayer

Pray your own prayer of thanksgiving and praise to God. Pray for His help on this lesson.

Memory

Spend a few minutes with both addition and subtraction flash cards (mix and match).

Lesson

You have learned how to add big numbers with two digits! Do you know what a digit is? *Digit* is another word for *number.* 1, 2, 3, 4, 5, 6, 7, 8, and 9 are all one-digit numbers. 39, 90, and 43 are some of the two-digit numbers.

Today we will learn how to subtract big numbers. We'll use your blocks. First, let's try this one: **68-18**. How do we do this? Start with six chunks of 10 and eight 1s. This makes 68!

68

Now subtract 18. That's one chunk of 10 and eight 1s.

-18

First, take away 8 single 1s from the 68. Do you have any left? No! Now take 1 chunk of 10 away from the 6 chunks of 10. How many chunks of 10 do you have left? You have 5 chunks left. This means 68 – 18 is 50. The answer is 50!

Now let's subtract these big numbers using **long subtraction**. Do you remember how we used long addition to add big numbers? Long subtraction is also like that!

First, we'll subtract the numbers on the right: 8 – 8 = 0. These are the 1s. Then we'll subtract the numbers on the left: 6 – 1 = 5. These are the chunks of 10. This is what it looks like:

10s	1s
6	8
– 1	8
5	0

Practice doing the following exercises using blocks or long subtraction:

68 – 35 ____	68 – 41 ____	68 – 26 ____

Subtraction gets harder with problems like this: 30 – 3. Let's use your blocks to try it. Remember, God will help you!

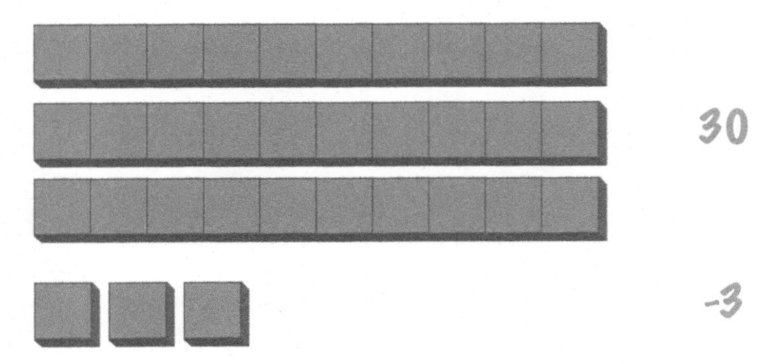

Look at the number 30. 30 has no single 1s. It only has 3 chunks of 10. So how would you take away the 3 single 1s? It's really easy. Do you remember how we turned 12 single 1s into a chunk of 10 and two single 1s? This time, we'll do the opposite. We'll take a chunk of 10 and turn it into 10 single 1s like this:

Subtract the 1s first. 10 – 3 = 7

Then subtract the chunks of 10s. 20 – 0 = 20

There's the answer! We have 2 chunks of 10 left, and 7 single 1s. That's 27!

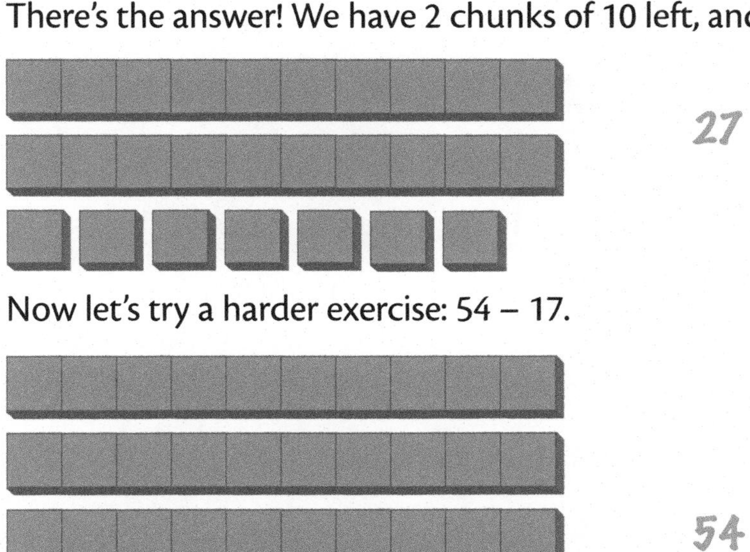

Now let's try a harder exercise: 54 – 17.

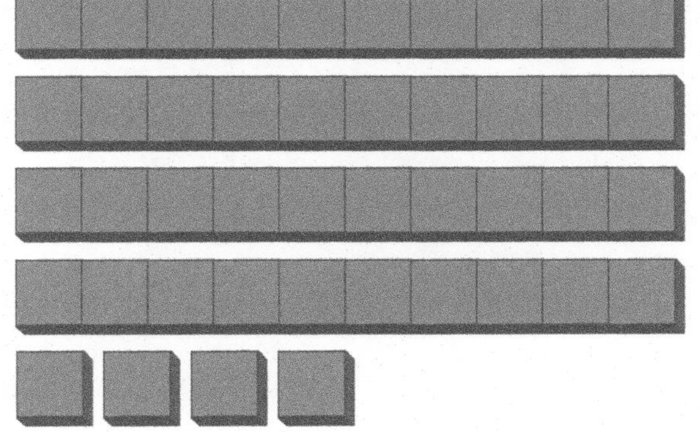

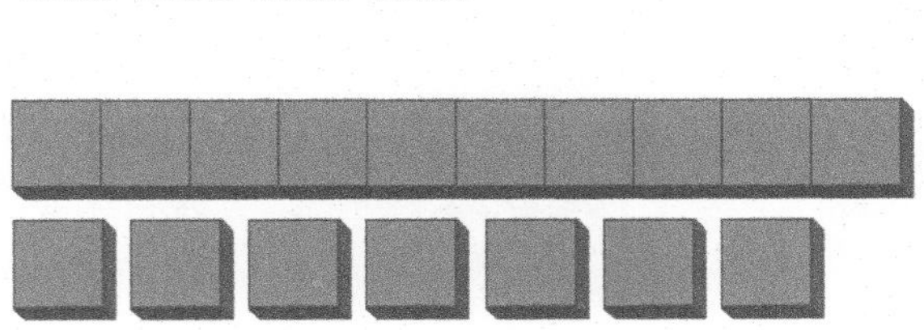

We don't have enough 1s to work with from 54. So now, we need to turn 1 of the chunks of 10 into 10 single blocks like this:

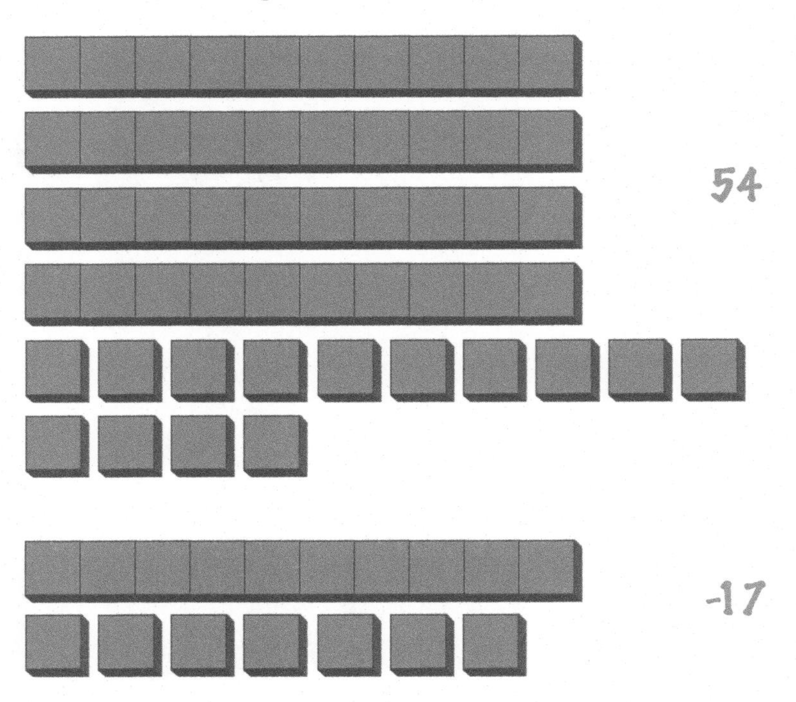

The 54 is broken up into 4 chunks of 10 and 14 single 1s. Now, we can subtract the 1s!

First subtract the singles: 14 – 7. How many 1s do we have? 7! Then subtract the chunks of 10: 4 - 1. How many chunks of 10 do we have now? 3! We have 3 chunks of 10 and 7 single 1s. So the answer is 37.

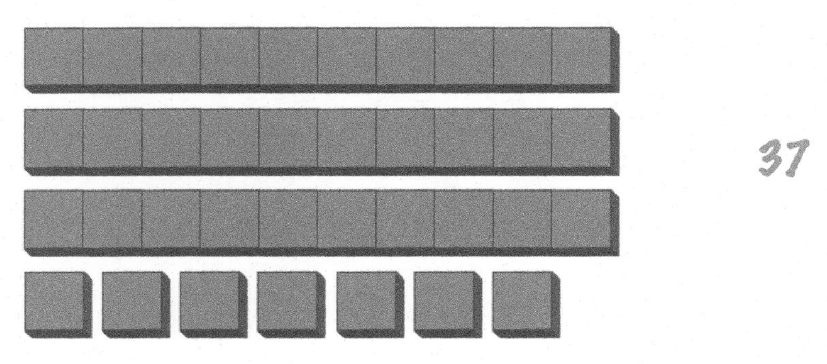

Here is the way we write this in long math:

$$\begin{array}{r} 54 \\ -\ 17 \\ \hline 37 \end{array}$$

Student Exercises

Add the 3 numbers in these exercises. Find the sum of the first two numbers. Cross the first two numbers out, and then add that sum to the last number for your final answer.

5 + 5 + 1 = 11	4 + 2 + 2 = ______
1 + 1 + 1 = ______	3 + 3 + 3 = ______
6 + 4 + 2 = ______	0 + 3 + 4 = ______
3 + 0 + 7 = ______	3 + 2 + 1 = ______

DAY 95 Practice

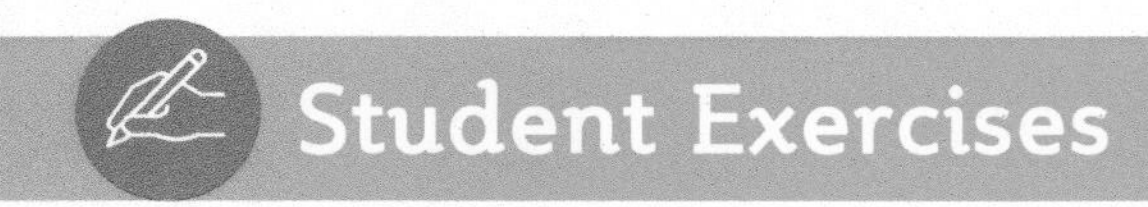

Student Exercises

Have you learned to subtract? Be thankful! God has given you a mind to learn these things. God is teaching you good things.

Subtract these numbers. Some are big. Some are small. You can use your memory, draw a picture, or use your blocks or coins.

11 − 5	8 − 3	12 − 8
15 − 7	28 − 7	83 − 43
66 − 35	49 − 26	36 − 14

Compare these sets. What shapes do the two sets have in common? Be careful: The shapes must also be the same color! Use crayons to draw the shapes that each set has in common. The shapes that are found in both sets will make your new set.

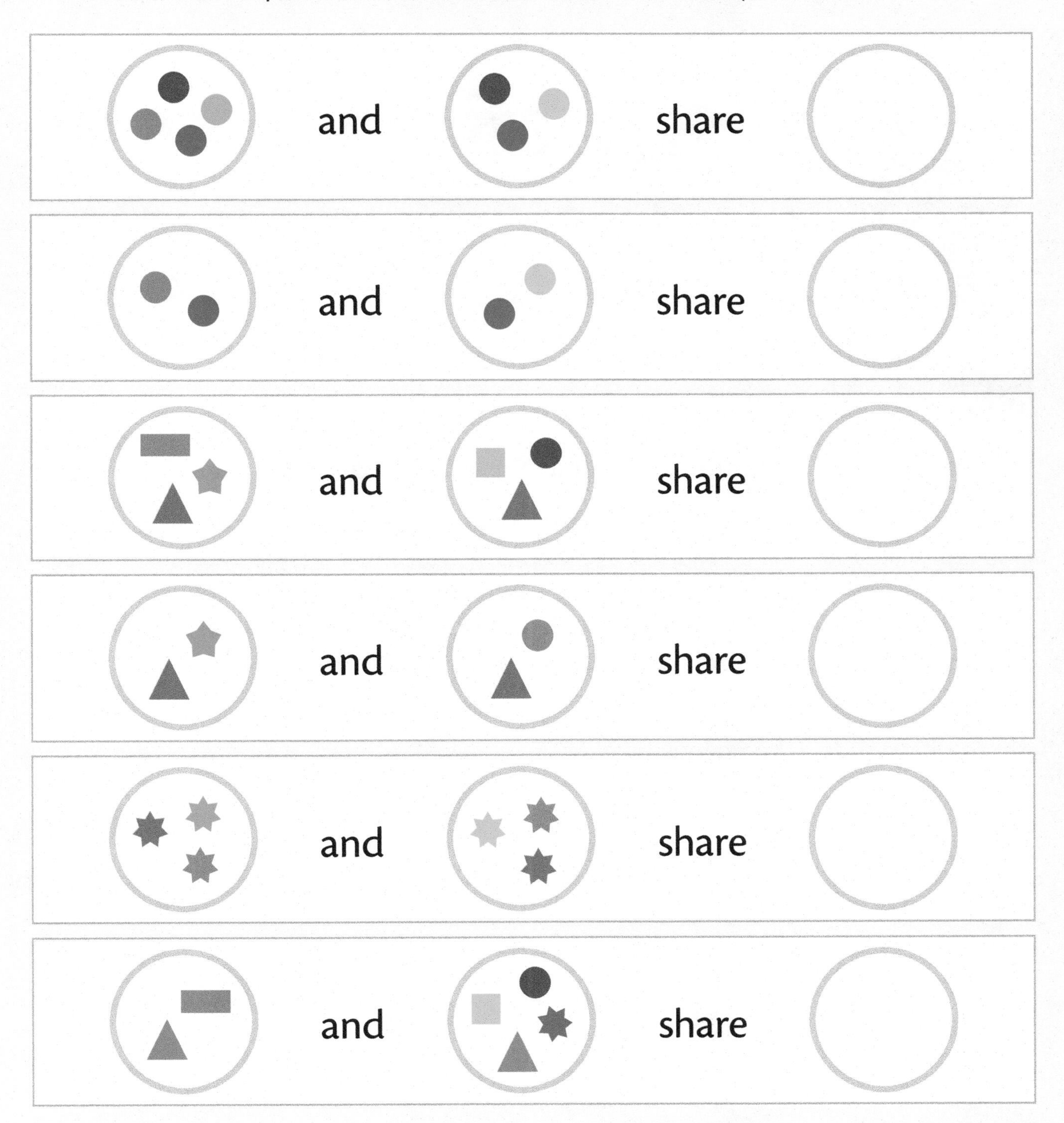

DAY 96 Playing with Patterns in Sudoku

This lesson further explores the use of math in games or culture, and is followed by one page of review exercises. This will require about 20 minutes of instruction from the parent/teacher.

Prayer

Pray your own prayer of thanksgiving and praise to God. Pray for His help on this lesson.

Memory

Spend a few minutes with both addition and subtraction flash cards (mix and match).

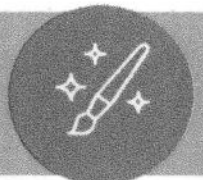

Activity (Extra Challenge)

> [God] has made everything beautiful in its time. (Ecclesiastes 3:11)

Why am I doing math? Maybe you are asking that question. These verses tell us that God has made everything beautiful. We should be happy in God. We should be excited about what He has made. We can be happy about using math in games. We should be glad to use math in our work. We use math to build houses. Today, we will use math to play a game. It is a game that is played by people all over the world. The game is called sudoku.

This is what a sudoku board looks like:

1	4	3	2
3	2	1	4
4	1	2	3
2	3	4	1

Which numbers can you see on the board? You can see 1, 2, 3, and 4. There are only three rules for this game:

1. Each board is made of sixteen small boxes. Look at the board below. Can you see the groups of four boxes in each corner? Each group of four boxes will have the numbers 1, 2, 3, and 4 in it. **Each number can only be used one time in each corner box.**
2. Each side-to-side row of boxes will have 1, 2, 3, and 4 in it. **Each number can only be used one time in each side-to-side row.**
3. Each up-and-down column of boxes will also have 1, 2, 3, and 4 in it. **Each number can only be used one time in each up-and-down column.**

Now we are going to play a sudoku game step by step.

4	3		
1	2	3	
		2	
2	1		

Step 1: Look at the first board on this page. Do you see the shaded corner group? It has all four numbers: 1, 2, 3, and 4!

Step 2: Now look at the shaded row of boxes below. What number is missing? Of course, the number 4 is missing. So we must write the number 4 in this square.

4	3		
1	2	3	
		2	
2	1		

4	3		
1	2	3	4
		2	
2	1		

Step 3: Look at the two shaded columns. Which numbers are missing here? That's right! 3 is missing in the first column. 4 is missing in the second column.

4	3		
1	2	3	4
		2	
2	1		

4	3		
1	2	3	4
3	4	2	
2	1		

Step 4: Notice, in the puzzle above, that there is now a row with 3, 4, and 2 in it. It needs a 1!

4	3		
1	2	3	4
3	4	2	1
2	1		

Step 5: Step 5: Now, in the puzzle above, let's find the missing numbers that go in the green-shaded corner group. Be careful. This one is harder. This corner is missing two

numbers: 3 and 4. But, where can the 3 go? It can't go on the left because there is already a 3 above it in the same column. That means the 3 belongs in the box on the right. What is the only number that can go in the box to the left of 3? That's right: 4!

4	3		
1	2	3	4
3	4	2	1
2	1	4	3

Step 6: Now you should be able to finish this puzzle. Go ahead and fill in the last two numbers.

Let's try a few more Sudoku puzzles. If you get stuck, your parent or teacher may be able to help you.

		2	1
	2	3	4
	3		
		4	3

4		2	1
2			3
3	4	1	

Student Exercises

Look at these sets. Circle the picture that is different from the others in each set. Can you explain why it is different?

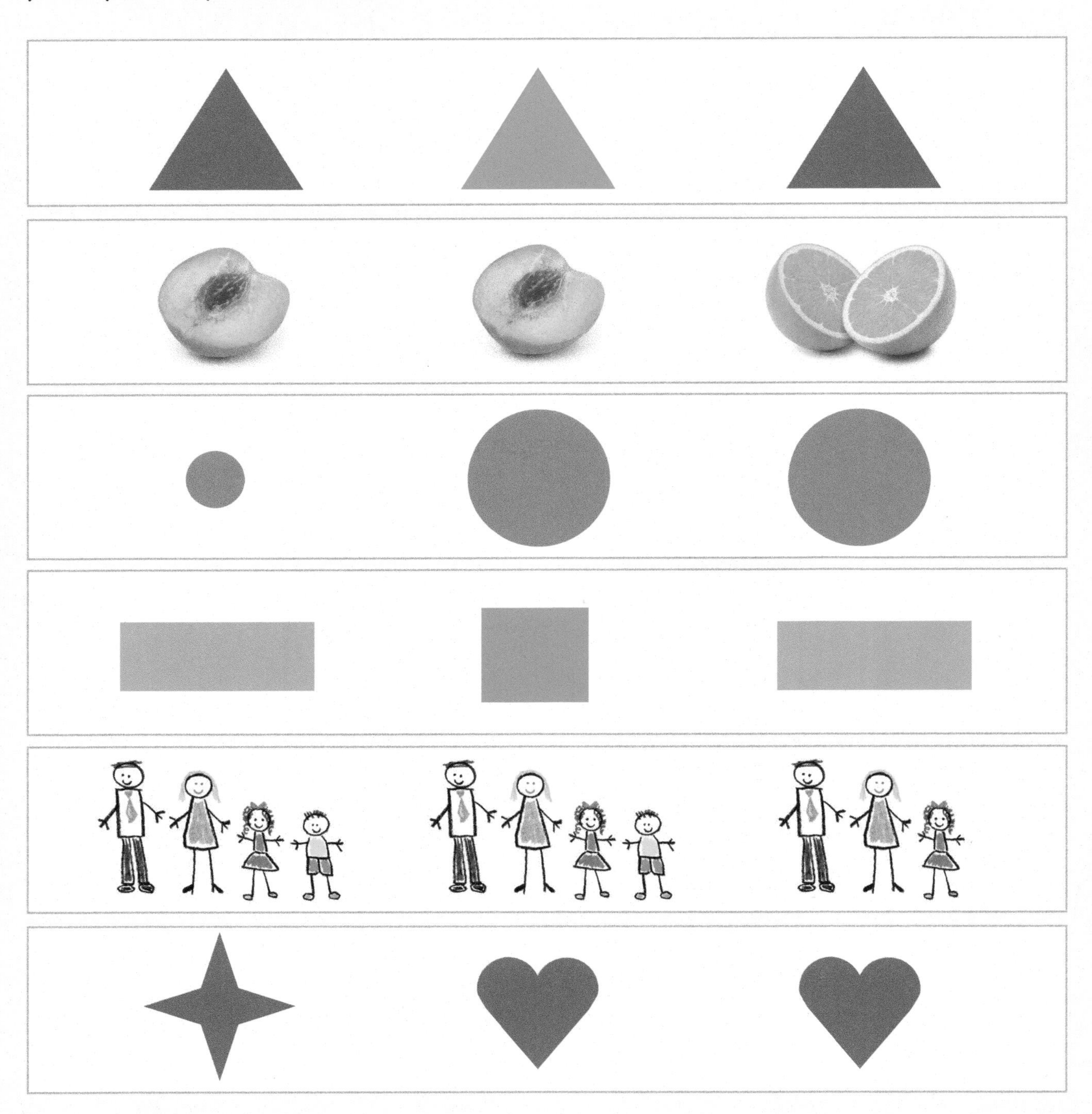

Student Exercises

God has made many ways for two smaller numbers to come together to make a bigger number. There are many right answers!

What smaller numbers can be used to make these bigger numbers? You can use your blocks or stones to make two sets if you need help. How many blocks or stones are in each smaller set? You're bringing these two smaller numbers together to make a bigger number. These are addition equations!

11 = _____ + _____	12 = _____ + _____
13 = _____ + _____	14 = _____ + _____
15 = _____ + _____	16 = _____ + _____
17 = _____ + _____	18 = _____ + _____
19 = _____ + _____	20 = _____ + _____

Fill in the numbers on the number line below. Start with 46. Then use this group of numbers to do the addition and subtraction exercises on the page. Remember, you will count to the right for addition. You will count backwards to the left for subtraction.

46 ____ ____ ____ ____ ____ ____ ____ ____ 55

46 + 4 = ____	46 + 5 = ____	46 + 6 = ____
54 − 4 = ____	54 − 5 = ____	54 − 6 = ____
49 + 2 = ____	48 + 0 = ____	52 − 3 = ____

Go Play with Bigger Numbers! DAY 98

This lesson integrates math into everyday life. This is an essential element to learning. The child is encouraged to apply God's patterns and wisdom to life in the home and community. Let's take a break from memory work and academic exercises, and identify ways in which to make math part of everyday life. The following are suggestions or examples, but other ideas may be added to the list.

Activity

Let's add or subtract bigger numbers every day! We can do math while . . .

1. Keeping track of money. How much money does Father or Mother spend on gas for the car driving to church? How much do they spend driving to church every week for one month? Write down how much money is spent every time they go to the gas station for a week (or a month). You may have to round the numbers up or down. Add the numbers together at the end of the week or month.

Or, you can keep track of how much money your family spends on groceries in a week. Write down how much money is spent every time you go to the grocery store. You may have to round the numbers up or down. Add the numbers together at the end of the week.

This exercise should be adapted to incorporate only numbers that would add up to less than 100.

2. Playing games and tracking points. You must add up points for Scrabble and Yahtzee. You'll use numbers bigger than 50! Add up the points as you go.

3. Filling up the pantry. Mother bought 35 potatoes. The family already ate 12 of them. How many are left? How many potatoes does Mother want to keep in the pantry? Use subtraction to figure out how many you need to buy at the store.

4. Making cookies. Jesus said, "Love your neighbor." How about making a batch of cookies for two or three neighbors. If you want to give 12 cookies to three different neighbors, how many cookies will you have to make?

$$12 + 12 + 12 = \_\_\_\_\_\_\_\_$$

When you have used math to serve others, thank God for teaching you about math. Pray that He would help you to find more ways to serve!

What is the longest trip you have ever taken?

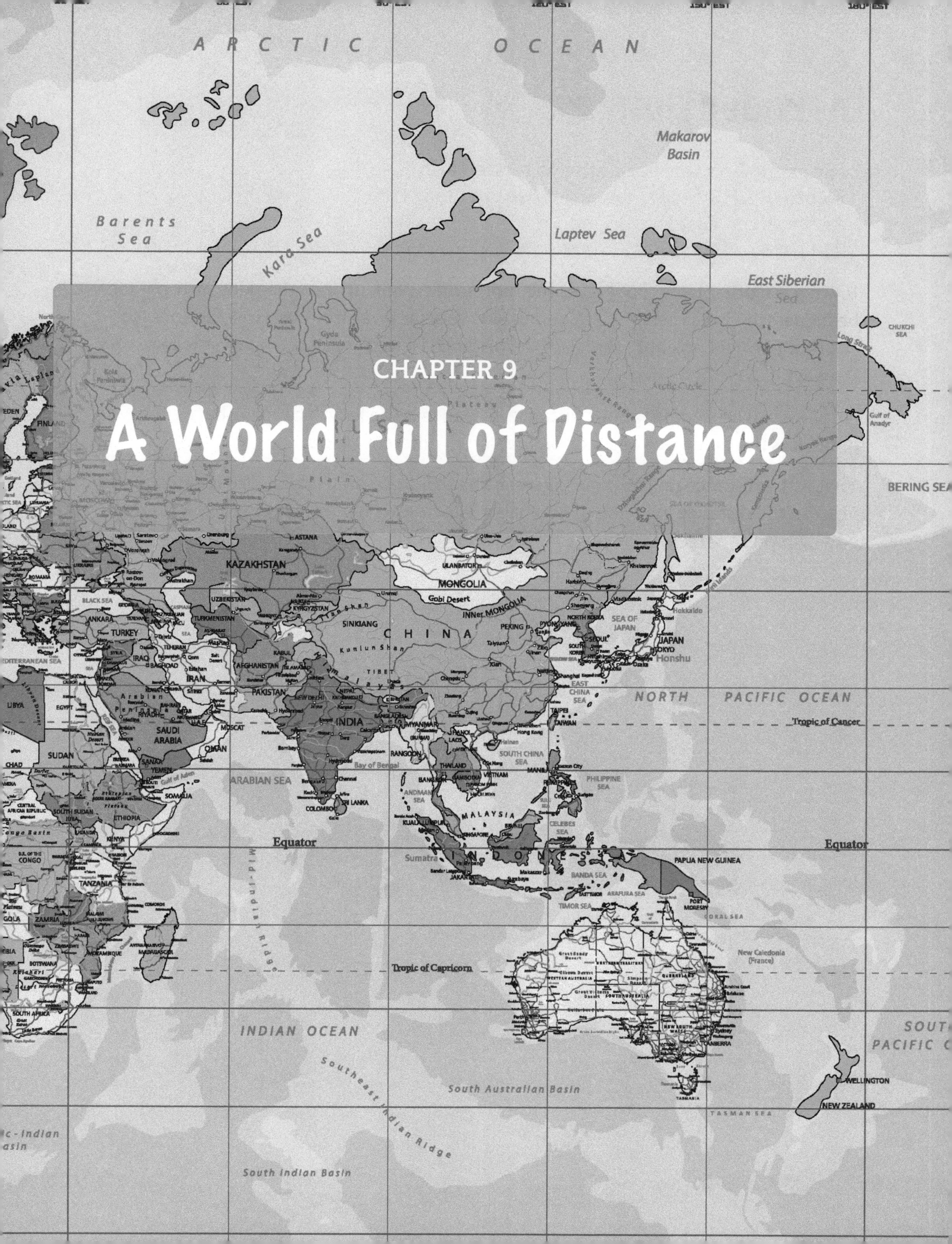

CHAPTER 9

A World Full of Distance

Introduction

> For as the heavens are high above the earth,
> So great is His mercy toward those who fear Him;
> As far as the east is from the west,
> So far has He removed our transgressions from us. (Psalm 103:11-12)

How big is God's mercy? Look over the mountains. Look up into the skies. You will see the bigness of God's mercy. He takes our sins very far away from us. How far does He take them away? Look at Psalm 103. The distance from the east to the west is very, very far away!

Today we will learn more about distance. Distance is the length of space between two things. It is the difference of space between two different places.

What is the distance on the page between the trunks of these trees? Use your ruler to measure the distance in inches. If you had a 3-inch-long hammock, could you pretend to hang it on the trees in the picture? Do you think the hammock would be too long?

Distances Around the House

DAY 99

This lesson explores distances in God's world, and is followed by one page of review exercises. This will require about 20 minutes of instruction from the parent/teacher.

Prayer

Our Father in Heaven, You have created a very big world full of short things and long things. You are big. Your mercy is so big for those who worship You. We praise You today and ask for Your help as we learn about Your world. Amen.

Memory

Spend a few minutes with both addition and subtraction flash cards (mix and match).

Let's explore God's world by measuring distances again! Today we'll measure distance by counting the steps you take between places.

Pick four places in the house or in the yard. Write a name for each place below:

1. ________________________________
2. ________________________________
3. ________________________________
4. ________________________________

For each activity below, start at the first place number. Walk to the second place number. Count the number of steps you take as you walk. Write down your number of steps:

- 1 to 2: ____________
- 2 to 3: ____________
- 3 to 4: ____________
- 4 to 1: ____________

Which was the longest distance you traveled? ___________

Which was the shortest? _________________

Which was the most fun for you? ______________________

Which was the most difficult? ___________________

God has laid out many interesting paths for us to use! For this last exercise, walk from place 2 to place 3 using different paths. Write down how many steps you took on each path:

- Path 1: __________
- Path 2: __________
- Path 3: __________

> A man's heart plans his way,
> But the LORD directs his steps. (Proverbs 16:9)

Student Exercises

Try finding the answers for these addition and subtraction exercises using your memory. You can also use your blocks, coins, or fingers if you need help.

You will have to use long addition and long subtraction for the bigger numbers. Remember, you need to add or subtract the numbers on the right first (the 1s). Then you will add or subtract the numbers on the left (the 10s).

$\begin{array}{r} 5 \\ +\,3 \\ \hline \end{array}$	$\begin{array}{r} 7 \\ +\,2 \\ \hline \end{array}$	$\begin{array}{r} 8 \\ +\,5 \\ \hline \end{array}$
$\begin{array}{r} 12 \\ +\;\;2 \\ \hline \end{array}$	$\begin{array}{r} 32 \\ +\,54 \\ \hline \end{array}$	$\begin{array}{r} 42 \\ +\;\;6 \\ \hline \end{array}$
$\begin{array}{r} 10 \\ -\;\;7 \\ \hline \end{array}$	$\begin{array}{r} 11 \\ -\;\;4 \\ \hline \end{array}$	$\begin{array}{r} 15 \\ -\;\;7 \\ \hline \end{array}$
$\begin{array}{r} 13 \\ -\;\;4 \\ \hline \end{array}$	$\begin{array}{r} 64 \\ -\,21 \\ \hline \end{array}$	$\begin{array}{r} 85 \\ -\,34 \\ \hline \end{array}$

DAY 100 **Practice**

Student Exercises

This is a new exercise. Find the distance between the two points on the number line. Pretend to jump from point to point. You will count how many jumps it takes to get from a starting point to an ending point.

1 2 3 4 5 6 7 8 9 10 11 12 13 14 15 16 17 18 19 20

1 and 6 ______	4 and 3 ______
2 and 5 ______	5 and 2 ______
3 and 4 ______	6 and 1 ______
10 and 15 ______	4 and 12 ______
20 and 16 ______	14 and 6 ______
3 and 10 ______	5 and 15 ______

Here is more practice with subtraction exercises. There are two different kinds of exercises here. Can you find both patterns?

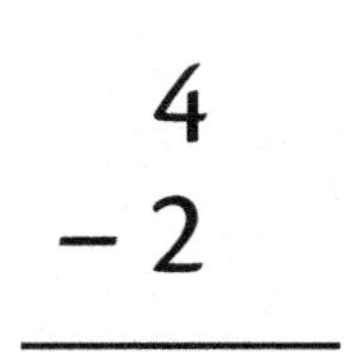

$$\begin{array}{r} 8 \\ -\ 4 \\ \hline \end{array}$$

$$\begin{array}{r} 9 \\ -\ 9 \\ \hline \end{array}$$

$$\begin{array}{r} 10 \\ -\ \ 5 \\ \hline \end{array}$$

$$\begin{array}{r} 5 \\ -\ 5 \\ \hline \end{array}$$

$$\begin{array}{r} 2 \\ -\ 1 \\ \hline \end{array}$$

$$\begin{array}{r} 1 \\ -\ 1 \\ \hline \end{array}$$

$$\begin{array}{r} 12 \\ -\ \ 6 \\ \hline \end{array}$$

$$\begin{array}{r} 6 \\ -\ 3 \\ \hline \end{array}$$

DAY 101 The Car and the Crow

This lesson explores distances in the real world by comparing different routes, and is followed by two pages of review exercises. This will require about 20 minutes of instruction from the parent/teacher.

Prayer

Pray your own prayer of thanksgiving and praise to God. Pray for His help on this lesson.

Memory

Spend a few minutes with both addition and subtraction flash cards (mix and match).

Lesson

How do you get from one place to another? Walking is one way. But there are also other ways to travel. People drive in cars. Crows fly. God gave wings to the crow. He gave minds to people. He gave minds to people so they can build cars.

Cars usually have to drive on roads. They cannot cut across fields, ponds, and rivers. They cannot drive through houses and buildings.

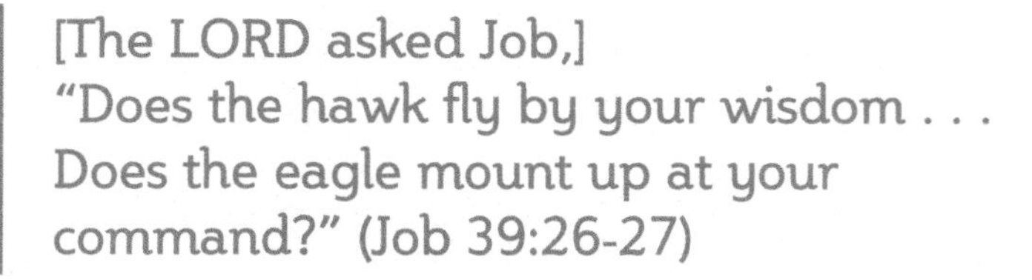

> [The LORD asked Job,]
> "Does the hawk fly by your wisdom . . .
> Does the eagle mount up at your
> command?" (Job 39:26-27)

Eagles, hawks, and crows can soar over the houses. They can fly over the fields, ponds, and rivers. They don't have to walk on the roads. They are free to fly wherever they want in the wide-open sky. They can fly straight from one place to another.

Activity

Let's explore this city! We're going to measure the distance between places. How far does the crow fly from one place to another? How far does the car drive? Let's find out!

Here's the map of the city:

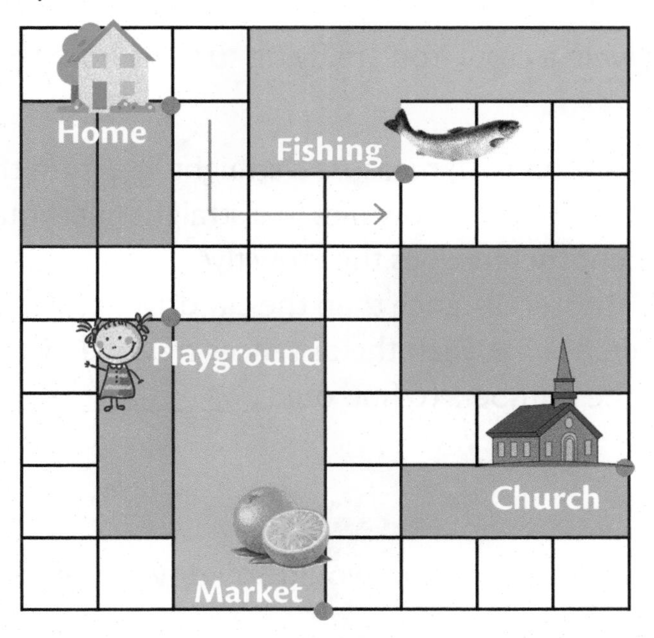

The green shapes are parks. The blue shapes are lakes. The black lines are roads. Special places are marked with red dots. Can you see the home? The playground? The market? And the church?

Imagine that your family lives in this city. Imagine that you and your parent drive the car from your home to the market to buy groceries. On the way home, you decide to stop at the playground. Soon we will find the distance from your house to the market. Then we will find the distance from the market to the playground. How could you figure this out?

The sides of the squares are each 1 centimeter long. Pretend that each centimeter is a city block. (A city block is a little longer than 300 feet, or about 100 meters.) Let's find out how many blocks it is to each place.

Here is one example for you. Let's go fishing today! Follow the red line from your house to the fishing spot. How many blocks would you have to walk? That's right! Four!

1. Find a way from your house to the market. Count the number of centimeters or blocks. How many centimeters or blocks would you have to drive to get there?

   ____

2. Find a way from the market to the playground. Count the number of centimeters or blocks. How many centimeters or blocks would you have to drive to get from the market to the playground? ___________

Now imagine that you were a crow. You are flying straight from the market to the playground.

1. Use your centimeter ruler to measure a straight line from the market to the playground. Remember, the crow flies in a straight line. You may have to round up. How many centimeters does the crow fly? ___________
2. The crow flies a shorter distance than the car drives to get to the playground. What is the difference between the distances traveled by the car and by the crow? Subtract the numbers to find out. ___________ – ___________ =

   ____

Now try these:

1. How many centimeters or blocks would you drive to get from your home to the church? ___________
2. How far would the crow have to fly (in centimeters)? ___________
3. Can you find a distance between two places that would be the same for the crow and the car? ____________________

Student Exercises

Round each number to the nearest 10. Sometimes you will round up. Sometimes you will round down. Do all the numbers need to be rounded?

92 → ______	51 → ______
86 → ______	47 → ______
75 → ______	35 → ______
69 → ______	24 → ______
55 → ______	4 → ______

God made some numbers bigger than other numbers. Put these numbers in the order God made for them!

Fill in each blank with the correct symbol: < (smaller than) or > (bigger than). Then read the math sentence. Remember, the baby shark eats the bigger number!

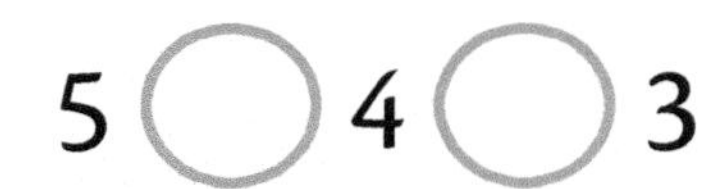

11 ◯ 7 ◯ 12

0 ◯ 8 ◯ 0

12 ◯ 22 ◯ 21

16 ◯ 6 ◯ 61

Practice DAY 102

Student Exercises

Find the distance in each exercise. Each child's house is marked by a red dot. If you're driving a car, count the number of city blocks (or centimeters) to find out how far you need to go. If you're flying like a crow, use your ruler to measure how far you need to fly. Each centimeter is one city block.

Amy
Candice
Bobby
Dan
Emily

Amy's house to Bobby's house by car: ___________

Amy's house to Candice's house by car: ___________

Bobby's house to Emily's house by car: ___________

Bobby's house to Emily's house by crow: ___________

Candice's house to Dan's house by crow: ___________

Dan's house to Candice's house by car: ___________

How many chunks of 10 do you need to make these numbers? How many singles (1s) do you need? Write the answers in the blanks.

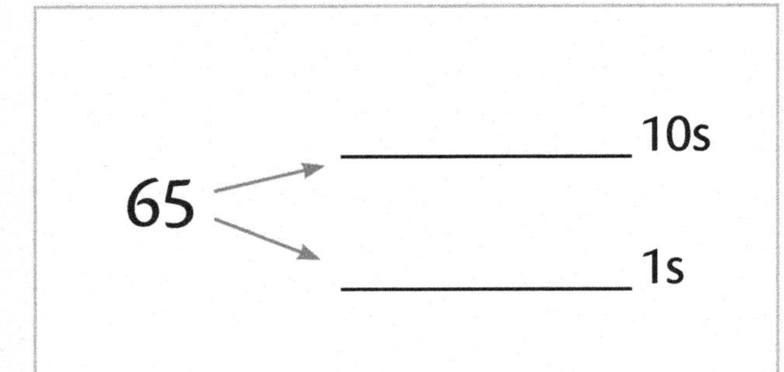

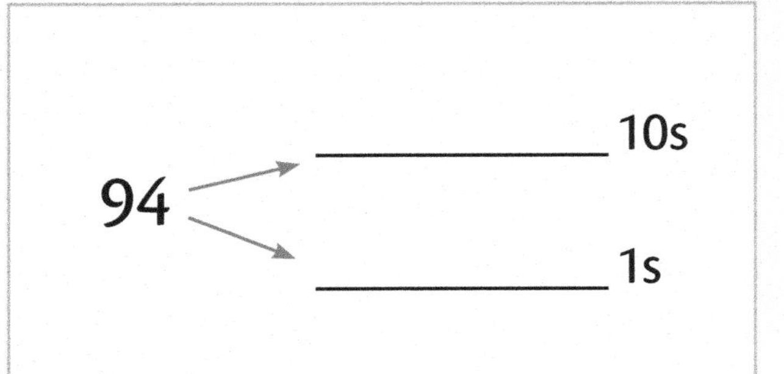

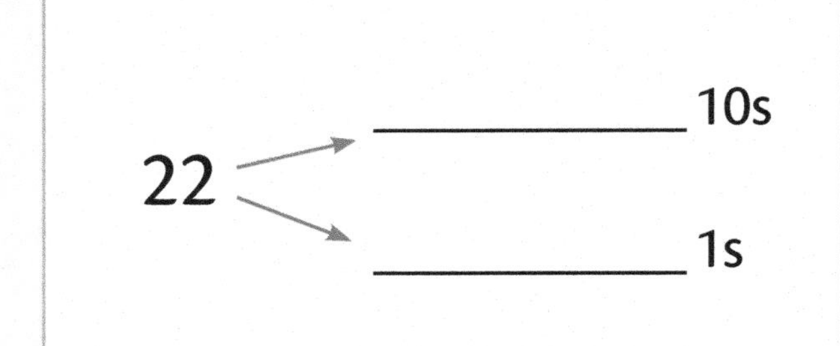

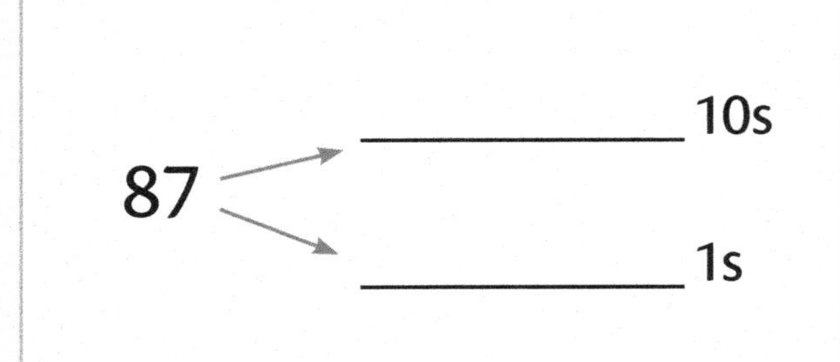

45 → ________ 10s
→ ________ 1s

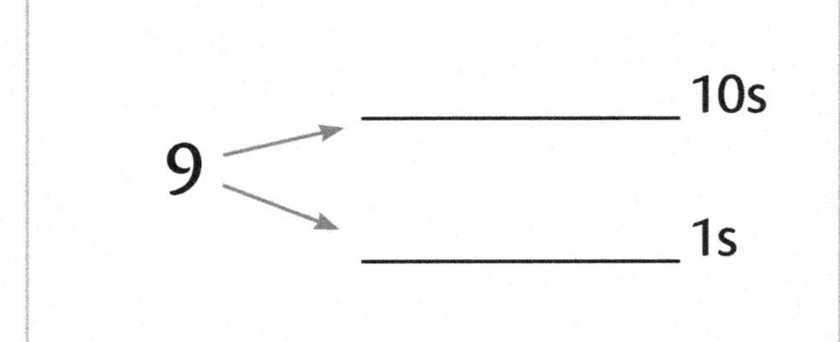

30 → ________ 10s
→ ________ 1s

43 → ________ 10s
→ ________ 1s

Measuring Weight DAY 103

This lesson introduces scales and weights. It is followed by two activities and two pages of exercises. This will require about 30 minutes of instruction from the parent/teacher.

Prayer

Pray your own prayer of thanksgiving and praise to God. Pray for His help on this lesson.

Memory

Spend a few minutes with both addition and subtraction flash cards (mix and match).

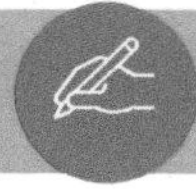

Lesson

> Honest weights and scales are the LORD's;
> All the weights in the bag are His work (Proverbs 16:11).

Today you will learn how to measure weight. Americans usually weigh stuff using ounces and pounds. Most other countries use grams and kilograms.

There are 16 ounces in 1 pound.

There are 1,000 grams in 1 kilogram.

If you dropped a bowling ball on your foot, you might break your toes. If you dropped a golf ball on your foot, you wouldn't break anything. Why not? The bowling ball is heavier. Most bowling balls weigh about 15 pounds (7 kilograms)! A golf ball isn't even 1 pound (1/2 kilogram)!

Actually, just so you know, 150 golf balls will weigh the same as 1 bowling ball!

Some scales compare the weight of one thing with another. It works a little bit like a teeter totter. Which is the heavier ball?

When your mom or dad buy fruit and vegetables at the store, they have to use a scale. It usually has a dial and numbers like this one.

Most food and vegetables are sold by the pound or the kilogram.

2 medium-sized oranges make up about 1 pound.

5 medium-sized oranges make up about 1 kilogram.

We can use different scales to check our own weight. Sometimes the scales have a pointer that points to your weight. Sometimes the scales give you a digital number. How much do these boys weigh? If you look carefully at the first scale, you can see the red line pointing between 20 and 30.This boy weighs 25 kilograms! The other boy is using a digital scale. A digital scale shows the weight with a number instead of using a pointer. This boy weighs 72.2 pounds.

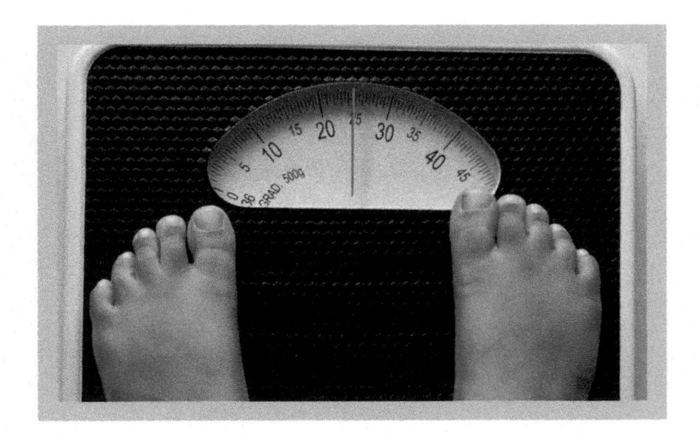

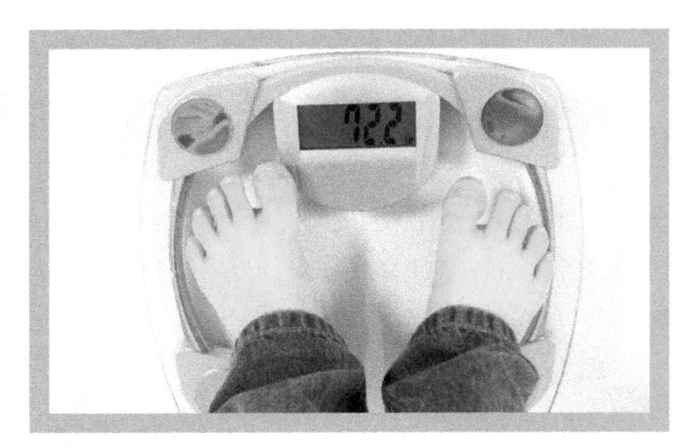

Student Exercises

Let's pretend to go to the store!

Your mom or dad has asked you to buy 10 oranges. There are 2 oranges in a pound. How many pounds do you need to buy?

____

1 pound

1 pound

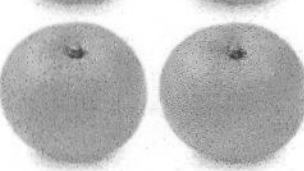

1 pound

1 pound

1 pound

The average cantaloupe weighs 3 pounds. Your mom wants 3 cantaloupes. How many pounds of cantaloupes will you buy?

____

3 pounds

3 pounds

3 pounds

The cantaloupes costs $1 a pound. That's $9.
The oranges cost $1 a pound. That's $5.
How much money will you need to buy the oranges and the cantaloupes?

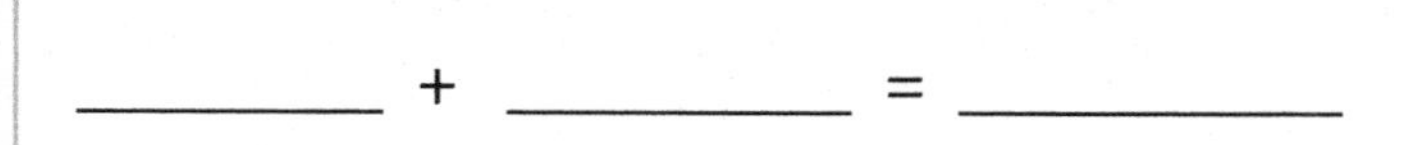

________ + __________ = __________

Avoid weighing full grown mastiffs.

Find a pet to weigh. It is hard to weigh a dog or cat. They don't stand still very long on the scale. Here's how to do it:

Weigh yourself on the scale.

Your weight ________________

Pick up your pet, and step on the scale again. Ask a helper to read the weight.

Your weight ________________

Weight of you with your pet ________________

Subtract your weight from the weight of you with your pet. The answer is your pet's weight!

God makes some things heavier than others. Human beings make some heavy things like cars. We make lighter things like balloons! In the exercises below, think about which thing weighs more than the other. Just like a teeter totter, the heavier side of a scale always goes down. The lighter side goes up. Circle the scale that's tipped the right way.

Extra Challenge

Here is a fun math game to play. It's called Wonder Squares! God made a wonderful pattern that you will discover in this game. The rules are simple. A wonder square has three sideways rows, three up-and-down columns, and two diagonals. That makes eight patterns. All three numbers in all 8 patterns are supposed to add up to the same number. Fill in the math square below. Make sure all eight patterns add up to the same number.

2	7	6	→15
9	5	1	→15
4	3	8	→15

↙15 ↓15 ↓15 ↓15 ↘15

1		2
2	1	3

Hint: Make sure all eight patterns of three numbers add up to 6. (2+1 + 3 = 6.)

5		7
	4	
1		3

Hint: Make sure all eight patterns of three numbers add up to 12. (5 + 4 + 3 = 12.)

DAY 104 Practice

Student Exercises

Now, let's pretend to weigh things using blocks instead of pounds or kilograms. The scales in this lesson can weigh things by balancing the two sides. When the blocks balance with the thing on the other side, we know the thing's weight. Count the blocks and write the number in the blank.

How many blocks equal the same weight as the orange?

____

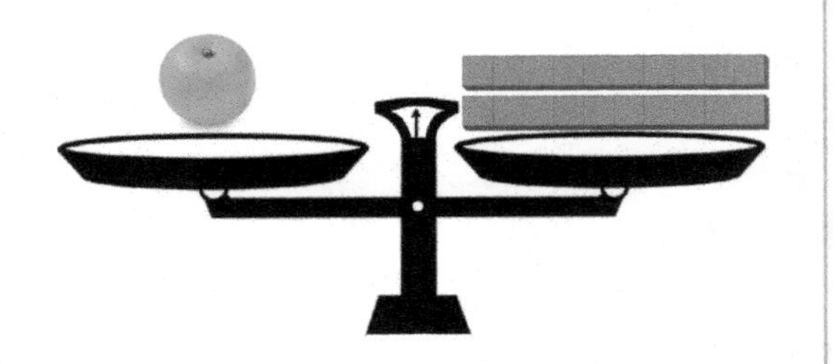

How many blocks equal the same weight as the ping pong ball?

____

How many blocks equal the same weight as these scissors?

____

How many blocks equal the same weight as this big bowling ball?

____

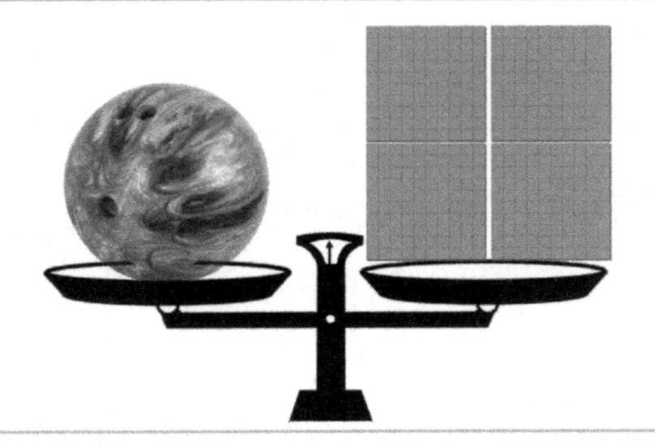

Try finding the answers for these addition and subtraction exercises using your memory. You can also use your blocks, coins, or fingers if you need help. You will have to use long addition and long subtraction for the bigger numbers.

$\begin{array}{r} 5 \\ +\ 5 \\ \hline \end{array}$	$\begin{array}{r} 4 \\ +\ 8 \\ \hline \end{array}$	$\begin{array}{r} 8 \\ +\ 6 \\ \hline \end{array}$
$\begin{array}{r} 7 \\ +\ 9 \\ \hline \end{array}$	$\begin{array}{r} 14 \\ +\ 13 \\ \hline \end{array}$	$\begin{array}{r} 26 \\ +\ 31 \\ \hline \end{array}$
$\begin{array}{r} 7 \\ -\ 5 \\ \hline \end{array}$	$\begin{array}{r} 6 \\ -\ 3 \\ \hline \end{array}$	$\begin{array}{r} 13 \\ -\ 7 \\ \hline \end{array}$
$\begin{array}{r} 16 \\ -\ 8 \\ \hline \end{array}$	$\begin{array}{r} 56 \\ -\ 35 \\ \hline \end{array}$	$\begin{array}{r} 87 \\ -\ 35 \\ \hline \end{array}$

DAY 105 Distance in the Bible

This lesson introduces the student to distances in biblical times, and is followed by one page of review exercises. This will require about 10 minutes of instruction from the parent/teacher.

Prayer

Pray your own prayer of thanksgiving and praise to God. Pray for His help on this lesson.

Memory

Spend a few minutes with both addition and subtraction flash cards (mix and match).

Lesson

Cars can go very fast. They can drive over very long distances. We measure how far we drive using miles or kilometers. A mile is a little bit longer than a kilometer.

Some people run for a mile. That's a long way! You can run a mile in 10 minutes. It takes about 30 to 40 minutes to walk a mile. But you can drive a mile in the car in 1 minute.

Activity

Ask your parent or teacher to help you with these questions:

- How many miles (or kilometers) is it from your church to your home? ________
- How many miles (or kilometers) is the closest hospital from your home? ________
- How many miles (or kilometers) is the closest grocery store from your home?

  ____

When Jesus came to the earth 2,000 years ago, there were no cars or buses. People walked a long way to get places. If they didn't walk, they had to ride an animal.

The next page shows a map of the country where Jesus lived. Read the following Bible verses and answer the questions. You can use the map and your centimeter ruler to figure out how far they walked. Each centimeter stands for 10 miles. Two centimeters is 20 miles. Four centimeters is 40 miles.

- Read Luke 2:1-7. How far did Mary and Joseph walk from Nazareth to Bethlehem? _________
- Read Matthew 4:12-17. How far did Jesus walk from Nazareth to Capernaum? _________
- Read 2 Samuel 5:1-5. How far did David walk from Hebron to Jerusalem? _________

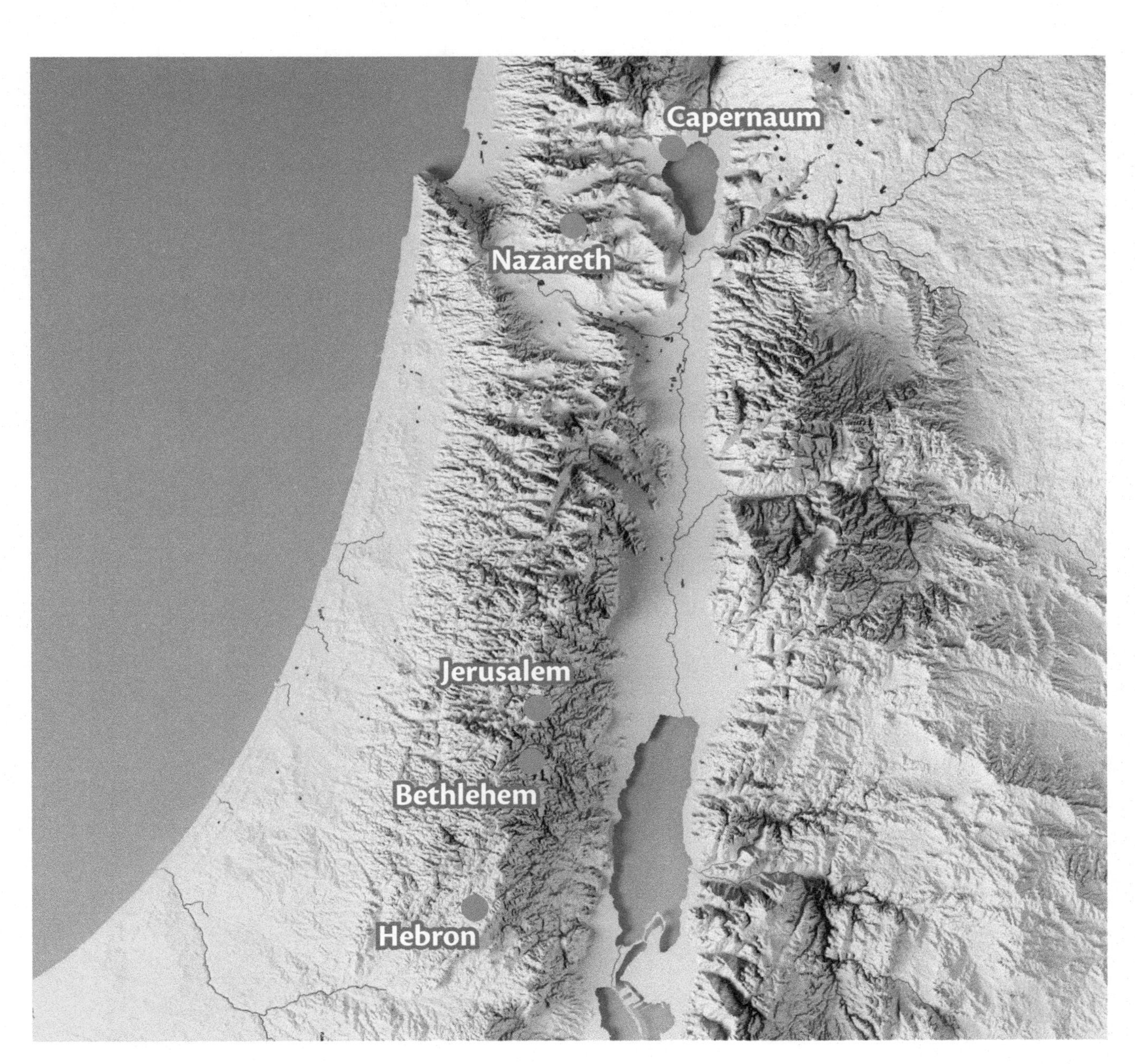

Student Exercises

Each exercise starts with two sets of numbers. Compare these sets. Which numbers do they have in common? Write these numbers in the space provided. This makes a new set of numbers! If they don't have any numbers in common, you will make an empty set.

Set {20, 30, 40} and {10, 20, 30} share = {________}

Set {45, 40, 35} and {35, 30, 25} share = {________}

Set {46, 18, 4} and {18, 4, 46} share = {________}

Set {0, 9, 99} and {8, 91, 98} share = {________}

Set {10, 20, 30, 40} and {20, 40, 60, 80} share = {________}

Set {13, 23, 33, 43} and {23, 32, 43, 52} share = {________}

Practice DAY 106

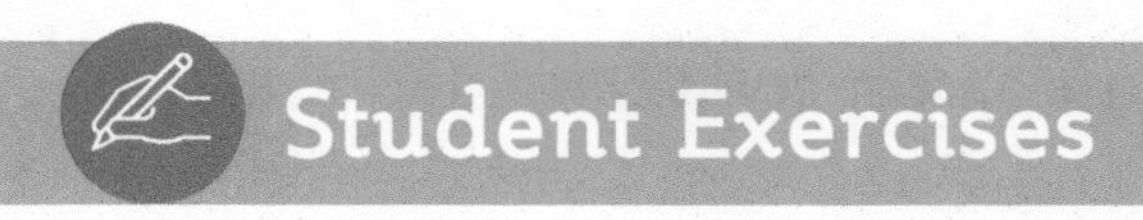

Student Exercises

Let's subtract bigger numbers using our blocks. When you subtract or take away blocks, try crossing out some of the blocks in the picture. Then count how many are left.

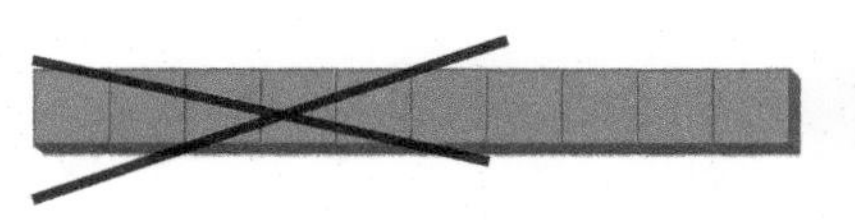

13 - 6 = __________

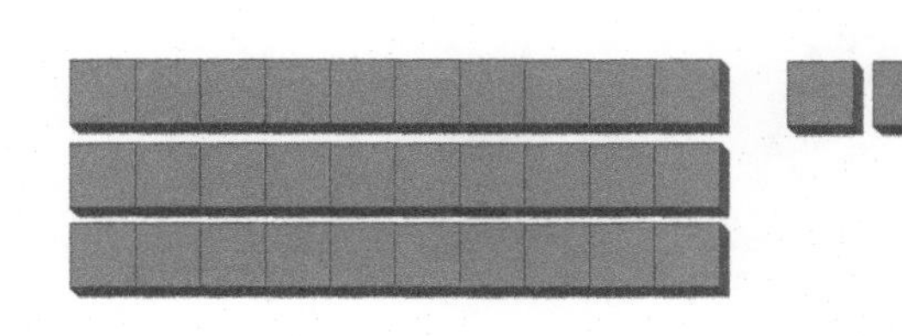

32 - 7 = __________

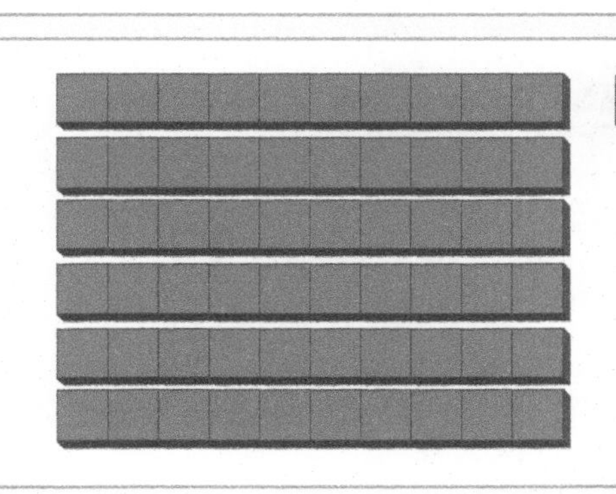

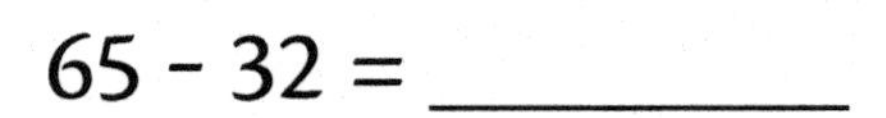

65 - 32 = __________

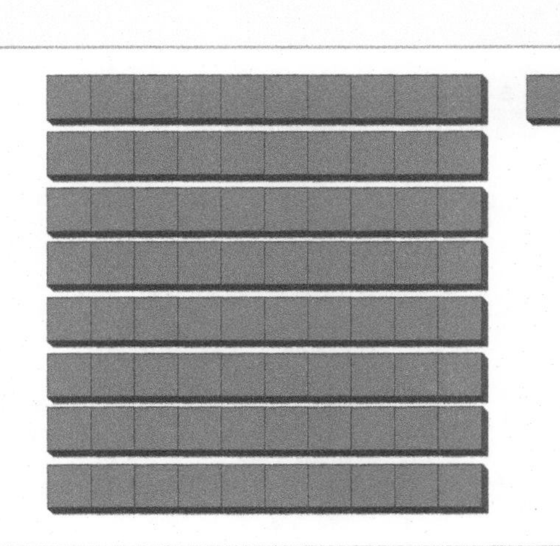

84 - 6 = __________

Let's practice some harder addition and subtraction exercises. Try to find the answers for these addition and subtraction exercises using your memory. You can also use your blocks or coins if you need help.

$\begin{array}{r} 6 \\ +\,7 \\ \hline \end{array}$	$\begin{array}{r} 6 \\ +\,9 \\ \hline \end{array}$	$\begin{array}{r} 8 \\ +\,5 \\ \hline \end{array}$
$\begin{array}{r} 6 \\ +\,8 \\ \hline \end{array}$	$\begin{array}{r} 17 \\ -\;\,6 \\ \hline \end{array}$	$\begin{array}{r} 17 \\ -\;\,9 \\ \hline \end{array}$
$\begin{array}{r} 18 \\ -\;\,8 \\ \hline \end{array}$	$\begin{array}{r} 18 \\ -\;\,9 \\ \hline \end{array}$	$\begin{array}{r} 14 \\ -\;\,6 \\ \hline \end{array}$

Practice DAY 107

Student Exercises

Find the distance in each exercise. Each child's house is marked by a red dot. If you're driving a car, count the number of city blocks (or centimeters) to find out how far you need to go. If you're flying like a crow, use your ruler to measure how far you need to fly. Each centimeter equals one city block.

Adam's house to Erik's house by car: __________

Adam's house to Dorothy's house by car: __________

Adam's house to Dorothy's house by crow: __________

Cody's house to Barbie's house by crow: __________

Cody's house to Barbie's house by car: __________

Barbie's house to Dorothy's house by car: __________

Let's think about this! We can add 2 bananas and 1 banana to get 3 bananas. We can add 2 apples and 2 apples to get 4 apples. But what about this?

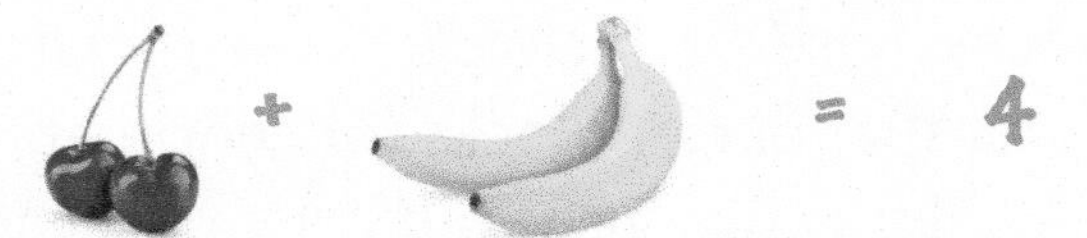

Why is this so confusing? God created different kinds of fruits, trees, and animals. We cannot treat them like they are the same or it would be confusing. We cannot add two different sets that do not have anything in common with each other.

> Then God said, "Let the earth bring forth grass, the herb that yields seed, and the fruit tree that yields fruit according to its kind, whose seed is in itself, on the earth"; and it was so. (Genesis 1:11)

Now . . . both of these sets do have something in common with each other. Both sets are made of fruit. Now we can add 2 banana fruits and 2 cherry fruits and get 4 fruits.

Each member of the following sets are fruits created by God.

Let's count fruit in these baskets, and add up the numbers!

Count only the apples in each basket. Write the numbers down and add them.

__________ + __________ = __________

Count only the oranges in each basket. Write the numbers down and add them.

__________ + __________ = __________

Count all the fruit in each basket. Write the numbers down and add them.

__________ + __________ = __________

Go Find Distances! DAY 108

This lesson integrates math into everyday life. This is an essential element to learning. The child is encouraged to apply God's patterns and wisdom to life in the home and community. Let's take a break from memory work and academic exercises, and identify ways in which to make math part of everyday life. The following are suggestions or examples, but other ideas may be added to the list.

Activity

> The plans of the diligent lead surely to plenty. (Proverbs 21:5)

You will need help from your parent/teacher with these activities. Let's think about two things today. First, we will think about order, and then we will think about distance. God wants us to make good use of our day. Plan your day wisely. What will you do first? Where will you go first?

A. Think about order. What will you do first? What will you do next? Plan the best way to do something before you start to do it.

1. Mow the lawn. What is the fastest way to mow a lawn? You don't want to miss any part of it!
2. Mop the kitchen floor. When you mop, you don't want to walk on the floor while it is wet. Where would you need to start? Where would you need to end? You want to be able to leave the kitchen when you're done!
3. Make a sandwich. What is the fastest way to make a sandwich?
4. Washing the dishes. Some dishes are easier to clean after they soak. Start soaking the pans in water first so they will be easier to clean later. What would you do next? Wash the cups and glasses. Then wash the dishes. Do the silverware next. Then, finish up with the pans. Wash, rinse, and dry.

B. Think about distance. Could you save time by going three places in one trip? When you can do three things together, you save time. You can also save money.

Let's say that you need to go to three places—the grocery store, the library, and the gas station. How far are these places from your house?

1. How many miles would you go if you took three trips? You'll have to add up the miles it takes to get there and to come home.

	Going there	Coming home
• How many miles is it from your house to the grocery store?	________	________
• How many miles is it from your house to the library?	________	________
• How many miles is it from your house to the gas station?	________	________
• Add these up. How many miles would you drive altogether?	________	

2. What if you went to all three places on the same trip?
    - First, plan your route. Which place is closest to home: The library, the gas station, or the grocery store? Which place is next? Which place should you visit last? Write them down in order:

    - Starting point: Home
    - Place 1: ____________
    - Place 2: ____________
    - Place 3: ____________
    - How many miles is it from home to place 1? ________
    - How many miles is it from place 1 to place 2? ________
    - How many miles is it from place 2 to place 3? ________
    - How many miles is it from place 3 back to home? ________
    - Add these up. How many miles would you drive altogether? ____________
3. Find the difference between your two sums. You will find how many miles you could save by combining car trips!

_______ – ________ = __________

Oh, the depth of the riches both of the wisdom
and knowledge of God! How unsearchable are
His judgments and His ways past finding out!

"For who has known the mind of the Lord?
Or who has become His counselor?"
"Or who has first given to Him
And it shall be repaid to him?"

For of Him and through Him and to Him are all
things, to whom be glory forever. Amen.

ROMANS 11:33-36

God made tulips to bloom, seasons to change, and windmills to spin.

CHAPTER 10

Exploring Worlds of Numbers

Introduction

This lesson introduces symmetry as seen in God's creation, and is followed by an activity and a page of review exercises. This will require about 20 minutes of instruction from the parent/teacher.

God has made a beautiful world! He didn't make everything the same. He made different kinds of animals. He made day and night. He made big numbers and small numbers. We have seen God's patterns with numbers and shapes. But now we will look at another pattern God planned for our world. We call this pattern **symmetry**. When God created the world, He said it was very good. He made it beautiful. Symmetry is beautiful.

Look at these two pictures. Which one is more beautiful? Why is this more beautiful?

Finding Symmetry DAY 109

Prayer

Father God, You made us to enjoy the beauty You have created. Show Yourself to us today as we seek You, and we will rejoice in You. Amen.

Memory

Spend a few minutes with both addition and subtraction flash cards (mix and match).

What is symmetry? One kind of symmetry is made when something has two sides that look exactly the same. You have found symmetry when one side of a picture looks exactly like the other side.

The best way to find symmetry in a picture is to draw a line down the middle of it. Does one side look exactly like the other side? The line through the middle of the butterfly shows symmetry. Draw a line down the middle of the other pictures on this page to find symmetry.

Now let's look at a person's face. Look at your own face in a mirror. Or look at another person's face. Draw an imaginary line down the middle of your nose. Both sides of your face look the same!

Sometimes a picture shows lots of symmetry—amazing symmetry. Look at the picture of the snowflake. You can draw an up-and-down line across it like we did with the butterfly. But you can also draw a side-to-side line. You will still see perfect symmetry above and below your line! Can you imagine slanted lines that still show symmetry? Yes! If you are having a hard time imagining lines, try turning the book sideways. The snowflake still looks the same! There are six different positions where you can turn the snowflake and see the same symmetry.

Activity

Let's look for symmetry around the house or in the yard! Maybe you can draw some examples of what you find. Draw a line down the middle of your picture. Does it have symmetry? Does it look the same on both sides?

Here are a few examples of symmetry that you may find at your house:

- A kitchen cabinet with two doors that look exactly alike.
- A flower with 6 petals; slowly spin it around! Does it look the same in six positions?
- A fan with 5 blades.
- A guitar with two sides that look exactly alike.
- A chair with two sides that look exactly alike.
- Hands held out in front of you.

Student Exercises

Add these numbers together to find a bigger number. Try to answer these from memory. You can use your little stones if you need help. Write the answer for the first six exercises on the line. Write the answer for the last six exercises under the line.

13 - 5 = ______	10 - 4 = ______
12 - 7 = ______	17 - 8 = ______
19 - 3 = ______	14 - 1 = ______

29 - 4 ____	73 - 52 ____	44 - 34 ____
48 - 35 ____	65 - 13 ____	38 - 23 ____

DAY 110 Practice

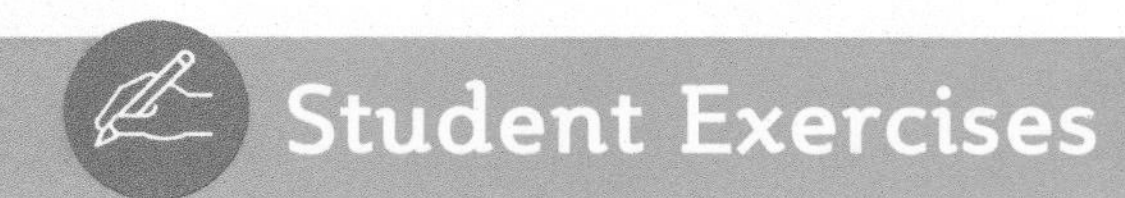

Let's do these story exercises!

Your parent gives you 21 carrot sticks for you and two friends. Everybody should get the same number of carrot sticks. But what is wrong with these groups?

How many carrot sticks should each child get? ________

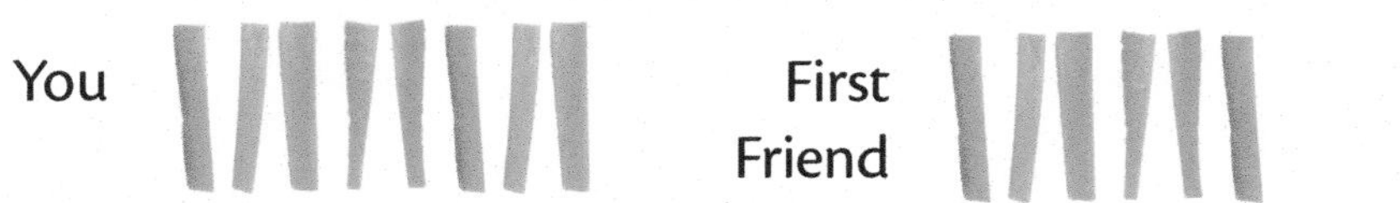

Your parents say you will be going by car to visit Grandmother. Let's pretend it is 30 miles away. How many miles will it take to drive to her house and back again?

$$\begin{array}{r} 30 \\ +\ 30 \\ \hline \end{array}$$

Sometimes people can drive 1 mile in 1 minute on a highway. That means that it would take the same number of minutes as miles to go between two places. If you and Grandmother both live along the highway, how many minutes would it take to go from your house to her house? ____________

How much older is one of your parents than you? How much older or younger are your brothers, sisters or friends than you are? Let's figure out the differences! You can use your blocks if necessary. Be sure to write the name of your friend or brother or sister in the blank line first.

Parent __________________________ __________ - __________ = __________

Name __________________________ __________ - __________ = __________

Name __________________________ __________ - __________ = __________

Name __________________________ __________ - __________ = __________

Count the number of sides for each of these shapes. Write the number of sides you counted **inside** the shape. Now count the number of corners for each of these shapes. Write the number on the **outside** of the shape.

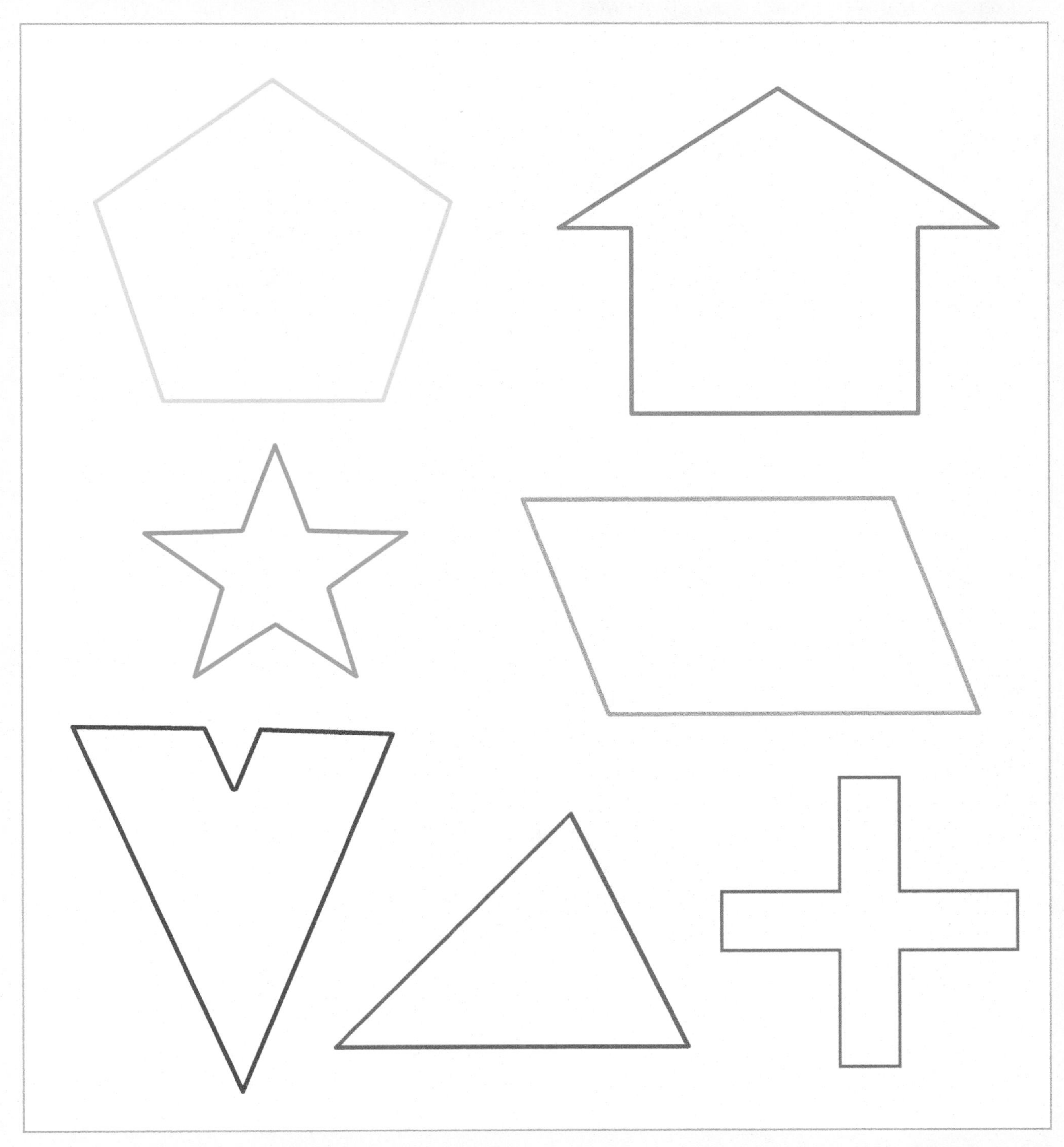

DAY 111 Naming by Numbering

The lesson explores symmetry using numbers, and is followed by two pages of new and review exercises. This will require about 20 minutes of instruction from the parent/teacher.

Prayer

Thank God for something you have learned. Ask Him to help you as you do this lesson. OR Pray your own prayer of thanksgiving and praise to God.

Memory

Spend a few minutes with both addition and subtraction flash cards (mix and match).

Lesson

> "You shall also make a lampstand of pure gold . . . Six branches shall come out of its sides: three branches of the lampstand out of one side, and three branches of the lampstand out of the other side." (Exodus 25:31-32)

God told Moses to build a place called a tabernacle. He also told him to make a lampstand like this one to go inside the tabernacle. How many candles do you see on the left? How many candles do you see on the right? Do you know why it has symmetry? There are the same number of candles on each side of the lampstand. If you drew an imaginary line down the middle of the lampstand, it would look the same on both sides. God wanted symmetry in the special lampstand for the tabernacle. We call this side-by-side symmetry.

God made the starfish to look like a star. You can use a number to name each leg. I chose to call the first leg "0." I've called the others 1, 2, 3, and 4.

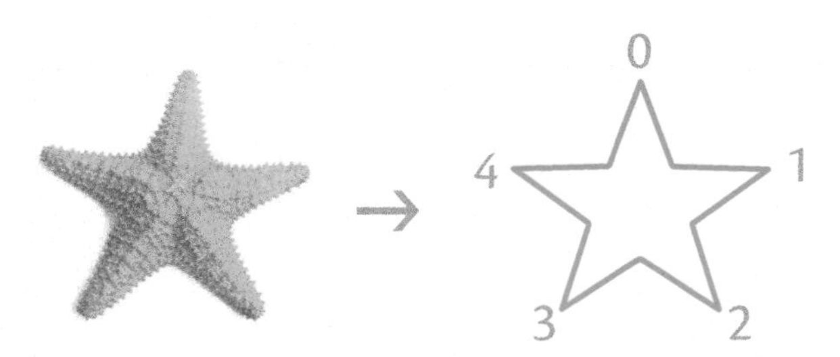

Sometimes numbers play a part in a group. A set of numbers might live together in a smaller world, like these four cars in a little toy train. The pieces of track are also connected to each other in their own group.

Let's build a group of numbers that live together in this shape with 6 sides. You can see it has 6 sides and 6 corners or points. We'll name each point with a group of numbers 0, 1, 2, 3, 4, and 5. When we name each point with a number, we are labeling all the parts of this shape that look exactly the same. How many numbers did we use to name each point? That's right! There are 6 numbers altogether. The size of the set is 6. There are six different spinning symmetries in this shape!

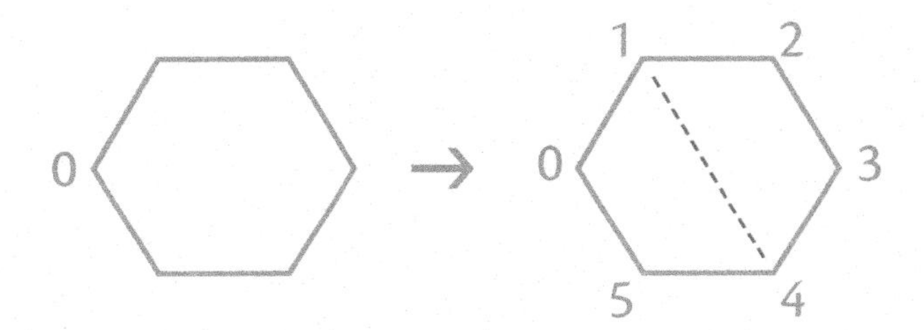

Now look at the line drawn in the middle of the shape. The line connects the numbers 1 and 4 which are exactly across from each other We say that 1 is opposite 4. Do you see how the parts on both sides of the line look exactly the same? The shape has side-by-side symmetry! How can we get two matching sides? We draw a line down the middle between two points on the shape that are opposite each other..

The numbers on the shape work together as a **group**.

1. Which numbers did we use to name corners of this shape?
2. Count how many numbers are used for this symmetry. That's the size of this group or set of numbers.
3. Each number has two neighbors. Who are the neighbors of 3?
4. Who are the neighbors of 5?
5. Count from 0 all the way around to 0 again. Touch each point as you go. Count 0 when you start and when you end. How many points did you touch? Remember, you'll count 0 twice.
6. What number is opposite 2?

Student Exercises

Use numbers to name each corner of the shapes below. Start with 0 for each shape. Think of these numbers as living together in a family.

The student should order the numbers in a clockwise direction, and 0 may be entered at any point.

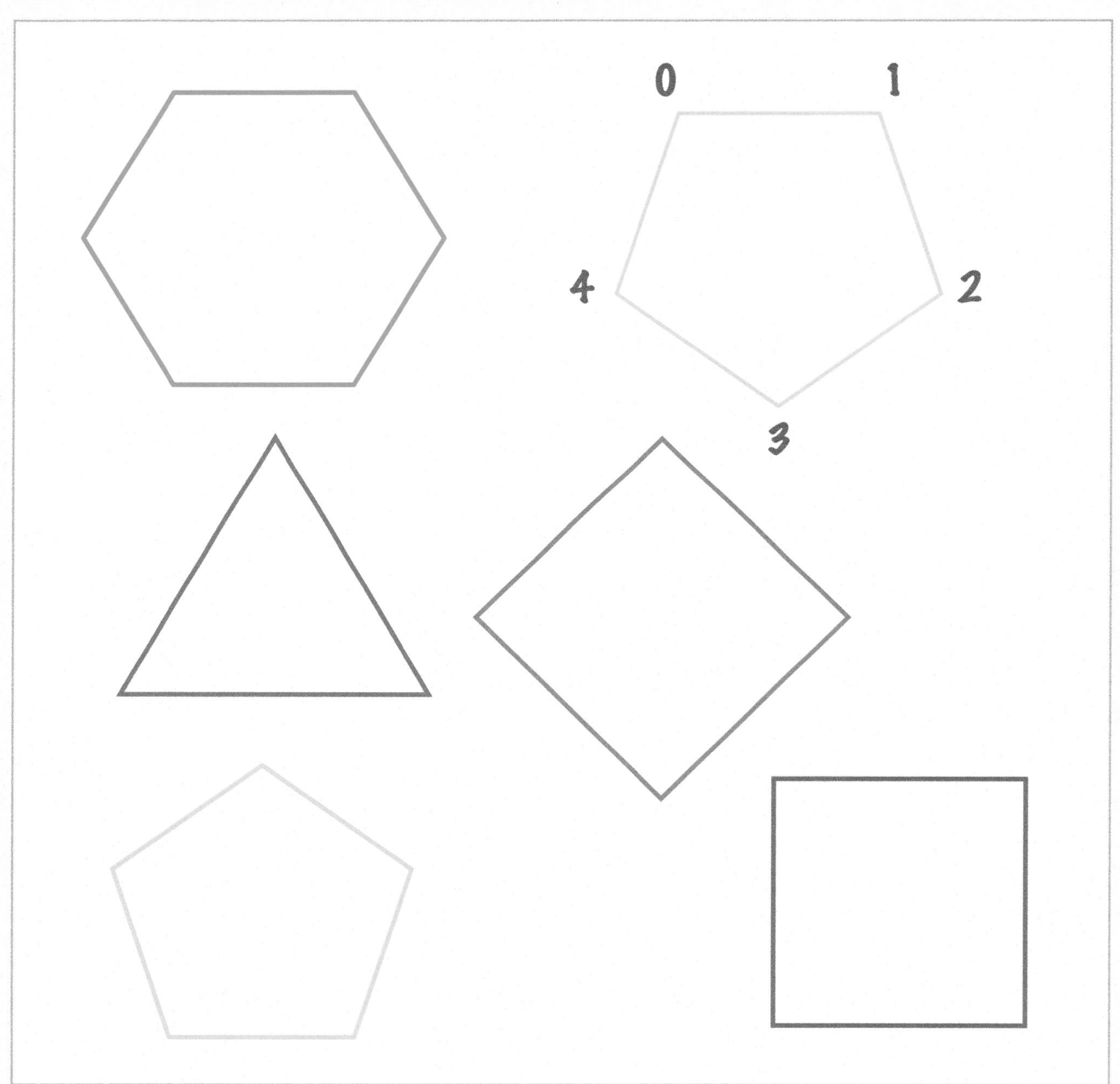

Write the numbers in the order God made for them. Sometimes you will have to count down (or backward) to find the missing number. Sometimes you will have to count up (or forward) to find the number that comes next. Are you getting better at counting up and down? If so, thank God that you are learning math! If it is still hard, ask God for help. Trust Him; He loves you and He will help you!

0, ________, ________

________, 4, ________

________, ________, 8

________, 6, ________, ________

________, ________, ________, 12

________, ________, 11, ________

3, ________, ________, ________, ________

________, ________, ________, ________, 10

DAY 112 Practice

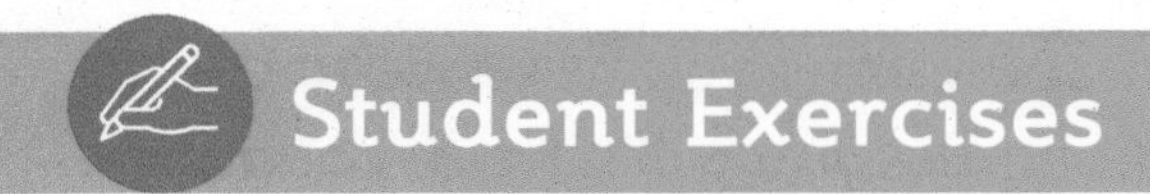

Let's draw some sets! The "Size" number tells you how many things to draw in each set. Can you think of different things to draw for each of these exercises?

You can use a separate sheet of paper if you need more room for your drawings.

Size 2 < Dad & Mom!

Size 4

Size 6

Size 8

Size 10

How can we split these bigger numbers into smaller numbers? Take out 5 stones for your first exercise. Split the 5 stones into two sets. Count the number of stones in each group. Write these numbers in the blanks. Finish by reading the addition equation you made.

There will be several right answers for these exercises. Try to do each exercise without using your stones, but you can use them if you need help.

5 = __2__ + __3__	5 = ______ + ______
8 = ______ + ______	8 = ______ + ______
7 = ______ + ______	7 = ______ + ______
12 = ______ + ______	12 = ______ + ______

DAY 113 Spinning Symmetry

This lesson explores symmetry using patterns God has made, and is followed by an activity and one page of review exercises. This will require about 20 minutes of instruction from the parent/teacher.

Prayer

Thank God for something you have learned. Ask Him to help you as you do this lesson. OR Pray your own prayer of thanksgiving and praise to God.

Memory

Spend a few minutes with both addition and subtraction flash cards (mix and match).

Extra Challenge

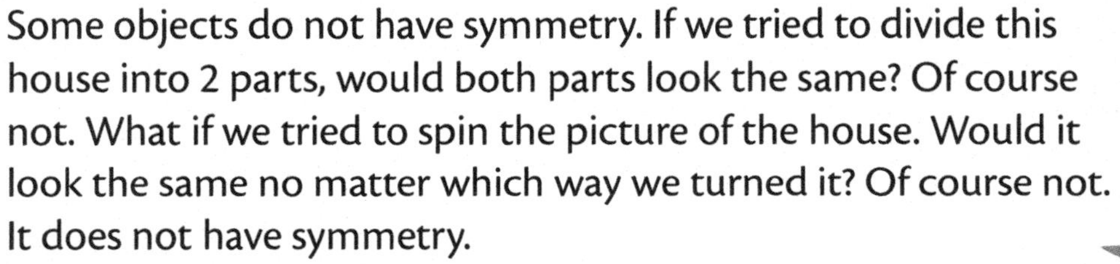

Some objects do not have symmetry. If we tried to divide this house into 2 parts, would both parts look the same? Of course not. What if we tried to spin the picture of the house. Would it look the same no matter which way we turned it? Of course not. It does not have symmetry.

Today, we will cut out two shapes that have beautiful spinning symmetry. These shapes look the same even when you turn them into four or six different positions. They have a pattern.

Let's find the patterns and symmetry of our shapes! Follow these steps to get your shapes ready:

1. Find the shapes on the next page. Each corner of these shapes are marked with a number. The first shape has a set of 4 numbers: 0, 1, 2, 3. The second shape has a set of 6 numbers: 0, 1, 2, 3, 4, 5.
2. Cut out the shapes.
3. Punch a small hole in the star for each shape.
4. Attach the shapes to a piece of cardstock or cardboard. Use brass fasteners.
5. Write "Wheel 1" on the shape with 4 corners. Write "Wheel 2" on the shape with 6 corners.

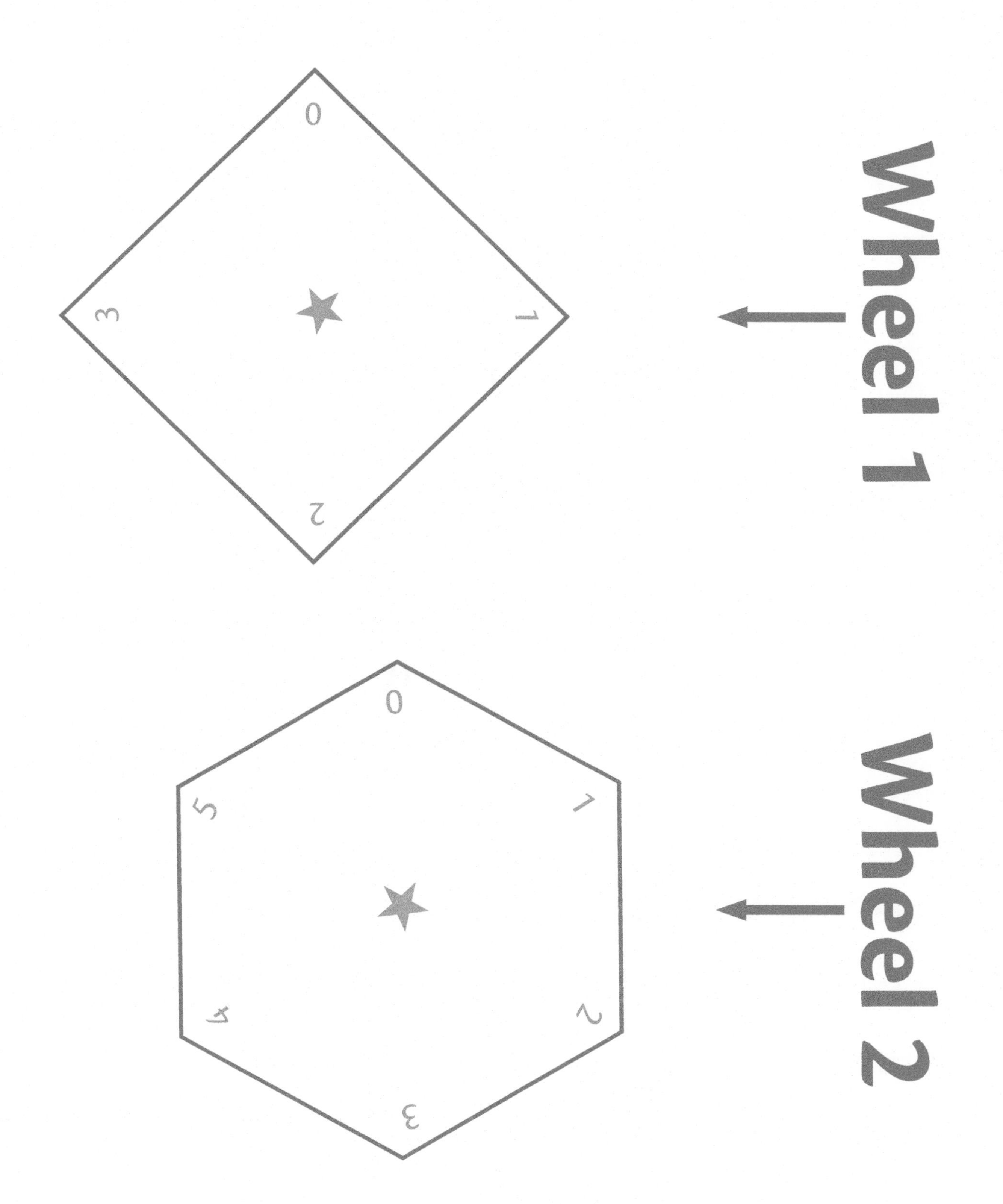
Wheel 1
0
1
2
3
Wheel 2
0
1
2
3
4
5

THIS PAGE
INTENTIONALLY LEFT BLANK

6. Draw a little arrow on the cardstock above each shape as shown on the cut-out page.

Now let's see the symmetry of our shapes. What would happen if you turned the picture of the house at the beginning of our lesson? You wouldn't see any spinning symmetry. The house would look like it was lying on its side. Or the house would be upside down. It would not look the same if you were to turn it upside down.

Let's watch the symmetry of Wheel 1. Line its 0 up with the arrow on the cardstock. Then turn the wheel upwards (counter-clockwise) until the 1 lines up with the arrow. Do you see how the shape looks the same as it did when the 0 was lined up with the arrow? Now line the 2 up with the arrow. Then line the 3 up with the arrow. This object has four symmetries. This means you can turn the shape to four different positions, and it will look the same in each position.

Repeat this with Wheel 2. How many symmetries does this wheel have? Hint: If you're not sure, count how many numbers you needed to name all the corners.

Let's look at one more pattern with our shapes.

Imagine that all God's numbers live in one very big group. This is the group we use when we count. When we count, we usually get bigger and bigger numbers. They go higher and higher.

Each of these wheels are given a set of numbers. Wheel 1 has numbers 0, 1, 2, and 3. Wheel 2 has numbers 1, 2, 3, 4, and 5. This set of numbers does not go higher than 5. This group of numbers live together in a smaller world, like cars in the little toy train.

Now let's play with some math groups!

We are going to spin the two wheels and count. Here's how it works:

Start with Wheel 1. Line the 1 up with the arrow. Turn the wheel upwards (counter-clockwise) 2 places or positions. What number is lined up with the arrow now?

Let's write this out like this.

$$1 \curvearrowleft 2 = 3$$

We can read this as "1 move 2 ends up at 3."

Let's try another one using Wheel 2. Line up the number 2 with the arrow. Turn the wheel upwards (counter-clockwise) 5 places. What number is lined up with the arrow now? Notice that we never count higher than 5. If we try to count higher than 5, we end up at 0 or even past it.

Let's write this out this way:

2 ↶ 5 : 1

Try these exercises with your wheels.

Using wheel 1:

1 ↶ 1: ____________

2 ↶ 2: ____________

3 ↶ 3: ____________

Using wheel 2:

3 ↶ 5: ____________

2 ↶ 4: ____________

1 ↶ 3: ____________

Student Exercises

Let's look at another kind of symmetry. Look at the number sentences below. The numbers on each side of the plus sign are trading places! Notice that you can add numbers in any order and the answer will always be the same. This works for addition. It does not work for subtraction.

We call this the commutative property.

$$3 + 2 = 2 + 3$$
$$5 = 5$$

Let's try a few more of these. Write the answer on the line below each addition exercise.

$10 + 5 = 5 + 10$

________ = ________

$10 + 13 = 13 + 10$

________ = ________

$10 + 20 = 20 + 10$

________ = ________

$6 + 8 = 8 + 6$

________ = ________

$22 + 7 = 7 + 22$

________ = ________

Let's subtract bigger numbers using our blocks. When you subtract or take away blocks, try crossing out some of the blocks in the picture. Then count how many are left.

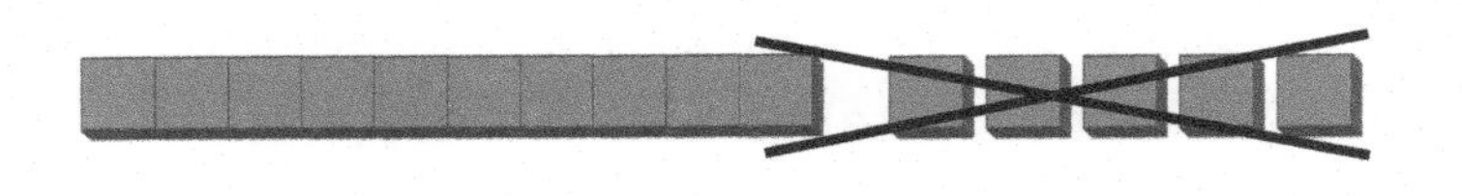

15 - 6 = ________

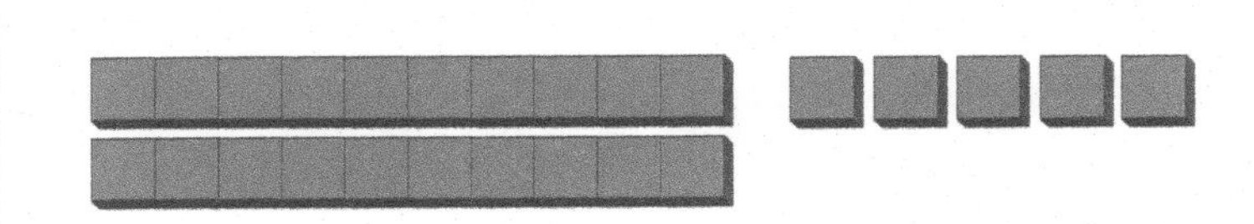

25 - 4 = ________

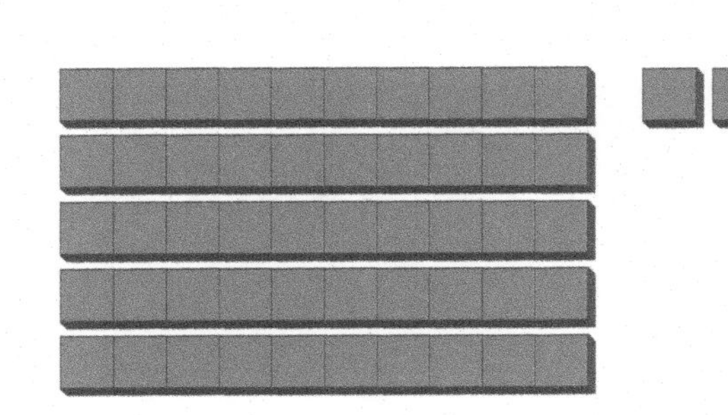

54 - 21 = ________

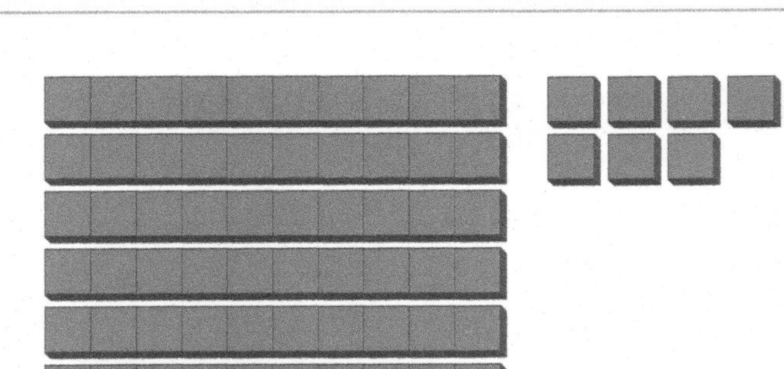

78 - 27 = ________

DAY 114 Practice

Extra Challenge

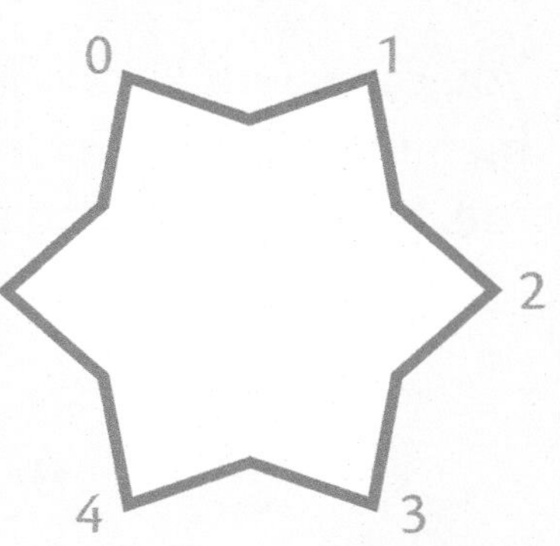

On Day 126, you made two symmetrical shapes into wheels. You turned the wheels to find out what number to end on. Now, instead of turning a wheel, you can get the same answer by moving your pencil around the shape.

This shape is symmetrical. It looks like a star. This group has 6 numbers: 0, 1, 2, 3, 4, and 5. Now let's count using your pencil.

For the first exercise, set your pencil on the number 0. Move your pencil up 2 places (clockwise). Make sure the numbers are getting bigger as you go! What number is your pencil pointing at now? We can read this as "0 move 2 stops at 2."

$0 \curvearrowright 2 =$ __2__	$1 \curvearrowright 3 =$ ______
$2 \curvearrowright 4 =$ ______	$3 \curvearrowright 5 =$ ______
$5 \curvearrowright 1 =$ ______	$4 \curvearrowright 2 =$ ______
$1 \curvearrowright 2 =$ ______	$3 \curvearrowright 2 =$ ______

You can count kittens in a litter. You can count family members on a couch. These are members of sets. God made kittens, and He made your family. God made numbers too. Now you can find numbers living together in sets. How many numbers are living together in these sets? Count the numbers in each set.

4 1 ________	0 ________
7, 8, 9 ________	11, 12 ________
0, 3, 6, 9, 12 ________	3, 4, 5, 6 ________
2 ________	5 ________
4, 5 ________	4, 6, 8 ________

Can you count these puppies and kittens? Let's count by adding the same number together a bunch of times! Count the kittens by 1s. Count the puppies by 2's. When you count the puppies this way you are skip counting!

1 + 1 + 1 = ________

1 + 1 + 1 + 1 = ________

1+ 1 + 1 + 1 + 1 = ________

There are 5 sets of ________ kittens.

2 + 2= ________

2 + 2 + 2 = ________

2 + 2 + 2 + 2 = ________

2 + 2 + 2 + 2 + 2 = ________

There are 5 sets of ________ puppies.

When we skip count by 2's it sounds like this: 2 ... 4 ... 6 ... 8 ... 10!

DAY 115 Symmetry in Psalm 67

This lesson explores symmetry in God's Word, and is followed by an activity and a page of review exercises. This will require about 20 minutes of instruction from the parent/teacher.

Lesson

Praise God! Have you seen how Jesus created a beautiful world using symmetry? Let's find some more symmetry!

Let's draw lines through the center of this triangle to make symmetry! There are three ways to do this. Take your pencil and follow the dotted lines on each of the three triangles below. Remember, where both sides of the picture look alike, there is side-by-side symmetry.

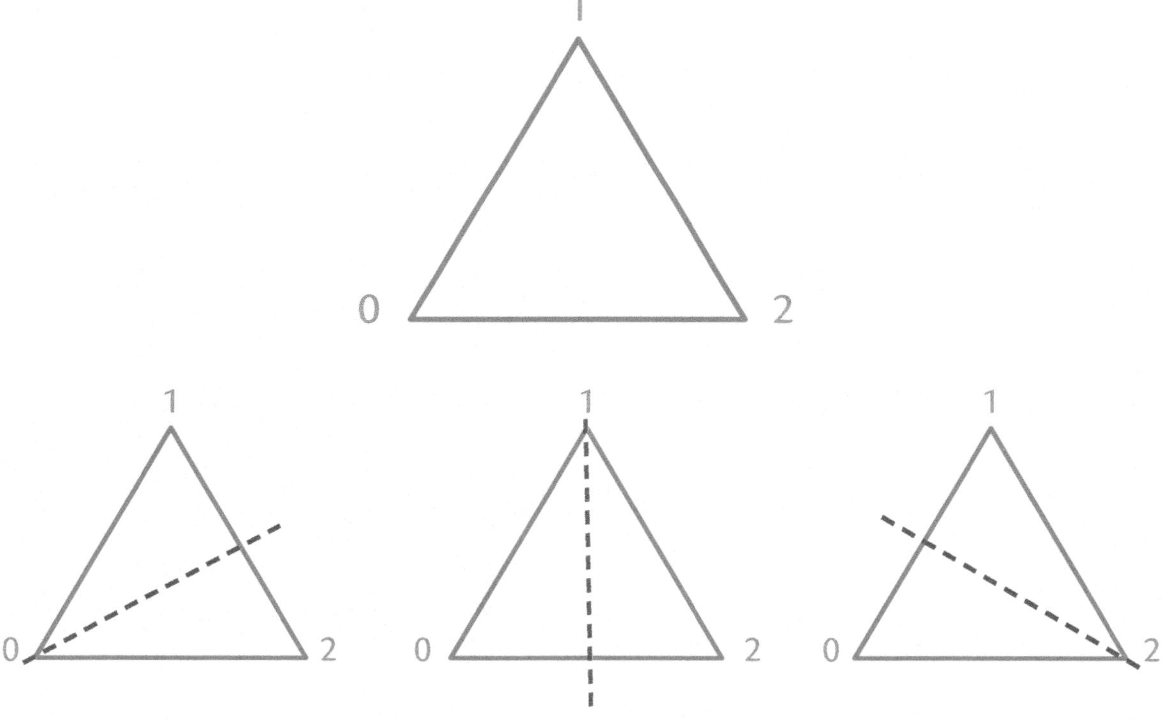

Do you see how both sides of each triangle look exactly the same? Can you see God's creative beauty in the triangle?

> [The LORD] has made the earth by His power,
> He has established the world by His wisdom. (Jeremiah 10:12)

Activity

God loves symmetry so much that we find symmetry in His Word. Let's read Psalm 67. This psalm tells us how people from all over the world are going to praise Jesus.

Yellow (God's Blessing)	God be merciful to us and bless us, And cause His face to shine upon us, Selah That Your way may be known on earth, Your salvation among all nations.
Orange (People Praise God)	Let the peoples praise You, O God; Let all the peoples praise You. Oh, let the nations be glad and sing for joy!
Red (God Rules)	For You shall judge the people righteously, And govern the nations on earth. Selah.
Orange (People Praise God)	Let the peoples praise You, O God; Let all the peoples praise You. Then the earth shall yield her increase;
Yellow (God's Blessing)	God, our own God, shall bless us. God shall bless us, And all the ends of the earth shall fear Him.

You can hear the symmetry in this poem when you read it. You can see the symmetry by coloring it. Lightly color each box with the correct colored pencil. You will need yellow, orange, and red pencils.

Do you see something beautiful about this pattern? It's like side-by-side symmetry, but it's up-and-down instead! Does this pattern help you to remember what you have read? God expresses beauty in so many different ways in His creation and in the church. That's what Psalm 27:4 tells us.

> ... I have desired of the LORD, to behold the beauty of the LORD,
> And to inquire in His temple. (Psalm 27:4)

Student Exercises

Use your ruler to draw a line showing the side-by-side symmetry in each picture. The same basic picture should be repeated on both sides of the line. Don't worry about the two sides being *exactly* the same!

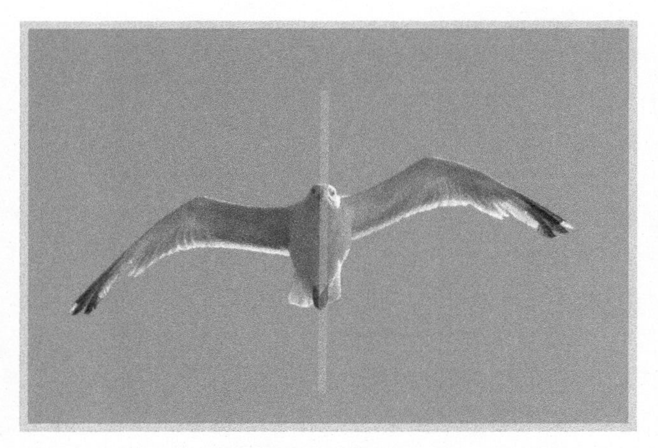

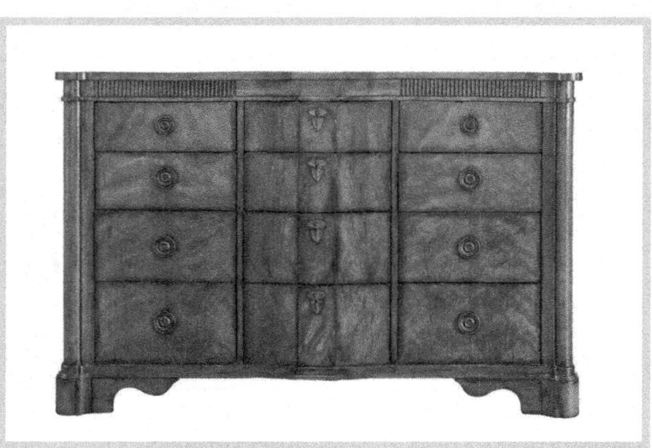

Practice DAY 116

Extra Challenge

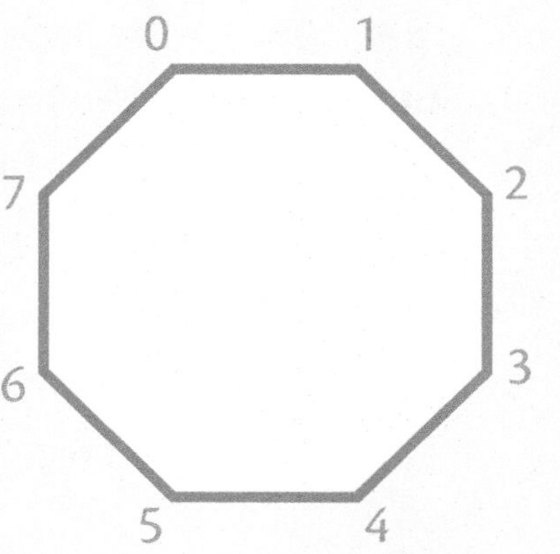

This shape is symmetrical. We call it an **octagon**. It has 8 sides. We've called its corners 0, 1, 2, 3, 4, 5, 6, and 7. This time you won't need to cut out the shape. Instead, you will count using your pencil.

For the first exercise, set your pencil on the number 2. Move your pencil forward 5 places (clockwise). The numbers should be getting bigger as you go. What number is your pencil pointing at now? We can read this as "2 move 5 ends up at 7."

2 ↷ 5 = 7	5 ↷ 2 = ______
1 ↷ 4 = ______	4 ↷ 1 = ______
5 ↷ 5 = ______	6 ↷ 6 = ______
6 ↷ 7 = ______	0 ↷ 0 = ______

Using your stones, let's see what small numbers can be split out of a bigger number. Take out 5 stones for the first exercise. Split the stones into two sets. Count the members of each set. Write these numbers in the blanks. Finish by reading the addition equation you made. Say something like, "Five equals one plus four."

There will be several right answers for these exercises. Try to do the remaining exercises without using your stones, but you can use them if you need help.

5 = ___1___ + ___4___	5 = ______ + ______
5 = ______ + ______	5 = ______ + ______
11 = ______ + ______	11 = ______ + ______
11 = ______ + ______	11 = ______ + ______

Let's skip count sets of puppies and piglets!

2+2 = ______

2 + 2 + 2 = ______

2 + 2 + 2 +2 = ______

2 + 2 + 2 + 2 + 2 = _______

2 + 2 + 2 +2 + 2 + 2 = _______

There are 6 sets of _____ puppies!

Skip counting the puppies sounds like this: 2 ... 4 ... 6 ... 8 ... 10 ... 12!

3 + 3 = _________

3 + 3 + 3 = ________

3 + 3 + 3 + 3 = ________

There are 4 sets of ______ piglets!

Skip counting the piglets sounds like this: 3 ... 6 ... 9 ... 12!

Practice DAY 117

Student Exercises

Add these numbers together to find the correct bigger number. Try to answer these from memory. You can use your little stones if you need help.

Write the answer for the first six exercises on the line. Write the answer for the last six exercises under the line.

$8 + 2 = \underline{\ 10\ }$	$8 + 3 = \underline{\qquad}$
$8 + 4 = \underline{\qquad}$	$2 + 5 = \underline{\qquad}$
$2 + 6 = \underline{\qquad}$	$2 + 7 = \underline{\qquad}$

$\begin{array}{r} 4 \\ +\ 3 \\ \hline \end{array}$	$\begin{array}{r} 3 \\ +\ 6 \\ \hline \end{array}$	$\begin{array}{r} 3 \\ +\ 8 \\ \hline \end{array}$
$\begin{array}{r} 10 \\ +\ \ 0 \\ \hline \end{array}$	$\begin{array}{r} 10 \\ +\ \ 1 \\ \hline \end{array}$	$\begin{array}{r} 10 \\ +\ \ 2 \\ \hline \end{array}$

Which of these 4 numbers in each set are even? Which of these numbers are odd? Write the even numbers in the circle marked **even**. Write the odd numbers of the set in the circle marked **odd**. Can you remember which are even and which are odd? If not, you can use your stones to help you. Remember, every even number can be paired into sets of 2 friends.

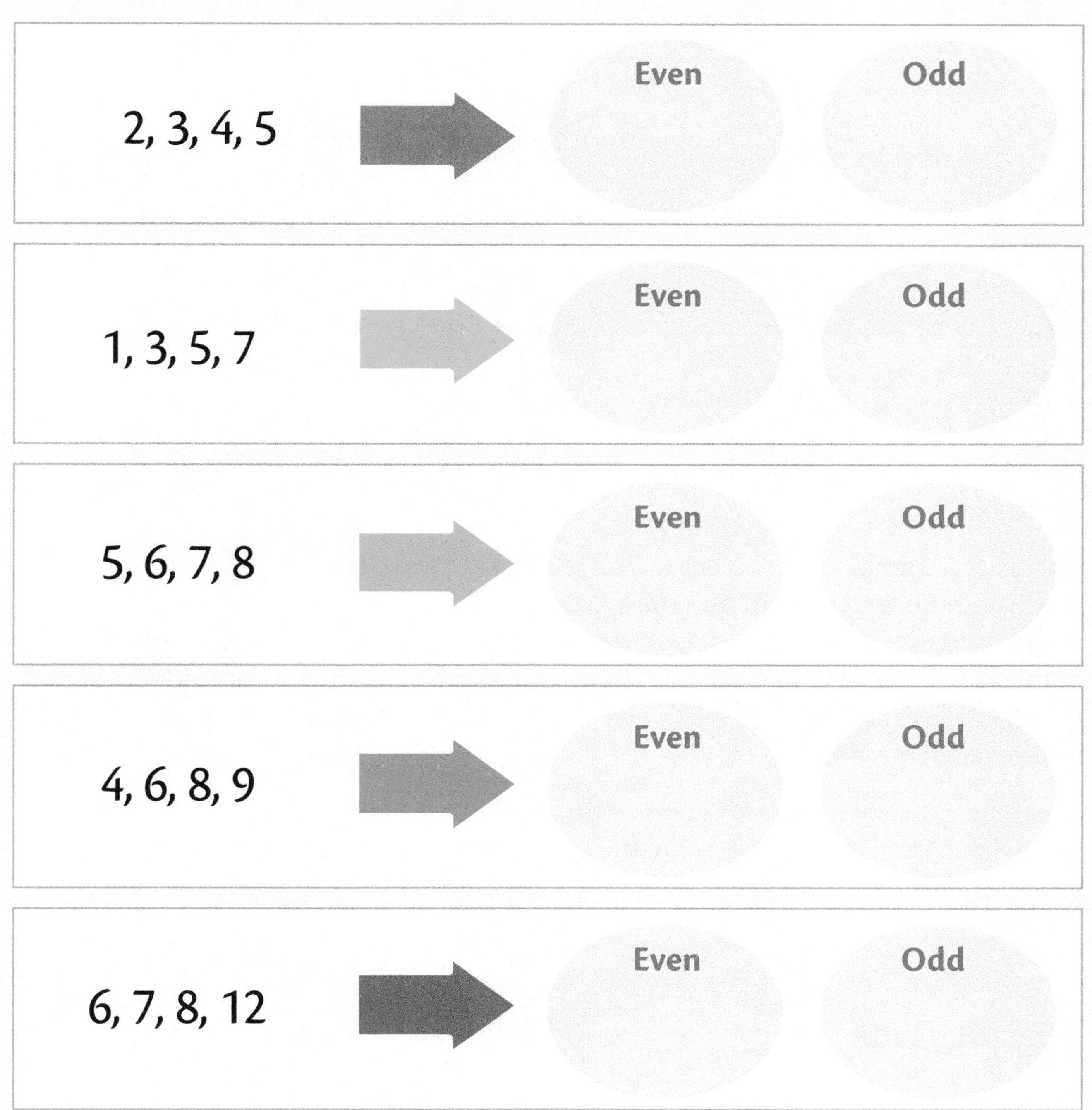

Crack the Code DAY 118

This lesson explores symmetry. The parent/teacher will read the codes, and the student will use the wheels from Lesson 126 to find the answers. The parent/teacher may wish to provide more clues than the ones given. This will require about 15 minutes of instruction from the parent/teacher.

Prayer

Thank God for something you have learned. Ask Him to help you as you do this lesson. OR Pray your own prayer of thanksgiving and praise to God.

Memory

Spend a few minutes with both addition and subtraction flash cards (mix and match).

Lesson

Have you ever seen a safe? You can keep important things locked inside a safe.

A safe has a strange key to unlock it. The key is made of wheels that have numbers on them. You must know the right secret numbers to use. Then you must turn the wheels to the right numbers in order to unlock the safe. The secret numbers are called a code. Today, we're going to play a game to crack the code for five different safes!

Your parent/teacher will read clues to help you find the right code. Use the clues and your wheels from Lesson 126 to find the codes. Write the codes in the blanks.

Safe 1: The numbers on Wheel 1 and Wheel 2 will be the same. The number on Wheel 2 is 3.

Wheel 1: Code __________ Wheel 2: Code __________

Safe 2: The numbers on Wheel 1 and Wheel 2 will be 2 more than 2.

Wheel 1: Code __________ Wheel 2: Code __________

Safe 3: The number on Wheel 1 will be the same as the number of eyes in your head. The number on Wheel 2 will be the same as the number of fingers on one of your hands.

Wheel 1: Code __________ Wheel 2: Code __________

Safe 4: The number on Wheel 1 will be the same as the number of sides in a triangle. The number on Wheel 2 will be the same as the number of sides in a square.

Wheel 1: Code __________ Wheel 2: Code __________

Safe 5: The numbers on both Wheels 1 and 2 will be 3 more than 3.

Wheel 1: Code __________ Wheel 2: Code __________

DAY 119 Practice

Student Exercises

Write numbers to name the points in each picture. Start with 1 and go up. You are counting spinning symmetries! Each picture will look the same no matter which way you look at it. God has made many things to look the same no matter which position you spin them to!

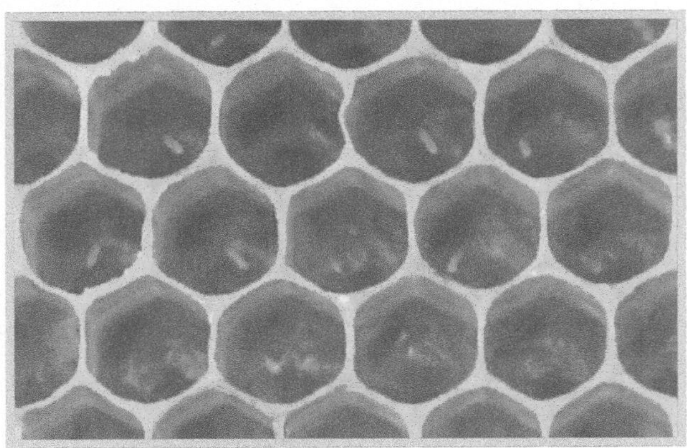

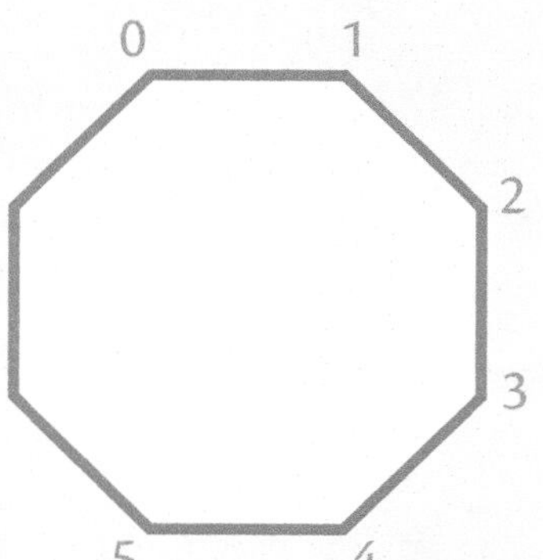

Let's find ways to end up at the number 0! We'll do the first exercise together:

What number is in the first blank below? 2! Put your pencil on the number 2. What number is in the second blank? 6! Move your pencil up (clockwise) 6 places from 2. What number did you get? 0! Let's try another exercise.

Choose another number. Write it in the first blank, then put your pencil on your number. Move your pencil up (clockwise) until you come to 0. How many places did you move? Write this number in the second blank.

2 ↷ 6 : 0	____ ↷ ____ : 0
____ ↷ ____ : 0	____ ↷ ____ : 0
____ ↷ ____ : 0	____ ↷ ____ : 0
____ ↷ ____ : 0	____ ↷ ____ : 0

Go Make Symmetry! DAY 120

This lesson integrates math into everyday life. This is an essential element to learning. The child is encouraged to apply God's patterns and wisdom to life in the home and community. Let's take a break from memory work and academic exercises, and identify ways in which to make math part of everyday life. The following are suggestions or examples, but other ideas may be added to the list.

Activity

God loves beauty and symmetry! He told Moses to make the tabernacle beautiful. Do you remember the lampstand that we learned about? It was symmetrical. God told Moses to make curtains of purple, blue, gold, and scarlet for the tabernacle. He also wanted Moses to make a special plate with 12 stones on it. The high priest would wear this plate in the tabernacle. Its stones were set up symmetrically. They were placed on the plate in an orderly way.

Let's make your home beautiful with symmetry! Now that you have learned about symmetry, you can use it to make things beautiful. With your parents' permission, find ways in which you can add more symmetry to the home.

When you are finished, praise God for the symmetry in the world. Thank Him for the patterns of symmetry you have discovered for yourself.

1. **On the wall.** Look at the pictures and other things hanging on the wall. Can you see a better way to arrange them to create more symmetry?
2. **In the yard.** Go outside and look at the porches and the yard. Could you make better symmetry with the design of the yard? Could you plant flowers, plants, and bushes outside in a symmetrical way? What could you add? What would you have to dig up and replant?
3. **Around the room.** Look at the decorations on the kitchen table. Study how the tables, chairs, bookshelves, and sofas are set up. Is there a better way to set up the furniture to add more beauty and symmetry? Maybe you need to clean things up a bit so you can see the symmetry.

Time waits for no man.

CHAPTER 11

A World Moving in Time

Introduction

> The days of our lives are seventy years;
> And if by reason of strength they are eighty years . . .
> So teach us to number our days,
> That we may gain a heart of wisdom. (Psalm 90:10, 12)

Time is God's gift to us. God gives us so many things, but every day He gives us time to do things.

God gave us ways to measure lengths and distances. He also gave us ways to measure time. We measure longer amounts of time in years, months, days, and hours. We measure shorter amounts of time in minutes and seconds.

What day is this today? What time is it now? These are things we want to know every day.

Age is one way of measuring time. How old are you? How many years have you lived since you were born?

How many months have you lived? How many days have you lived? Maybe you have not thought about this. But God wants us to number our days.

Here's a little challenge for you. How many seven-year periods will you live through to get to 70? If you are 7 years old, and you live to be 70, how many seven-year periods do you have left?

Your mother is older than you. Your grandfather is older than you and your mother. That means that your grandfather has lived longer. He has spent more time on Earth than you.

Today we will learn how to measure time!

Finding Time DAY 121

This lesson introduces time as God's gift and something that is measured. It is followed by one page of review exercises. This will require about 20 minutes of instruction from the parent/teacher.

Prayer

Our Father, we know that You have always been alive. You were in the beginning with the first man and woman on Earth. You are here with us now. You will be with us forever. Help us to learn about time today. Help us to be wise. Amen.

Memory

Spend a few minutes with both addition and subtraction flash cards (mix and match).

Lesson

How many things have happened to you since yesterday? It takes time for these things to happen. Imagine time as if it were a string. The things that happen to you are like beads on the string. God gives you time. He gives you the beads that go on the string for your life.

The most basic measurement for time is a second. One way to measure seconds is to count how many times you can say "Mississippi." Let's say this together:

1 Mississippi

2 Mississippi

3 Mississippi

4 Mississippi

That's about 4 seconds.

Activity

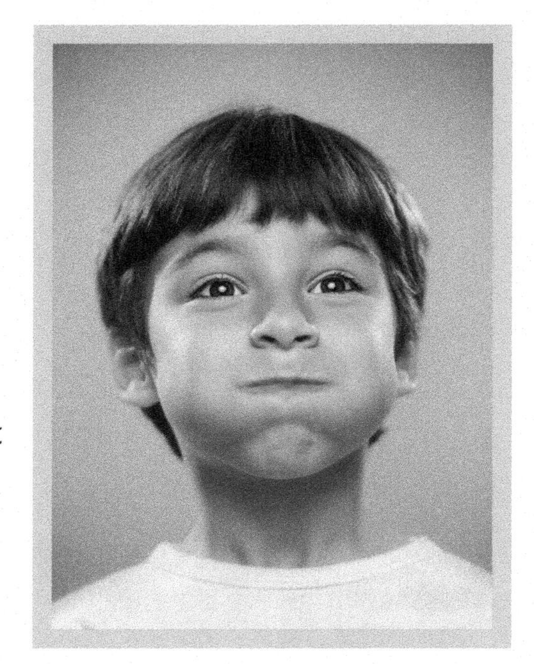

How quickly do seconds go by? Try these activities to help you understand time.

How many seconds can you hold your breath? Hold your breath while your parent or teacher counts Mississippi's.

____

How many seconds can you stand totally still without blinking?

____

How many seconds can you balance on one foot?

____

A minute is 60 seconds. It is about 60 Mississippi's. Ask your parent or teacher to time you while you count 60 Mississippi's.

1 Mississippi

2 Mississippi

3 Mississippi

. . .

58 Mississippi

59 Mississippi

60 Mississippi!

How close did you get? If you came within 10 seconds, that's pretty good!

Student Exercises

Try finding the answers for these addition and subtraction exercises using your memory. You can also use your blocks, coins, or fingers if you need help.

You will have to use long addition and long subtraction for the bigger numbers. Remember, you need to add or subtract the numbers on the right first (the 1s). Then you will add or subtract the numbers on the left (the 10s).

$\begin{array}{r} 3 \\ +\ 6 \\ \hline \end{array}$	$\begin{array}{r} 10 \\ +\ 3 \\ \hline \end{array}$	$\begin{array}{r} 7 \\ +7 \\ \hline \end{array}$
$\begin{array}{r} 14 \\ +\ 62 \\ \hline \end{array}$	$\begin{array}{r} 74 \\ +\ 24 \\ \hline \end{array}$	$\begin{array}{r} 2 \\ +1 \\ \hline \end{array}$
$\begin{array}{r} 11 \\ -\ 8 \\ \hline \end{array}$	$\begin{array}{r} 10 \\ -\ 4 \\ \hline \end{array}$	$\begin{array}{r} 74 \\ -\ 52 \\ \hline \end{array}$
$\begin{array}{r} 96 \\ -\ 83 \\ \hline \end{array}$	$\begin{array}{r} 67 \\ -\ 25 \\ \hline \end{array}$	$\begin{array}{r} 58 \\ -\ 21 \\ \hline \end{array}$

DAY 122 Practice

Student Exercises

Find the distance in each exercise. Each child's house is marked by a red dot. If you're driving a car, count the number of city blocks (or centimeters) to find out how far you need to go. Use the bridges (black lines) to cross the river. If you're flying like a crow, use your ruler to measure how far you need to fly. Each centimeter is one city block.

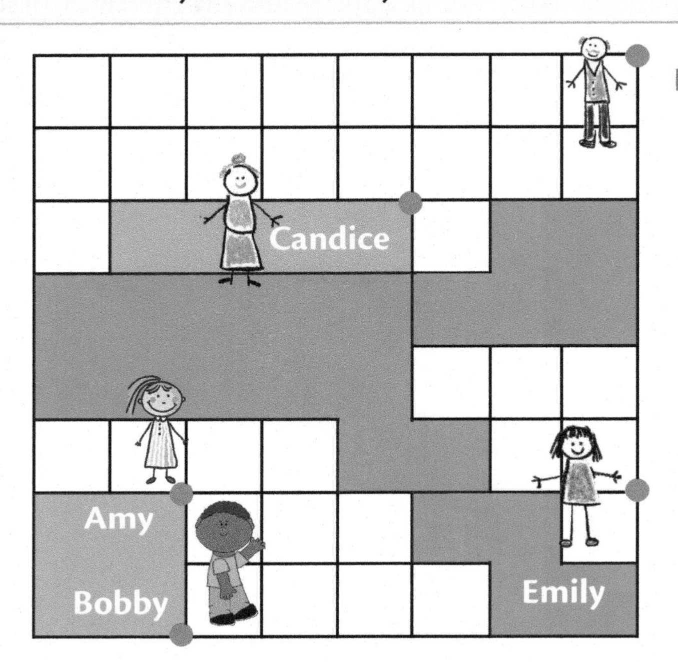

Amy's house to Bobby's house by car: ________

Amy's house to Candice's house by car: ________

Bobby's house to Emily's house by car: ________

Amy's house to Candice's house by crow: ________

Candice's house to Emily's house by crow: ________

Dan's house to Candice's house by car: ________

Dan's house to Amy's house by car: ________

Let's practice some harder addition and subtraction exercises. Try finding the answers to these addition and subtraction exercises using your memory. You may also use your blocks, stones, or coins if you need help.

$\begin{array}{r} 5 \\ +8 \\ \hline \end{array}$	$\begin{array}{r} 5 \\ +9 \\ \hline \end{array}$	$\begin{array}{r} 9 \\ +4 \\ \hline \end{array}$
$\begin{array}{r} 9 \\ +6 \\ \hline \end{array}$	$\begin{array}{r} 7 \\ +0 \\ \hline \end{array}$	$\begin{array}{r} 7 \\ +8 \\ \hline \end{array}$
$\begin{array}{r} 17 \\ -\ 7 \\ \hline \end{array}$	$\begin{array}{r} 17 \\ -\ 8 \\ \hline \end{array}$	$\begin{array}{r} 16 \\ -\ 7 \\ \hline \end{array}$
$\begin{array}{r} 16 \\ -\ 8 \\ \hline \end{array}$	$\begin{array}{r} 15 \\ -\ 7 \\ \hline \end{array}$	$\begin{array}{r} 15 \\ -\ 8 \\ \hline \end{array}$

DAY 123 Reading a Clock

This lesson introduces the student to an analog clock, and is followed by two pages of review exercises. This will require about 30 minutes of instruction from the parent/teacher.

Prayer

Pray your own prayer of thanksgiving and praise to God. Pray for His help on this lesson.

Memory

Spend a few minutes with both addition and subtraction flash cards (mix and match).

Lesson

> Then God said, "Let there be light"; and there was light. And God saw the light, that it was good; and God divided the light from the darkness. God called the light Day, and the darkness He called Night. So the evening and the morning were the first day. (Genesis 1:3-5)

We have learned how to measure time with seconds and minutes. We can also measure time with **hours**.

God gave us a day with two parts. Do you know what they are? Hint: Read Genesis 1:3-5 again if you're not sure. That's right: the daytime and the nighttime.

There are 60 seconds in 1 minute. There are also 60 minutes in 1 hour. A whole **day** is 24 hours long. Let's think about what this means for you.

What can you do in 1 second? ____

What can you do in 1 minute? ____

What can you do in 1 hour? ____

We use a clock or a watch to measure time in hours and minutes. The clock can help us know two things:

1. How long it takes me to do something.
2. What time it is now.

The sun is up by about 7:00 or 8:00 in the morning. The sun goes down in the evening at about 7:00 or 8:00. These times may be different depending on where you live on the earth. These times also change throughout the year. .

If it was 3:00 in the afternoon, you could go outside and play in the light. But what if it was 3:00 in the morning? If your mother told you that it was 9:00 at night, would this be a good time to go outside to play? No! It is nice to know the time of day. That's why we have clocks and watches.

Now, let's learn about the clock! A clock has three important parts:

1. The face
2. The hour hand
3. The minute hand

The face of the clock has twelve numbers. Count them. Start near the top with 1.

The hour hand is the short arrow. The minute hand is the long arrow.

Let's learn how to read the time. We read the short hand first. Where is it pointing? That's right! It's pointing at 1. We know the time is near "1 o'clock."

Now where is the minute hand pointing? It's pointing at 4. How do we read the minutes? Look at this clock. The minutes are written in blue. What blue number is the minute hand pointing at? It's pointing at the number 20, or 20 minutes.

The short hand (the hour hand) is pointing at 1. The long hand (the minute hand) is pointing at 20. What time does this clock show? We say, "It's one twenty." We write it this way:

1:20

Now look at this clock. The hour hand (the short hand) is pointing at 7. The minute hand (the long hand) is pointing at 12. It's pointing at 0 minutes. We say, "It's 7 o'clock!" And we write it this way:

7:00

Activity

Make your own clock! Cut along the dotted line to cut out the clock on the next page. You can make your own arrows for the short hand and the long hand. Attach them with a brass fastener.

1. Parent/teacher: set the hour and minute hands to a certain position for the child to tell the time. Repeat at least four times.
2. Parent/teacher: suggest a time and have the child set the positions of the hour and minute hands. Repeat at least four times.

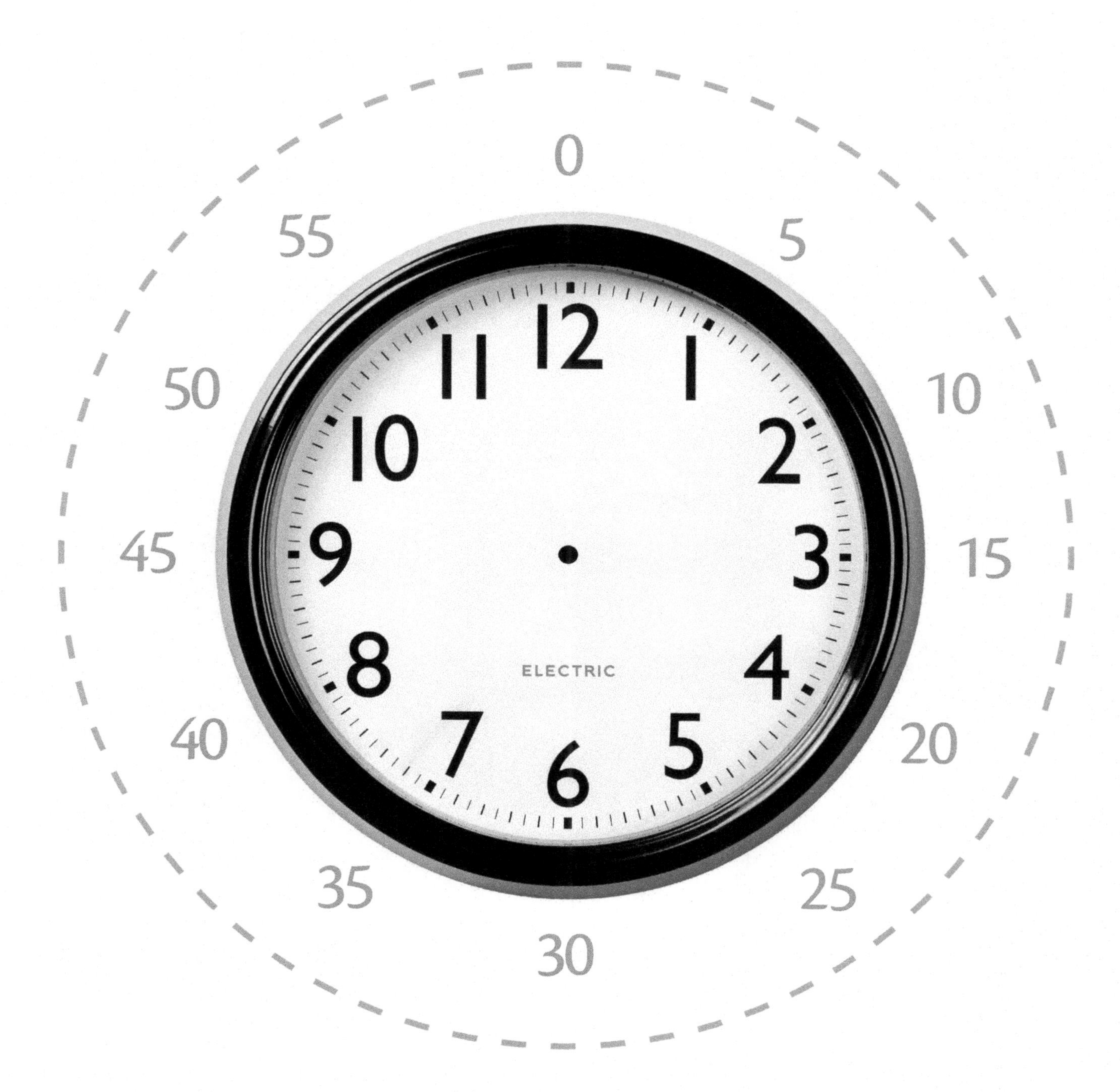
0
5
10
15
20
25
30
35
40
45
50
55
12
1
2
3
4
5
6
7
8
9
10
11
ELECTRIC

THIS PAGE
INTENTIONALLY LEFT BLANK

Student Exercises

Count forwards and backwards to write the numbers in the order God made for them. You may use your set of blocks (1s and 10s) to help you if you would like.

God made many numbers. This is just a small part of all of God's numbers!

24, ________, ________

________, 28, ________

________, ________, 32

________, 30, ________, ________

________, ________, 33, ________

________, ________, 37

53, ________, ________, ________

________, ________, ________, ________, 71

Round each of these numbers to the nearest 10. Sometimes you might have to round up. Sometimes you might have to round down. Remember, 5 is always rounded up.

9 → ______	2 → ______
45 → ______	25 → ______
76 → ______	38 → ______
67 → ______	51 → ______
94 → ______	89 → ______

Practice DAY 124

Student Exercises

Let's read and write time! For the first six clocks, read the time. Write it on the blank line under each clock. For the last six clocks, draw the short hand (the hour hand) and the long hand (the minute hand) on each clock to show the right time.

1:45

12:30

8:10

6:30

3:45

9:00

Let's compare some numbers using subtraction! Maybe your friend is faster than you. Or maybe your brother is taller than you. Whether you have more or less than someone or whether you are faster or slower, always be happy for your friends and brothers and sisters. Do not be jealous or angry.

> Let nothing be done through selfish ambition or conceit, but in lowliness of mind let each esteem others better than himself. (Philippians 2:3)

Pretend that you collected the eggs in the first picture from your chicken house. Pretend your friend collected the eggs in the second picture from her chicken house. How many more eggs did your friend collect?

Your Eggs

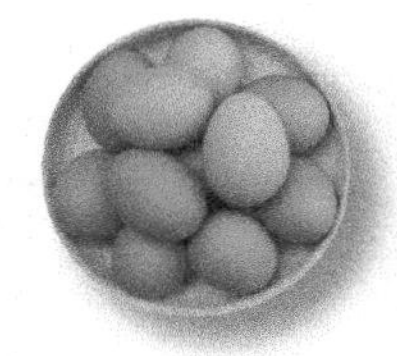

Your Friend's Eggs

________ – ________ = ________

Pretend that you ran a 5k race (5 kilometers or about 3 miles). You ran the race in 38 minutes. Your racing buddy ran it in 33 minutes. How many more minutes did it take you to run the race?

________ – ________ = ________

Let's compare running speeds! You will be one runner. Find someone else who would like to run. Find a place to run and mark a starting line and a finish line. Find an adult who can time you both. Let your friend run first and write below how many seconds it took your friend to reach the finish line. Write how many seconds it took you to run. How much faster or slower are you than your friend?

________ – ________ = ________

Slower Speed   Faster Speed

DAY 125 Reading a Calendar

This lesson introduces longer measurements of time, and is followed by two pages of review exercises. This will require about 20 minutes of instruction from the parent/teacher.

Prayer

Pray your own prayer of thanksgiving and praise to God. Pray for His help on this lesson.

Memory

Spend a few minutes with both addition and subtraction flash cards (mix and match).

Lesson

> "For in six days the LORD made the heavens and the earth, the sea, and all that is in them, and rested the seventh day. Therefore the LORD blessed the Sabbath day and hallowed it." (Exodus 20:11)

God gave us the **week** when He made the world in six days. He gives us six days to work. He gives us one day to rest. So, we can work six days in a row—Monday through Saturday. We usually rest on Sunday. We are not supposed to work more than six days in a row. God set the example for us at the beginning. He made the world in six days. He rested on Day 7.

Do you know these longer measurements of time?

- There are 24 hours in 1 **day**.
- There are 7 days in 1 **week**.
- There are about 4 weeks (or 30 days) in 1 **month**.
- There are 12 months in 1 **year**.

> Christ died for our sins according to the Scriptures, and that He was buried, and that He rose again the third day according to the Scriptures . . . (1 Corinthians 15:3-4)

We remember certain important things every week.

On Fridays, we remember that Jesus died on the cross for our sins on a Friday. On Saturdays, we remember that Jesus was buried. Then, on the third day (Sunday), Jesus rose from the dead. That's why on Sunday, we say, "Christ is risen today!"

We can keep track of the days of the week using a calendar. We can also keep track of the days of the month using a calendar. Some months have more days, and some months have less days. Here's a little poem to help you remember how many days are in the months of the year:

> Thirty days hath September, April, June, and November.
> All the rest have 31.
> But February's 28.
> The leap year comes once in four
> It gives February one day more.

February usually has 28 days. But every four years, February gets 29 days.

FEBRUARY

Sunday	Monday	Tuesday	Wednesday	Thursday	Friday	Saturday
				1	2	3
4	5	6	7	8	9	10
11	12	13	14	15	16	17
18	19	20	21	22	23	24
25	26	27	28			

Let's talk about time using this calendar.

- Read the days of the week at the top of the calendar.

- What is the first day of the week?
- We can look at the top of the calendar, above any numbered date, and see which day of the week that date will be. What day of the week is the 14th or the 28th?
- Does every month always start on Thursday?
- When is your birthday? What day of the week were you born on?
- What is the date one full week after February 5th?
- What is the day of the week 10 days after Wednesday the 7th?

Read the following Bible verses. Talk about Jesus's death and resurrection. What does this mean for us?

> Now when the Sabbath was past, Mary Magdalene, Mary the mother of James, and Salome bought spices, that they might come and anoint Him. Very early in the morning, on the first day of the week, they came to the tomb when the sun had risen. And they said among themselves, "Who will roll away the stone from the door of the tomb for us?" But when they looked up, they saw that the stone had been rolled away—for it was very large. And entering the tomb, they saw a young man clothed in a long white robe sitting on the right side; and they were alarmed.
> But he said to them, "Do not be alarmed. You seek Jesus of Nazareth, who was crucified. He is risen! He is not here. See the place where they laid Him." (Mark 16:1-6)

Student Exercises

Find the missing numbers! Try to find the missing pieces in these equations. You can use your little stones from the math toolbox.

For the addition exercises, ask: "How many stones do you have to add to get the bigger number?" Hint: What number is smaller? Start with that number. How many stones do you need to add to the small number to get the big number?

Now let's look at the first subtraction exercise, "What minus 2 equals 7?" Ask, "How many stones do I need to start with?" First look at the 7 stones you end with. Add the 2 stones that have been taken away. How many do you have? 9! If you start with 9 stones, and take away 2, you will have 7 left.

What plus 4 equals 5?	______ + 4 = 5
What plus 3 equals 6?	______ + 3 = 6
What minus 2 equals 7?	______ − 2 = 7
What minus 1 equals 6?	______ − 1 = 6
What plus 1 equals 10?	______ + 1 = 10
What plus 2 equals 8?	______ + 2 = 8
What minus 3 equals 9?	______ − 3 = 9

What time is it now? For the first six clocks, read the time. Write it on the blank line under each clock. For the last six clocks, draw the short hand (the hour hand) and the long hand (the minute hand) on each clock to show the right time.

3:25

11:50

1:15

2:00

9:30

8:00

DAY 126 Practice

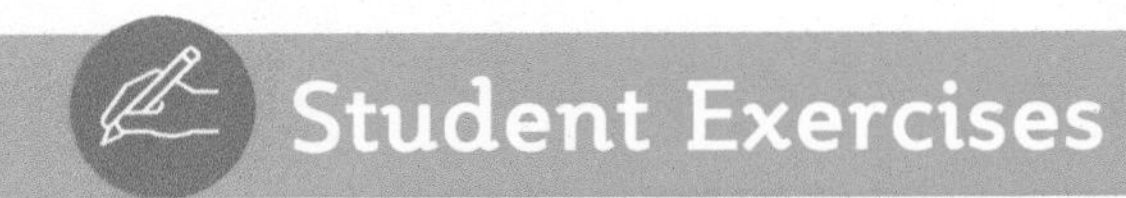

Student Exercises

Plan an imaginary month! A calendar helps you plan what you need to do. It helps you to remember the things you plan. Let's mark the date for each event listed. Use colored pencils to shade the dates on the calendar. Use the color each event is listed with.

Sunday	Monday	Tuesday	Wednesday	Thursday	Friday	Saturday
		1	2	3	4	5
6	7	8	9	10	11	12
13	14	15	16	17	18	19
20	21	22	23	24	25	26
27	28	29	30	31		

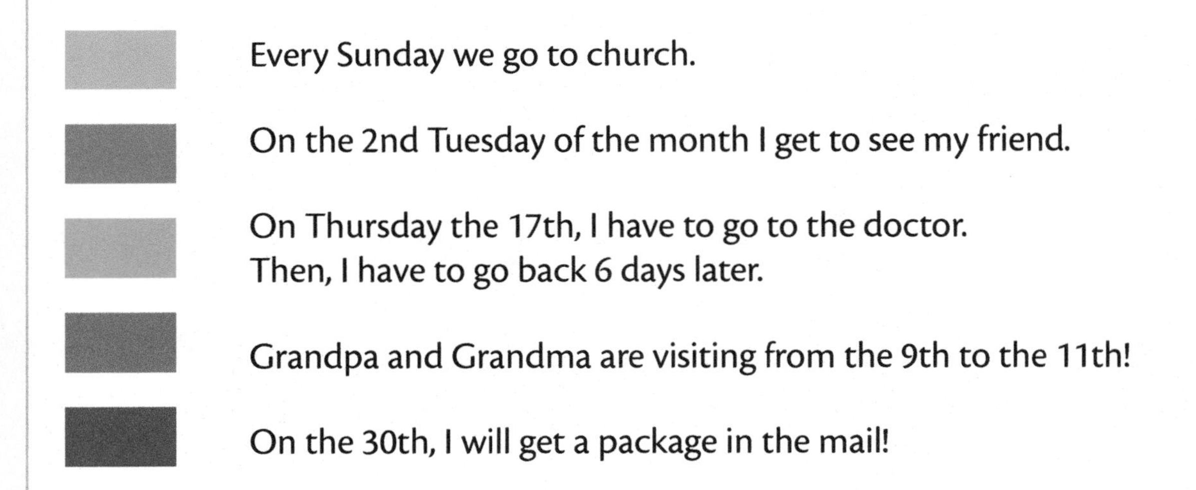

Every Sunday we go to church.

On the 2nd Tuesday of the month I get to see my friend.

On Thursday the 17th, I have to go to the doctor.
Then, I have to go back 6 days later.

Grandpa and Grandma are visiting from the 9th to the 11th!

On the 30th, I will get a package in the mail!

Try finding the answers for these addition and subtraction exercises using your memory. You can also use your blocks, coins, or fingers if you need help. You will have to use long addition and long subtraction for the bigger numbers.

$\begin{array}{r} 7 \\ +4 \\ \hline \end{array}$	$\begin{array}{r} 7 \\ +6 \\ \hline \end{array}$	$\begin{array}{r} 7 \\ +5 \\ \hline \end{array}$
$\begin{array}{r} 7 \\ +7 \\ \hline \end{array}$	$\begin{array}{r} 24 \\ +42 \\ \hline \end{array}$	$\begin{array}{r} 35 \\ +53 \\ \hline \end{array}$
$\begin{array}{r} 14 \\ -5 \\ \hline \end{array}$	$\begin{array}{r} 14 \\ -6 \\ \hline \end{array}$	$\begin{array}{r} 16 \\ -8 \\ \hline \end{array}$
$\begin{array}{r} 16 \\ -9 \\ \hline \end{array}$	$\begin{array}{r} 58 \\ -26 \\ \hline \end{array}$	$\begin{array}{r} 78 \\ -36 \\ \hline \end{array}$

When you read time on a clock like this, you may see the long minute hand pointing at a number. When the minute hand moves from the number 12 to the number 1, it takes 5 minutes. When it moves from 12 to 2, it takes 10 minutes. When it moves from 12 to 3, that's 15 minutes. When the minute hand has moved all the way around to the 12 again, it has gone 60 minutes. To tell time on this kind of clock, you will have to memorize the blue numbers below. It will help if you know how to skip count by 5's!

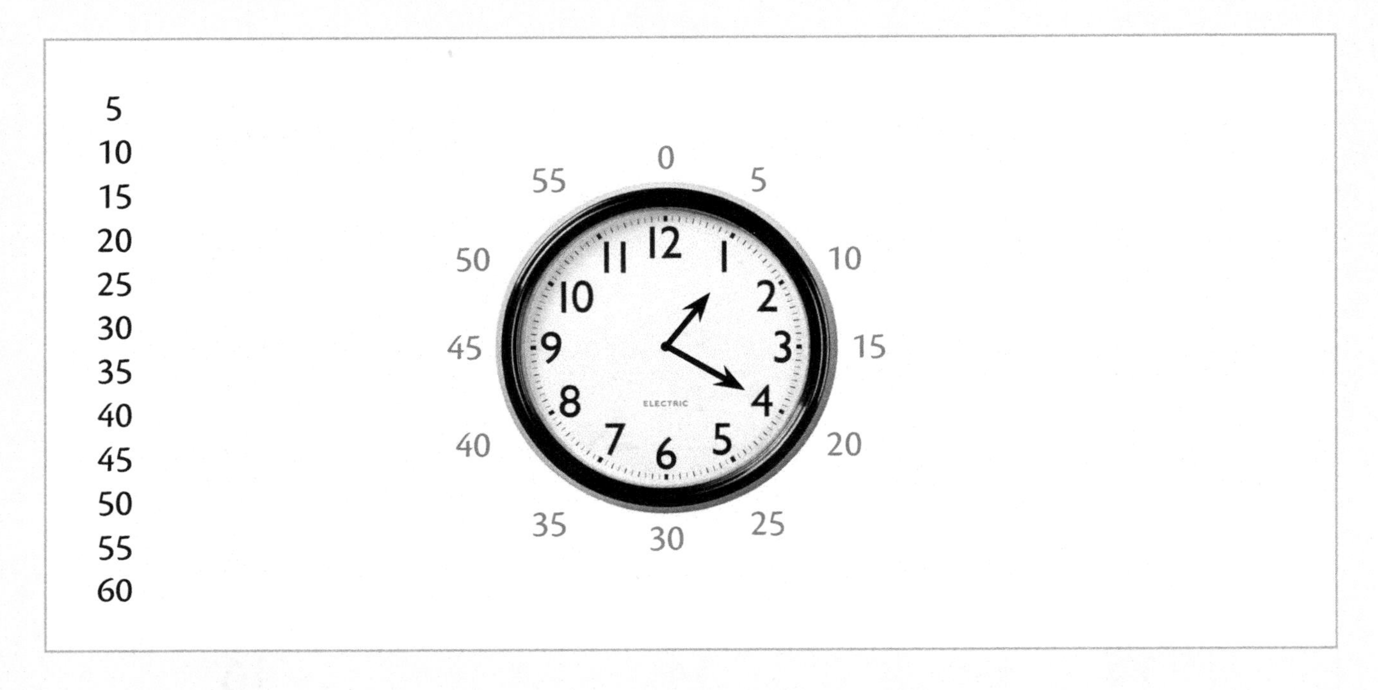

Now let's see if you can skip count by 2's all the way to 30. How many ants can you see? Count these ants in pairs!

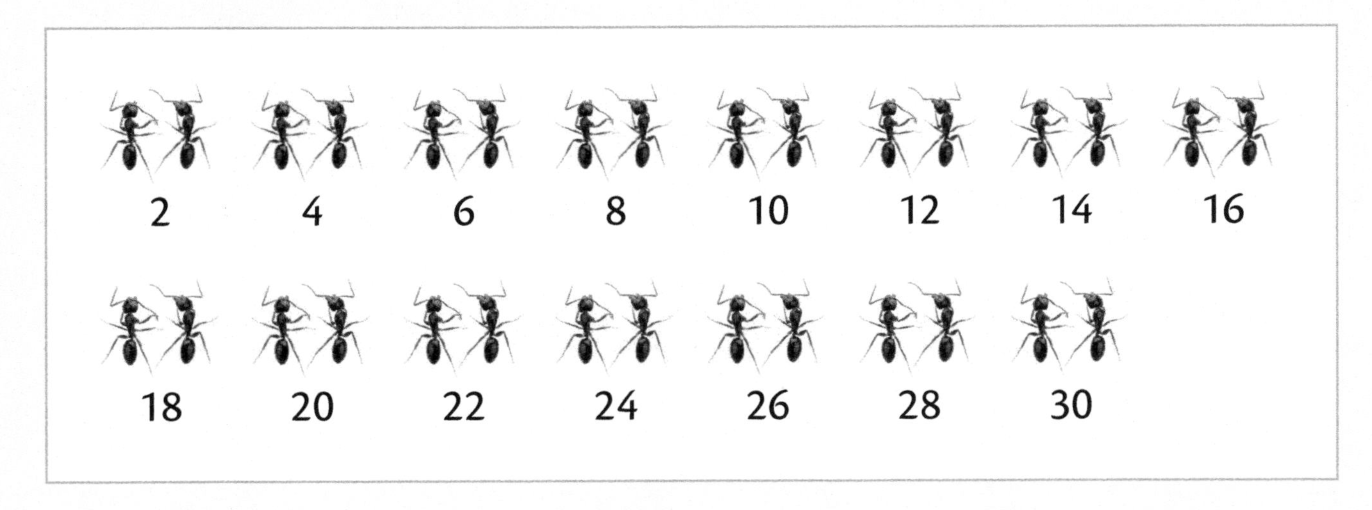

DAY 127 Groups and Time

This lesson introduces the student to the sequences of days and years, and is followed by two pages of review exercises. This will require about 20 minutes of instruction from the parent/teacher.

Prayer

Pray your own prayer of thanksgiving and praise to God. Pray for His help on this lesson.

Memory

Spend a few minutes with both addition and subtraction flash cards (mix and match).

Lesson

> Then the LORD said in His heart,
> "While the earth remains,
> Seedtime and harvest,
> Cold and heat,
> Winter and summer
> And day and night
> Shall not cease." (Genesis 8:21-22)

Do you remember who Noah was? God told Noah to build an ark because God was going to flood the whole world. After the flood, the Lord God promised He would never flood the whole earth again. He also promised two more things to Noah:

He promised that the sun would rise and go down every day. This would give us day and night every day of our lives. The order repeats every day.

The Lord also promised that there would be winter and summer every year. Since then, the earth has never missed a summer or a winter. These seasons repeat every single year of our lives.

God made days and years for us. Let's talk about how we measure days and years some more.

We measure time using different groups of numbers. We can measure time on a clock. A clock uses two groups of numbers. It uses numbers 1 to 60 to measure minutes. This is one group of numbers.

We measure hours using another group of numbers. We use numbers 1 to 12. So the clock keeps track of two groups of numbers at the same time!

Did you know that both hands of the clock start at the top of the clock and spin around? They always come around to 12 again, just where they started!

The minute hand and the hour hand both go around the clock. The minute hand moves faster than the hour hand. Each time the minute hand goes all the way around the clock, a new hour starts. Do you remember that a day has 24 hours? A clock only shows 12 hours. But 12 + 12 = 24! That means the hour hand must go around the clock two times in a whole day. From lunchtime today until lunchtime tomorrow, the little hand goes around the clock twice!

Do you know when a day starts? The day starts at midnight or 12:00. That's the middle of the night when you are sleeping. That's when Saturday becomes Sunday, or Tuesday becomes Wednesday. When a new day starts, the hour hand must go around twice before that day ends.

Let's practice making times on the clock you cut out in a previous lesson.

Point both hands at 12. This will be the 12:00 that means midnight, not lunchtime. Spin the minute hand (the long hand) forward so it points to 15 minutes (or the number 3). That's 12:15. Now spin the minute hand to show 35 minutes (or number 7). That's 12:35. Then spin the minute hand to 50. It will be pointing at the number 10. That's 12:50. Spin it forward again until it comes to 12. You've made it through a whole 60 minutes, or 1 hour!

But now you have to move the hour hand to 1. The minute hand made it all the way around the clock. A new hour has started. It's 1:00 in the morning. You're still asleep.

If you want to talk about AM/PM with your student now, you can. For now we will simply read the clock, and later on in book 2 we will talk about the complexity of military vs. standard time.

Spin the minute hand around the clock one more time. Change the hour hand to 2. Now it's 2:00 in the morning. Can you spin the minute hand around again? Then set the hour hand to 3:00 in the morning. What time do you usually wake up? Put the hour and minute hand at the time you wake up.

A new hour starts each time the minute hand goes all the way around the clock. The minute hand goes around the clock 24 times in one day.

Now let's look at the year. We use a calendar to help us measure time in a year.

The year uses another group of numbers to measure time. It uses the numbers 1 to 12 to count the months of the year. Every year the months repeat themselves in the same order as the year before. Can you name the months of the year? Here they are:

1. January
2. February
3. March
4. April
5. May
6. June
7. July
8. August
9. September
10. October
11. November
12. December

December is the last month of the year. After December, we start over again with January.

We keep track of the years from the time that Jesus was born. Every time we start a new year, we add 1 to the year. If this is the year 2023, then next year will be 2024. It has been about 2,023 years since Jesus was born in Bethlehem. Next year it will be about 2,024 years since Jesus was born. In 2045, how many years will it be since Jesus was born?

Student Exercises

God made some numbers bigger than other numbers. Put these numbers in the order God made for them!

Fill in each blank with the correct symbol: < (smaller than) or > (bigger than). Then read the math sentence. Remember, the baby shark eats the bigger number!

10 ◯ 7 ◯ 2	7 ◯ 11 ◯ 6
18 ◯ 9 ◯ 14	20 ◯ 40 ◯ 85
0 ◯ 2 ◯ 0	60 ◯ 55 ◯ 50
14 ◯ 8 ◯ 6	12 ◯ 21 ◯ 12

Try finding the answers for these addition and subtraction exercises using your memory. You can also use your blocks, coins, or fingers if you need help. You will have to use long addition and long subtraction for the bigger numbers.

$$\begin{array}{r} 4 \\ +\ 4 \\ \hline \end{array}$$

$$\begin{array}{r} 5 \\ +\ 3 \\ \hline \end{array}$$

$$\begin{array}{r} 6 \\ +4 \\ \hline \end{array}$$

$$\begin{array}{r} 2 \\ +\ 8 \\ \hline \end{array}$$

$$\begin{array}{r} 5 \\ +\ 6 \\ \hline \end{array}$$

$$\begin{array}{r} 15 \\ +\ 50 \\ \hline \end{array}$$

$$\begin{array}{r} 9 \\ -\ 4 \\ \hline \end{array}$$

$$\begin{array}{r} 12 \\ -\ 5 \\ \hline \end{array}$$

$$\begin{array}{r} 26 \\ -\ 3 \\ \hline \end{array}$$

$$\begin{array}{r} 13 \\ -\ 2 \\ \hline \end{array}$$

$$\begin{array}{r} 24 \\ -\ 12 \\ \hline \end{array}$$

$$\begin{array}{r} 31 \\ +\ 41 \\ \hline \end{array}$$

DAY 128 Practice

Student Exercises

The clock is a group of numbers in a spinning symmetrical shape. But we don't imagine the clock turning like we imagined turning star shapes. Instead, we turn the hands! We can start the hands at one number and move a certain number of places to another number. Let's do this with your cut-out clock from a previous lesson.

What happens when you go to church? The service starts at a certain time. It ends at a certain time. That's true of different things like parties, doctor's appointments, and work projects.

These events start at a certain time. Let's find out how long they last! For the first exercise, set your clock to 5 o'clock. Then, use the minute hand to make 5 hours go by. Don't forget to move the hour hand too! What number did the hour hand stop at? What time is this? Write the time on the line like this: 10:00.

A party starts at 5:00 and it is 5 hours long. __________

A play time starts at 10 and is 3 hours long: __________

A drive to camp starts at 9 and is 6 hours long: __________

A river cruise starts at 8 and is 10 hours long: __________

A church service starts at 3 and is 3 hours long: __________

A dentist visit starts at 11 and is 1 hour long: __________

The calendar works like a spinning symmetry too. The days all come around again and repeat the next week. The months of a year also come around again and repeat the next year! Answer the following questions about this calendar.

Sunday	Monday	Tuesday	Wednesday	Thursday	Friday	Saturday
		1	2	3	4	5
6	76	8	9	10	11	12
13	14	15	16	17	18	19
20	21	22	23	24	25	26
27	28	29	30	31		

What day of the week comes 2 days after Friday?

____

What day of the week comes 10 days after Wednesday?

____

What day of the week comes 5 days after Sunday?

____

What day of the week comes 7 days after Tuesday?

____

How many weeks are there between the 8th and the 15th? Remember, a week is 7 days long.

____

How many weeks are there between the 17th and the 31st?

____

How many weeks are between the 13th and the 20th?

____

DAY 129 Waiting for God

This lesson applies Scripture to the idea of periods of time, and is followed by one page of review exercises. This will require about 20 minutes of instruction from the parent/teacher.

Prayer

Pray your own prayer of thanksgiving and praise to God. Pray for His help on this lesson.

Memory

Spend a few minutes with both addition and subtraction flash cards (mix and match).

Lesson

Have you ever wanted something very badly? Was it hard to wait? Maybe you were planning to visit Grandma and Grandpa. But there was only a special time when you would go. Maybe you had to wait many weeks for the special time to come. It wasn't going to happen any sooner or any later than the special time.

God has a special time for everything. He has a special time for summer. He has a special time for winter. He had a special time for you to be born. You were born at the exact time He wanted you to be.

God wants us to wait on Him for His special time. Sometimes we have to wait a long time. After summer is over, we have to wait through fall, winter, and spring before summer comes again. But God always brings summer. Remember how He promised that every year will have summer and winter?

When we wait for God, He always comes. He always helps us. And He will help us wait for Him.

Read the following verses. Answer the questions.

1. Jonah 1:17: How many days was Jonah in the belly of the fish? How long did he have to wait for God to rescue him? ___________________________
2. John 11:17: How long was Lazarus in the grave before Jesus came to help?
____
3. Mark 5:25-34: How long did the woman have the bleeding sickness before Jesus healed her? ___________________________
4. Genesis 12:1-4; Genesis 21:5: How old was Abraham when he was promised that he would have a child or children? ___________________________
5. How old was Abraham when his son Isaac was born?
____
6. How many years did Abraham have to wait until Isaac was born? (You may need help to find this answer.) ___________________________

> Wait on the LORD;
> Be of good courage,
> And He shall strengthen your heart;
> Wait, I say, on the LORD! (Psalm 27:14)

Student Exercises

The clock uses two groups of numbers to measure time. The calendar uses a group of numbers to measure time. The seasons also use a group of numbers to measure time! The seasons use numbers 1 to 4. Can you name the four seasons?

Let's think about a whole year. Answer the questions below.

What is the 3rd month after March? ____________

What is the 10th month after June? ____________

What is the 5th month after November? ____________

What is the 2nd month after May? ____________

What is the 8th month after January? ____________

What is the next month (1st month) after December? ____________

> "While the earth remains,
> Seedtime and harvest,
> Cold and heat,
> Winter and summer,
> And day and night
> Shall not cease."
> (Genesis 8:22)

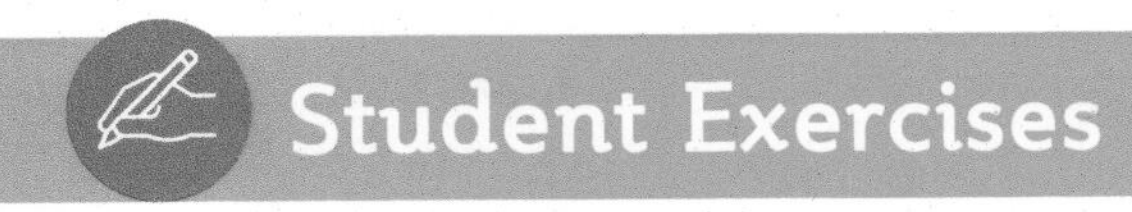

Student Exercises

Each exercise starts with two sets of numbers. Compare these sets. Which numbers do they have in common? Write these numbers in the space provided. This makes a new set of numbers! If they don't have any numbers in common, you will make an empty set.

Set {1,3,5,7} and {1,5,9,11} share = {________}

Set {2,4,6,8} and {8,2,6,4} share = {________}

Set {1,3,6,9} and {2,8,12,16} share = {________}

Set {1,2,3,4} and {2,3,4,5} share = {________}

Set {5,6,7,8} and {8,9,10,11} share = {________}

Set {1,3,6,12,15} and {1,4,7,11,15} share = {________}

Let's have fun with two more sudoku puzzles. Remember, every group of four boxes must have 1, 2, 3, and 4 in it. But each number can only be used once in each group.

Every horizontal (side-by-side) line of four squares and every vertical (up and down) line of four squares must also have 1, 2, 3, and 4 in it.

3		4	
	1	3	
2	3		4
1		2	3

1			2
3	2		1
	3	1	4

Go Add Money...Go Tell Time DAY 131

This lesson integrates math into everyday life. This is an essential element to learning. The child is encouraged to apply God's patterns and wisdom to life in the home and community. Let's take a break from memory work and academic exercises, and identify ways in which to make math part of everyday life. The following are suggestions or examples, but other ideas may be added to the list. This lesson involves two sections: an exercise in telling time and an exercise in adding money.

Activity 1

 ***Depending on the currency used, you may have to adjust the value (and currency symbols) for the products in these exercises.***

> Come now, you who say, "Today or tomorrow we will go to such and such a city, spend a year there, buy and sell, and make a profit [money]"...Instead you ought to say, "If the Lord wills, we shall live and do this or that." (James 4:13,15)

How do you use math? Buying things is one important way. Selling things is another important way. God wants us to buy and sell. This verse says we must remember that God is the one who helps us with money.

The Bible also says that God doesn't want us to steal. We steal if we take things that don't belong to us. God wants us to be honest so we don't steal. We must learn good math so we can be honest when we buy and sell things. Let's do some exercises with money!

1. You have 95 cents. You will buy an ice cream cone for 50 cents. You will buy a little bag of chips for 30 cents. How much money will you have left?

   First, add the amount of money you will spend to buy the ice cream and the chips.

   ____________ + ____________ = ______________

   How much money do will have left? Subtract the total you will spend on the ice cream and chips from 95 cents.

   ____________ – ____________ = ______________

2. You have 85 cents. You want to buy a bag of chips for yourself and for your sister. Each bag of chips costs 30 cents. How much money will you have left?

   First, add the amount of money you will spend to buy the two bags of chips.

   How much money will you have left? Subtract the total you will spend on the two bags of chips from 85 cents.

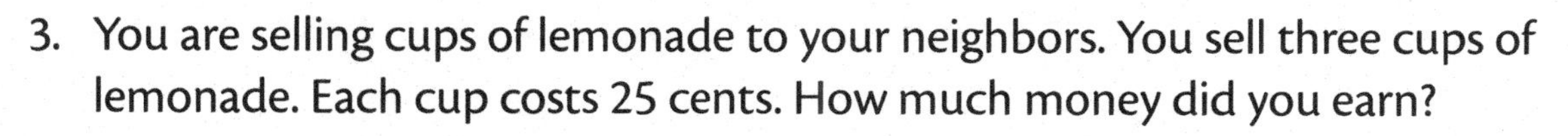

3. You are selling cups of lemonade to your neighbors. You sell three cups of lemonade. Each cup costs 25 cents. How much money did you earn?

   Remember, start by adding the first two numbers. Then finish by adding their sum and the third number. Use your colored blocks to figure out the first sum. Use long addition for the second sum.

   ____________ + ____________ + ____________ = ______________

4. Let's say you paid your mom 40 cents for the lemons and sugar. How much money do you have left? Hint: Subtract 40 cents from total amount of money you made in #3.

   ____________ – 40 = ______________

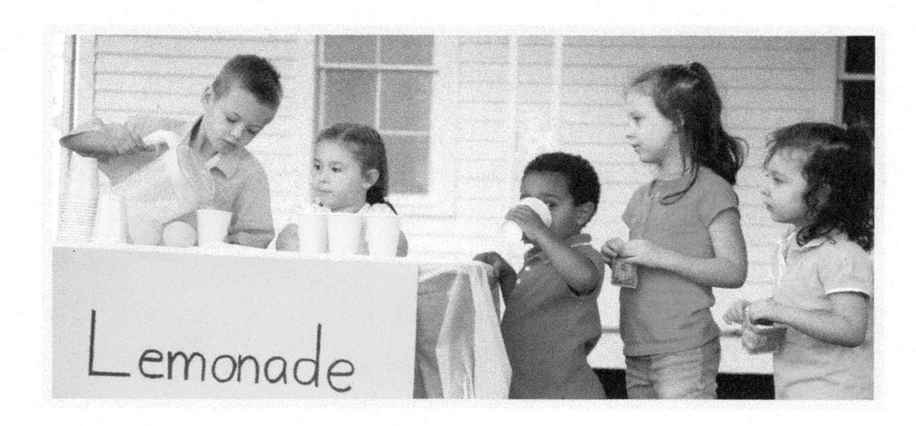

Activity 2

> To everything there is a season,
> A time for every purpose under heaven . . . (Ecclesiastes 3:1)

How we spend our time is important. Sometimes we go slowly. Sometimes we go quickly. Sometimes we may spend more time talking with our friends. This is how we build friendships. Sometimes we spend more time thinking about God's Word and praying. Sometimes we do our chores quickly. Sometimes we need to do our chores more slowly. But we always want to do a good job with the things God has given us to do.

We can use a watch, a clock, or a timer to measure time. Here are some activities where you can measure time. Be sure that your parent/teacher is supervising these activities.

1. **Running and exercise.** Exercise is good for your body. God wants us to take good care of our bodies. Measure how long it takes you and your brother, sister, or friend to run around the block (or down the driveway and back). Time yourself every week for a while. Can you beat your last time?
2. **Cooking.** We use timers to help us cook food. If you leave food in the oven or on the stove too long, what happens? You will burn the food. If you don't cook the food long enough, what happens? You won't be able to eat it. **Cook some food by following a recipe.** Set the timer for the right amount of time. You want the food to be just right when it's eaten .
3. **Drying clothes or dishes.** How long does it take for dishes to dry? How long does it take clothes to dry on the line? Use a timer or a clock to find out how long it takes.
4. **Doing your chores.** How long does it take you to do your chores and do them well? Use a timer or a clock to see how long it takes. Maybe you can work a little faster the next time you do your chores.
5. **Planning ahead.** Now that you can read a clock and a calendar, you can plan ahead for tomorrow or next week. What are you going to do tomorrow? Is there an appointment at the dentist or doctor that you want to remember? Be sure to put that on the calendar. Is there something you will need to get done before a certain time? Plan out your day to help you use your time well.

Let's do some thinking! On the last four exercises below, you will think of your own numbers to add or subtract! (Choose from 5, 10, 15, 20, 25, 30, 35, 40) Then think of the answer!

30 + 50 ____ ☐	30 + 30 ____ ☐	25 + 25 ____ ☐
☐ − 40 ____ ☐	95 − ☐ ____ ☐	☐ − 85 ____ ☐
25 + ☐ ____ ☐		

You've grown so much! Let's grow some more!

CHAPTER 12

More of God's Big Numbers

Introduction

Welcome to your last chapter of your very first math book! God's patterns are pretty neat, aren't they? These are God's ideas. This is God's math. He gave it to us. There are so many good uses for math. Have you found math useful? Have you learned to be thankful for math?

So far, you have learned about counting from 0 to 99. You have learned to add and subtract. You have learned to measure things. Hopefully, you are learning to love God and serve others too with math.

In this chapter, we are going to learn about adding and subtracting bigger numbers.

0 to 999 DAY 132

This lesson introduces triple-digit math using manipulative (blocks), and is followed by one page of new exercises. This will require about 30 minutes of instruction from the parent/teacher.

Prayer

Heavenly Father, thank You for giving us math. But, most of all we thank You for forgiving our many sins. They are more than we can count. Thank You for Jesus who died for our sins. Help us to be faithful with the things You give us to do. Amen.

Memory

Spend a few minutes with both addition and subtraction flash cards (mix and match).

Lesson

So far, you have learned to count to 99. Maybe there are 99 people now coming to your church. But what if 1 more person would come? How many people would be coming to your church? Let's find out.

Do you remember what we did when we counted from 1 to 10?

1, 2, 3, 4, 5, 6, 7, 8, 9 . . .

These numbers all take up one space. We call this space the 1s place.

What comes next? What is the next number? Of course it is 10. We have to use another **place** when we go to the number 10. This new place goes in front of the 1s place. We call it the **10s place**. But what happens when that 1 more person comes to church?

90, 91, 92, 93, 94, 95, 96, 97, 98, 99 . . .

Add one more number to 99, and we get 100! We need to use another place. This new place goes in front of the 10s place. It is called the **100's place**.

The number 100 can also be called 1 chunk of 100. 90 is 9 chunks of 10. 100 is 10 chunks of 10.

What does a group of 100 people look like? The picture shows 100 people. What a beautiful creation of God! He made men and women, boys and girls. They all look so different!

Now let's practice breaking a big number into chunks. Get your set of blocks out. They include single 1s, chunks of 10, and chunks of 100. We're going to use the big block of 100 for this lesson. Let's look at the big number 167. Imagine there are 167 people at church today. That's 1 chunk of 100, 6 chunks of 10, and 7 single 1s.

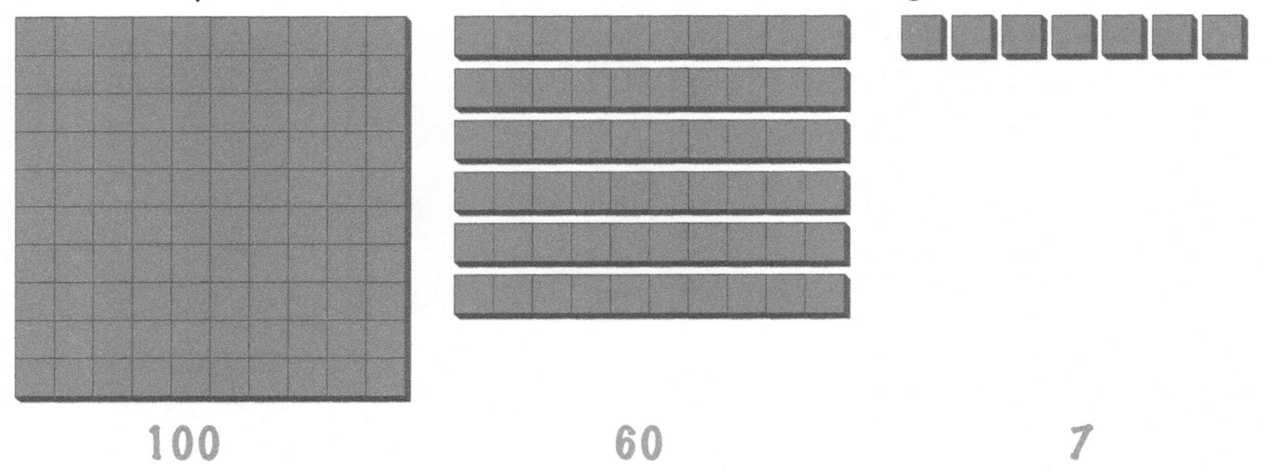

Let's look at the number 205. That would be a lot of people coming to church. How many single 1s? How many chunks of 10s? How many chunks of hundreds in this number? That's right! 2 chunks of 100s, no chunks of 10s, and 5 singles 1s. Let's build 205 using blocks.

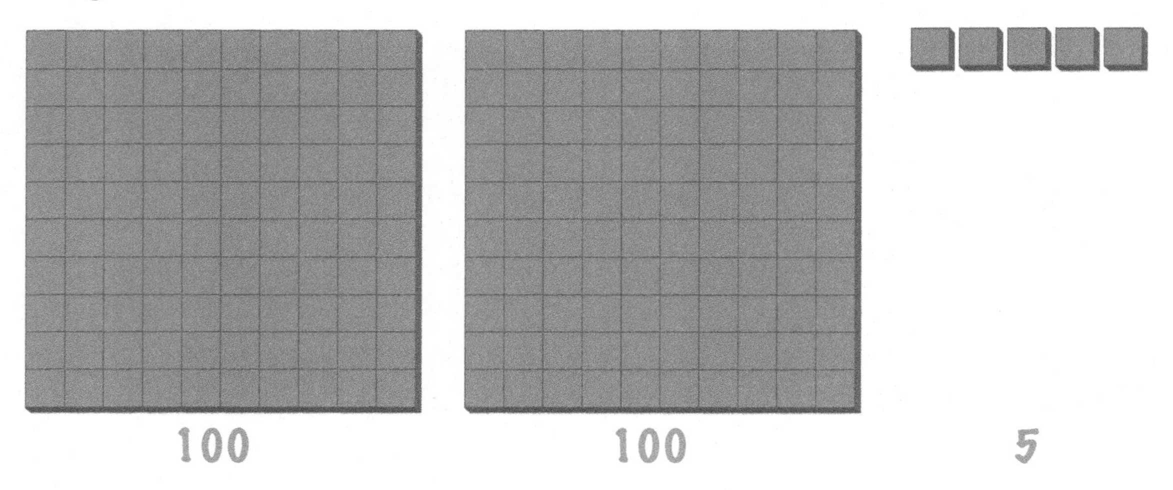

Here's another way to look at these big numbers. Some big jets have 100 seats. They can hold 100 people. Some buses have only 10 seats for 10 people. Motorcycles have a single seat for 1 person. Let's say there are 100 people in each jet airplane. There are 10 people in each bus. There's only 1 person on each motorcycle. How many people do we have altogether? Fill in the chart on the next page.

Airplanes	Buses	Motorcycles	How many people?

Let's say these big numbers out loud. You can read these numbers the long way and the short way. First, we will read the numbers the long way:

145 One hundred forty-five

283 Two hundred eighty-three

Let's read these numbers using the short way.

125 One twenty-five

741 Seven forty-one

500 Five hundred

Here's one more way to think of these numbers.

100 people live in these apartment buildings. 10 people live in these houses. 1 person lives in each little tent. How many people live in this area? That's right! 245 people live in this area!

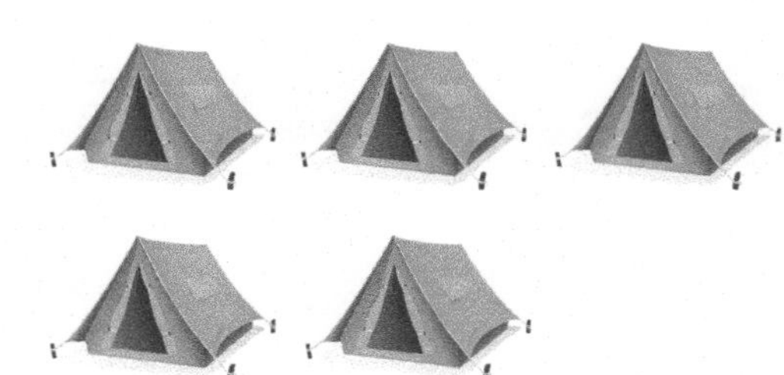

Student Exercises

How many chunks of 100 do you need to make these numbers? How many chunks of 10 do you need? How many singles (1s) do you need? Write the answers in the blanks.

Can you imagine these numbers as people riding on planes, buses, and motorcycles? Or can you imagine them living in apartments, houses, and tents?

123 → ______ 100's, ______ 10s, ______ 1s

453 → ______ 100's, ______ 10s, ______ 1s

745 → ______ 100's, ______ 10s, ______ 1s

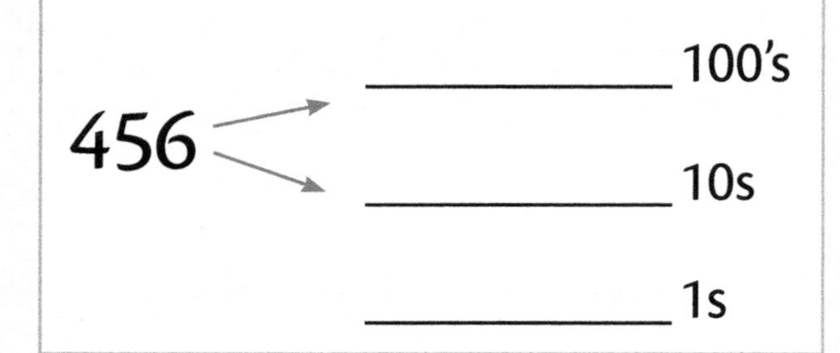

456 → ______ 100's, ______ 10s, ______ 1s

366 → ______ 100's, ______ 10s, ______ 1s

145 → ______ 100's, ______ 10s, ______ 1s

45 → ______ 100's, ______ 10s, ______ 1s

7 → ______ 100's, ______ 10s, ______ 1s

Student Exercises

Measure the sides of each shape in inches (in). How long is each side? Add all the sides together to find the length around the whole shape. Use your string to measure any shapes with curves. Write your answer inside each shape.

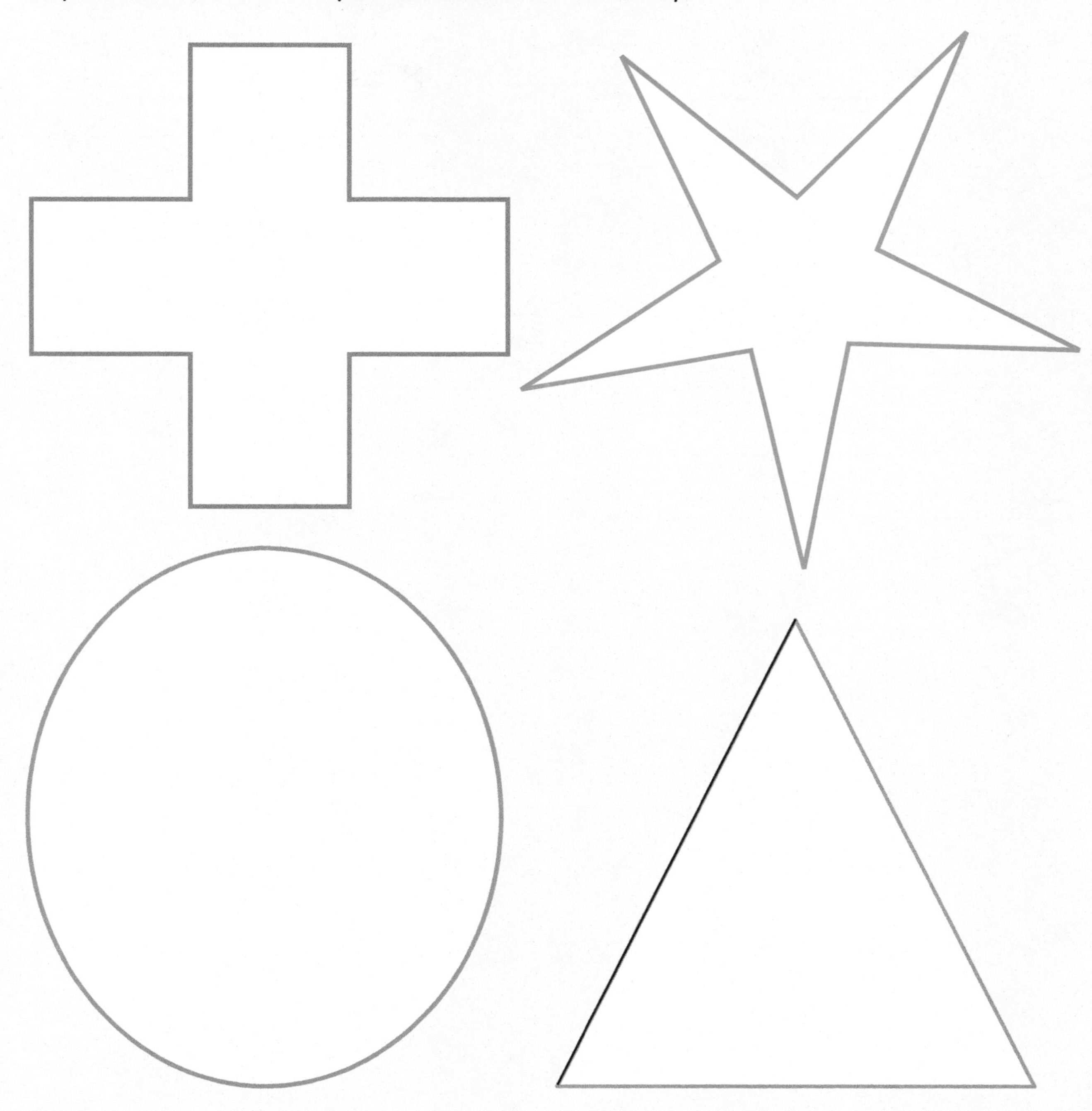

God has made many ways for two smaller numbers to come together to make a bigger number. There are many right answers! Find five different ways to break the big number 8 into two smaller numbers. You can use your blocks or stones to make two sets if you need help! (For this exercise, 2 + 6 is the same as 6 + 2.)

8 = ____ + ____

8 = ____ + ____

8 = ____ + ____

8 = ____ + ____

8 = ____ + ____

Let's count single 1s, chunks of 10, and chunks of 100s. Fill in the table with the number of 100's, the number of 10s, and the number of 1s.

Write the number of 1s, 10s, and 100s in each box. Put the number of 100s blocks in the 100s box. Put the number of 10s blocks in the 10s box. Put the number of ones blocks in the 1s box. Then, write the full number in the bigger box.

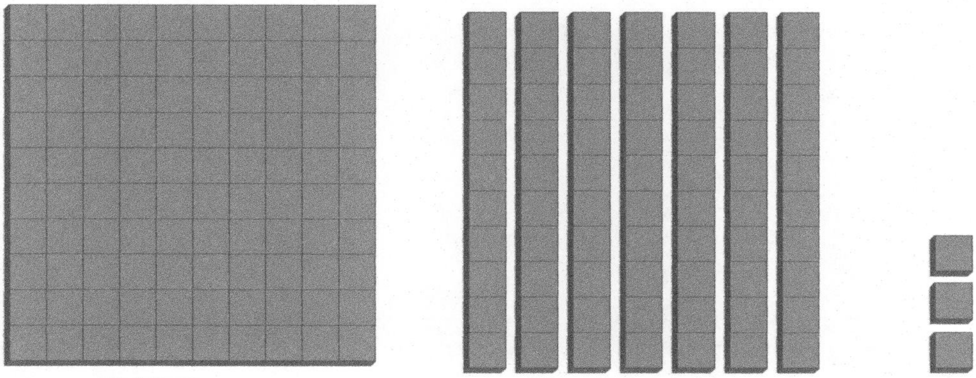

100s: 10s: 1s:

Total:

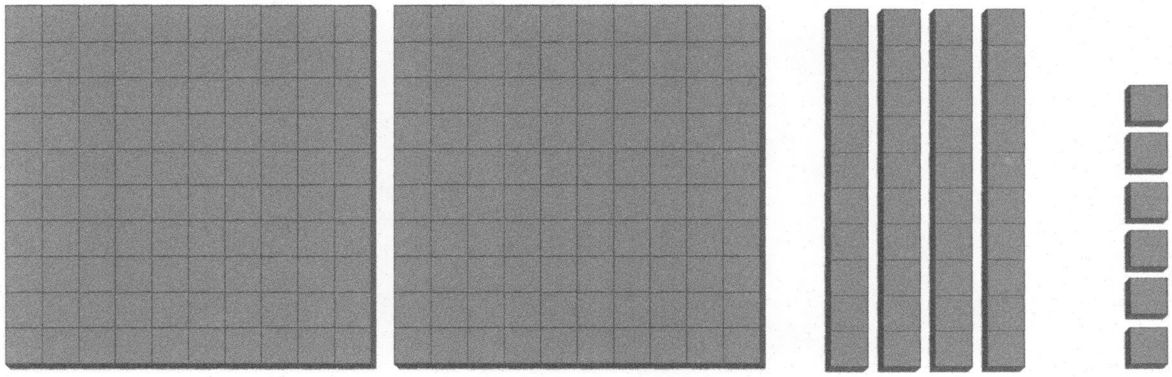

100s: 10s: 1s:

Total:

100s: 10s: 1s:

Total:

> Jesus said to them, "Bring some of the fish which you have just caught." Simon Peter went up and dragged the net to land, full of large fish, one hundred and fifty-three; and although there were so many, the net was not broken. (John 21:10-11)

How many fish did Peter catch? Write this number in the boxes.

100s: 10s: 1s: Total:

Ordering Big Numbers DAY 134

This lesson compares bigger and smaller three-digit numbers, and is followed by two pages of review exercises. This will require about 20 minutes of instruction from the parent/teacher.

Prayer

Pray your own prayer of thanksgiving and praise to God. Pray for His help on this lesson.

Memory

Spend a few minutes with both addition and subtraction flash cards (mix and match).

Lesson

You have just learned about big numbers. Let's do more math with them! How do they come together in addition? How do they come apart using subtraction? Today we will look at the order God has for these numbers.

You have already learned that some numbers are bigger than other numbers. A 10-year-old boy is older than a 7-year-old boy. A 17-year-old girl is older than a 13-year-old girl.

Take a look at these two numbers. Why is 17 bigger than 13?

13

17

Now can you see the difference? The number 17 is bigger than 13. It has more single 1s.

Compare 65 and 69. Which is bigger? Both have 6 chunks of 10. But 65 only has 5 single 1s, while 69 has 9 single 1s.

Bigger numbers can be a little harder to compare. First, think about how many chunks of 100 are needed to make each of the numbers below. Do you see that 642 is made up of six chunks of 100? That's because 6 is in the 100's place. Do you see that 842 is made of eight chunks of 100? It has 8 in the 100's place. That means that 842 is bigger than 642. It has more chunks of 100.

642 < 842

Now look at the numbers 738 and 768. Both numbers have seven chunks of 100. So which is bigger? When this happens, we need to think about the chunks of 10 next. Look at the numbers in the 10s place. 738 only has 3 chunks of 10. 768 has 6 chunks of 10. That means that 768 is bigger than 738.

738 < 768

Let's try another one. Look at the numbers 415 and 418. They're also big numbers! They both have 4 chunks of 100. But they also both have 1 chunk of 10. Now what should we do? That's right—we look at the singles next! 415 has only 5 single 1s. 418 has 8 single 1s. That means that 418 is bigger than 415.

415 < 418

Which of these numbers are bigger? Fill in the circle with the sign "<" (smaller than) or ">" (bigger than). Remember, the baby shark eats the bigger number!

349 ◯ 350	700 ◯ 70	305 ◯ 53
632 ◯ 631	299 ◯ 300	158 ◯ 185
0 ◯ 100	462 ◯ 362	125 ◯ 625

Student Exercises

Find the distance in each exercise. Each child's house is marked by a red dot. If you're driving a car, count the number of city blocks (or centimeters) to find out how far you need to go. If you're flying like a crow, use your ruler to measure how far you need to fly. Each centimeter equals one city block.

Candice
Amy
Bobby
Dan
Emily

Amy's house to Bobby's house by car: ________

Amy's house to Bobby's house by crow: ________

Amy's house to Candice's house by car: ________

Bobby's house to Emily's house by car: ________

Emily's house to Candice's house by crow: ________

Candice's house to Dan's house by crow: ________

Dan's house to Emily's house by crow: ________

Here's a fun game! Count by twos to go from your house to where your church meets! Start with 2. The next number will be 2 more. It will be 4. The next will be 6. Each number will be 2 more than the one before it.

2	3	7	12	15	43	76	19
5	4	9	11	31	59	17	11
12	16	6	13	14	18	49	19
23	12	8	7	12	22	55	42
9	19	16	10	19	14	54	11
15	39	9	20	16	33	14	21
12	32	34	18	54	24	26	12
27	22	19	20	22	13	38	28

Practice DAY 135

Student Exercises

Here are some big numbers! Which number is bigger? Which number is smaller? Fill in the blank with the correct sign: < (smaller than) or > (bigger than). Then read the math sentence. Remember, the baby shark eats the bigger number!

123 ◯ 153	537 ◯ 523
824 ◯ 273	54 ◯ 142
899 ◯ 900	345 ◯ 348
349 ◯ 450	400 ◯ 399
634 ◯ 632	34 ◯ 0

What time is it? For the first six clocks, read the time. Write it on the blank line under each clock. For the last six clocks, draw the short hand (the hour hand) and the long hand (the minute hand) on each clock to show the right time.

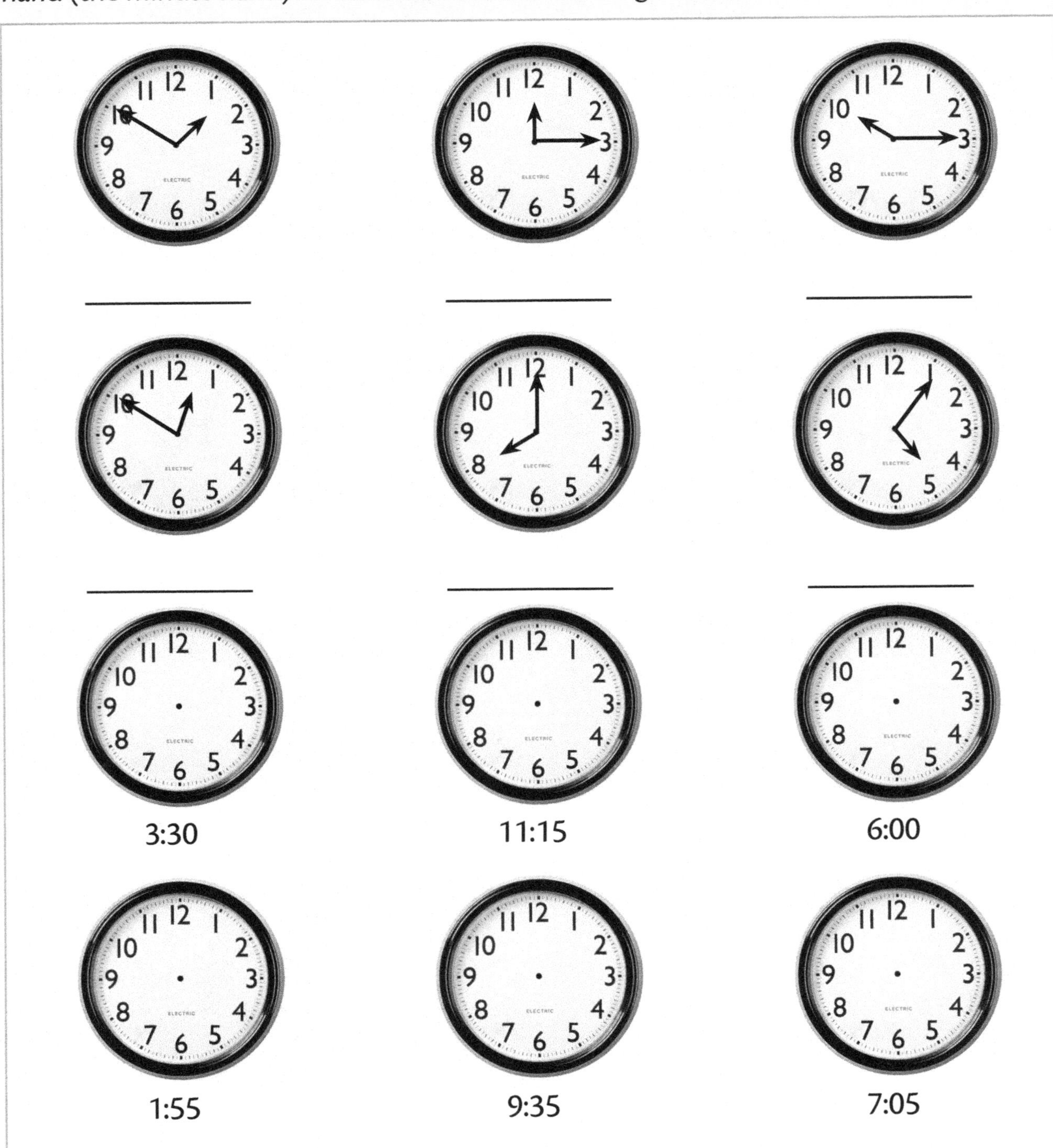

Bigger Numbers in the Scriptures DAY 136

This lesson explores larger numbers contained in Scripture, and is followed by one page of review exercises. This will require about 10 minutes of instruction from the parent/teacher.

Prayer

Pray your own prayer of thanksgiving and praise to God. Pray for His help on this lesson.

Memory

Spend a few minutes with both addition and subtraction flash cards (mix and match).

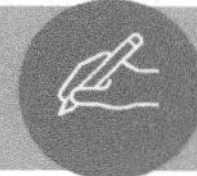

Lesson

Before Noah's flood, people lived much longer than they do today. Most people do not live to be over 100 years old today. You can read Genesis 5 to learn about the people who lived before the flood.

Let's compare their ages to yours. Compare their ages to your mother or father. Compare their ages to one of your grandfathers or grandmothers. Who lived longer?

Break each number up into chunks of 100, chunks of 10', and single 1s.

	How many...		
	100s	10s	1s
How old are you? ________	________	________	________
How old is your mother (or father)? ________	________	________	________
How old is your grandfather/mother? ________	________	________	________
How old was Adam when he died? ________	________	________	________
How old was Seth when he died? ________	________	________	________
How old was Enoch when God took him? ________	________	________	________
How old was Methuselah when he died? ________	________	________	________

Student Exercises

Look at these sets. Circle the picture that is different from the others in each set. Can you explain why it is different?

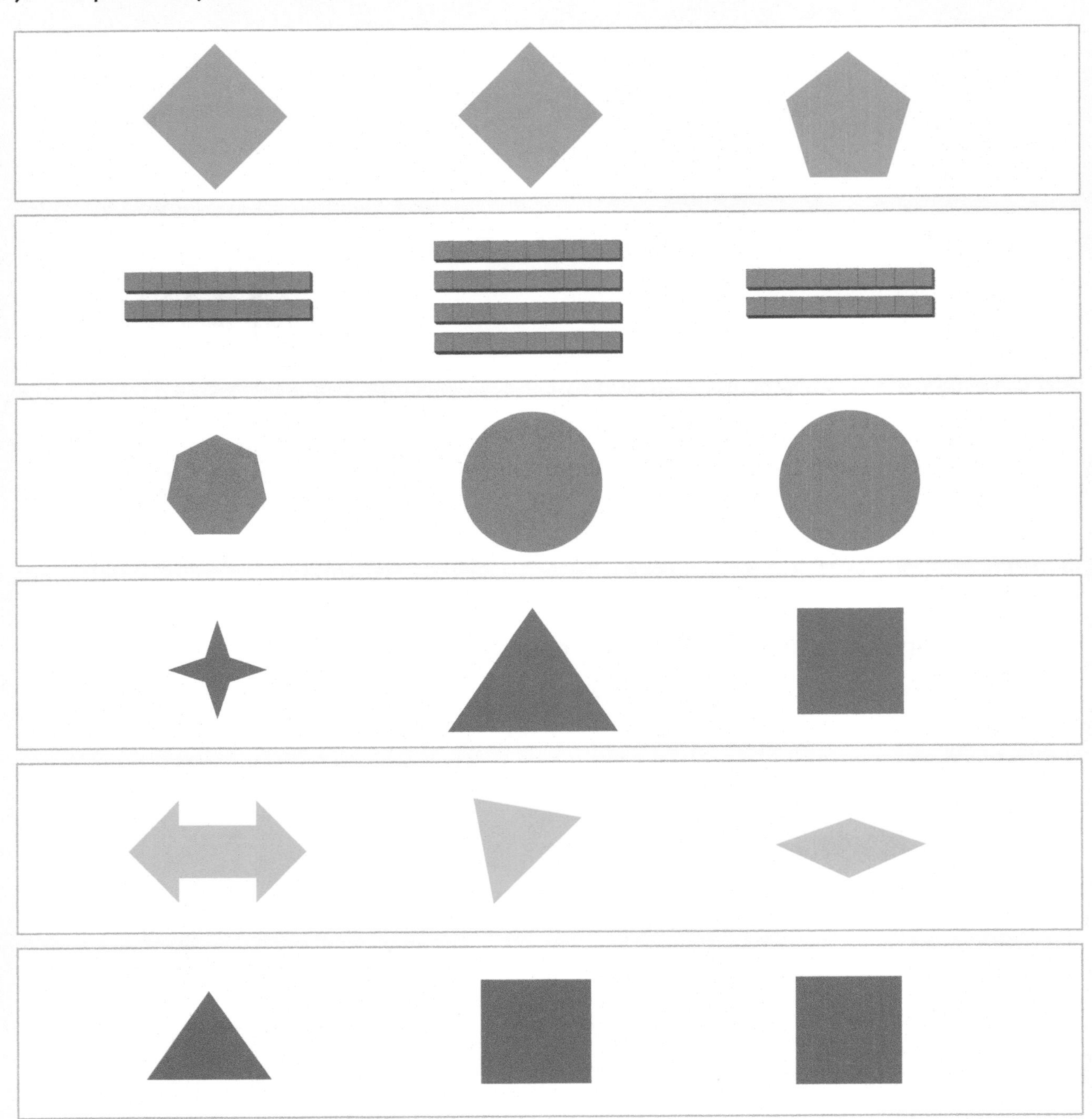

Student Exercises

Find the missing numbers! Try to find the missing pieces in these equations. You can use your stones to help you. For the addition exercises, ask: "How many stones do you have to add to the small number to get the bigger number?"

Now let's look at the subtraction exercise, "What minus 1 equals 5?" Ask, "How many stones do I need to start with? I need to take away one stone. And I need to have five stones left. So how many stones should I start with?" The best way to figure this out is to set out five stones. Add one stone. You will take this away later. How many stones do you have now? Six! If you start with six stones, and take one away, you will get five stones.

What plus 1 equals 3?	2 + 1 = 3
What plus 2 equals 4?	______ + 2 = 4
What minus 1 equals 5?	______ − 1 = 5
What minus 5 equals 6?	______ − 5 = 6
What plus 3 equals 7?	______ + 3 = 7
What plus 6 equals 8?	______ + 6 = 8
What minus 1 equals 9?	______ − 1 = 9

How many chunks of 100 do you need to make these numbers? How many chunks of 10? How many singles (1s)? Can you imagine them as people riding on planes, buses, and motorcycles?

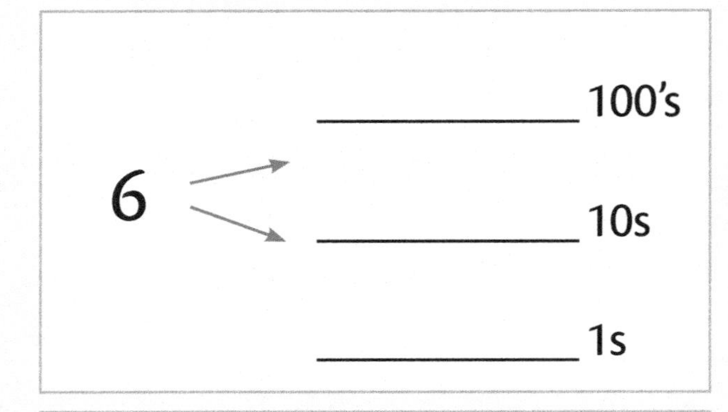

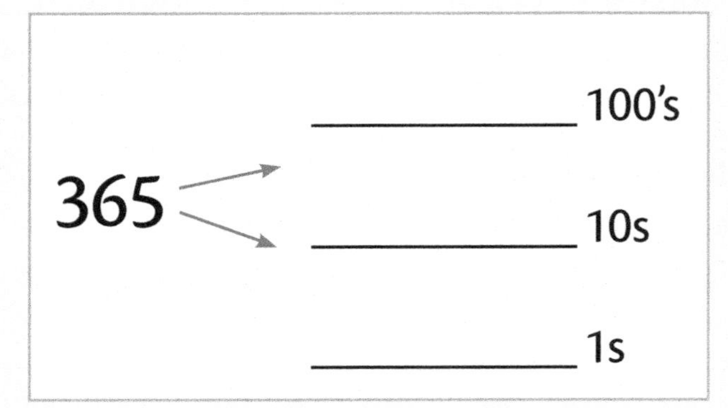

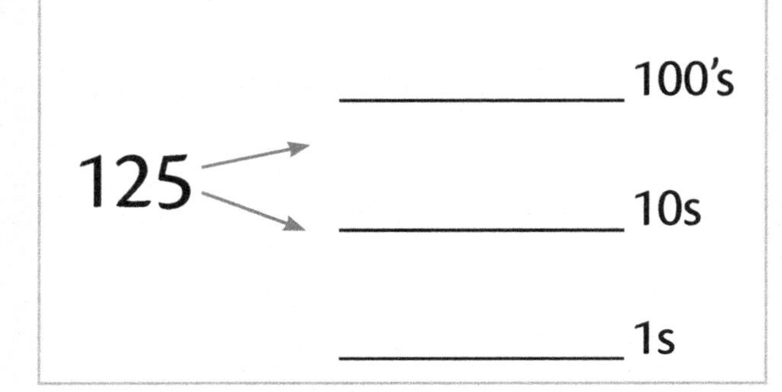

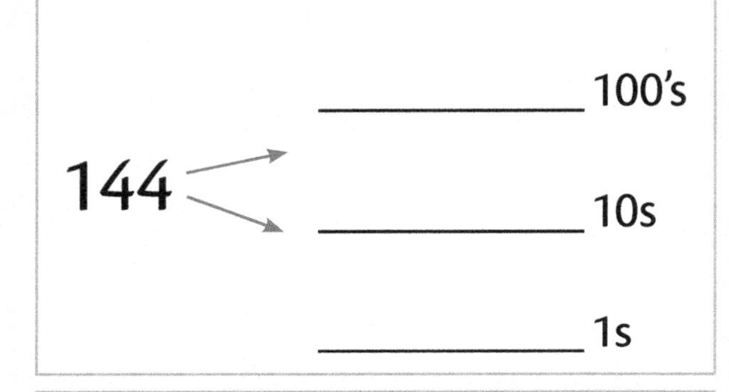

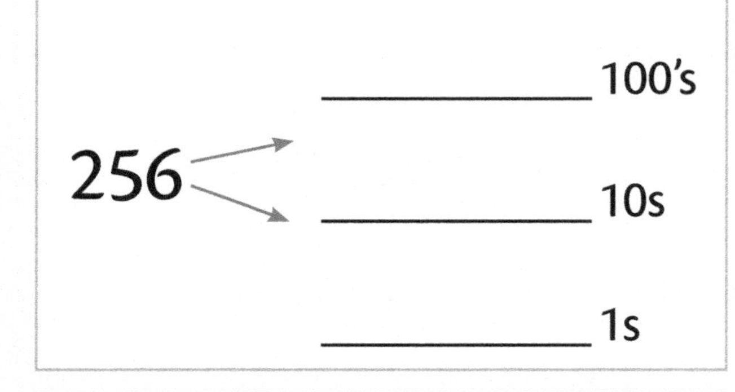

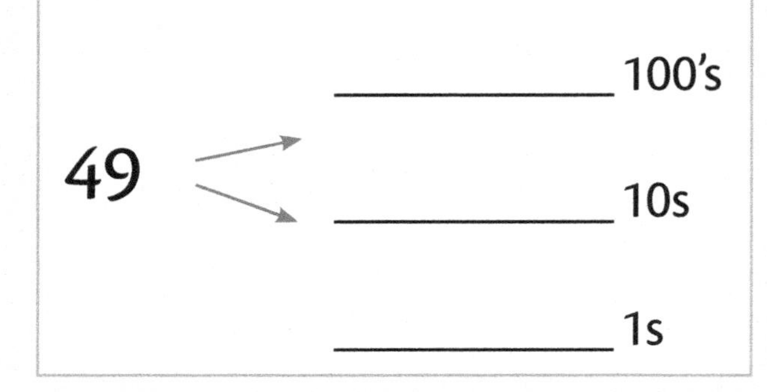

555

_______ 100's

_______ 10s

_______ 1s

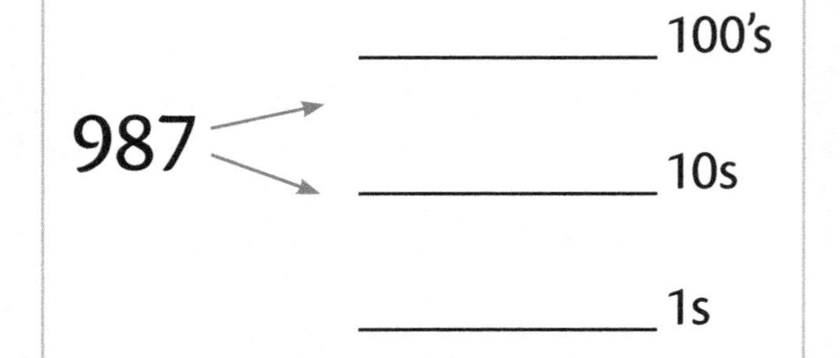

Adding and Subtracting Big Numbers DAY 138

This lesson introduces long addition and long subtraction for three-digit numbers, and is followed by two pages of review exercises. This will require about 30 minutes of instruction from the parent/teacher.

Prayer

Pray your own prayer of thanksgiving and praise to God. Pray for His help on this lesson.

Memory

Spend a few minutes with both addition and subtraction flash cards (mix and match).

Lesson

Today we will add and subtract big numbers. We'll learn about adding and subtracting numbers bigger than 100. Let's add using our blocks first. Then we'll do the math on paper.

Let's add 235 and 142. Get out two sets of blocks. Set them up like they are on the next page. Do you have 2 chunks of 100, three chunks of 10, and five single 1s to make 235? Do you have 1 chunk of 100, 4 chunks of 10, and two single 1s to make 142?

Now let's combine them. Start by adding the 1s first. Then combine the 10s. Finish by adding the 100's. How many chunks of 100, chunks of 10, and single 1s are there now?

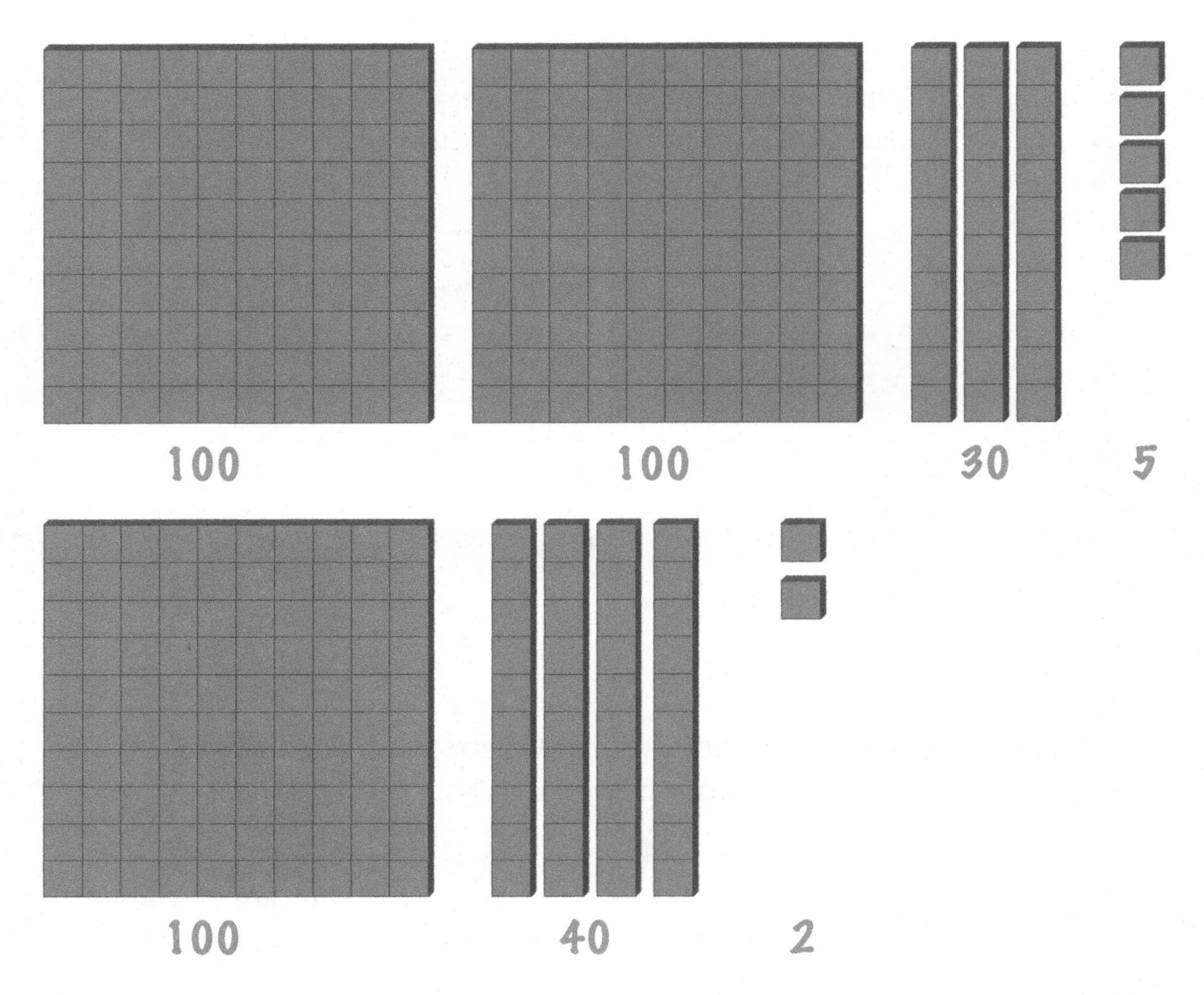

That's right! You have 3 chunks of 100, 7 chunks of 10, and 7 single 1s. This makes the number 377!

Let's add these numbers on paper now! Remember, you must add the 1s first. Then you add the chunks of 10. Then you add the chunks of 100. We will get 7 single 1s, 7 chunks of 10, and 3 chunks of 100. This makes . . . 377! This is how we add these big numbers in the long form:

235 +142 7	235 +142 77	235 +142 377

Now let's try long subtraction! We'll subtract one big number from another big number using your blocks: 273-112. Get out 273 blocks. They should look like this:

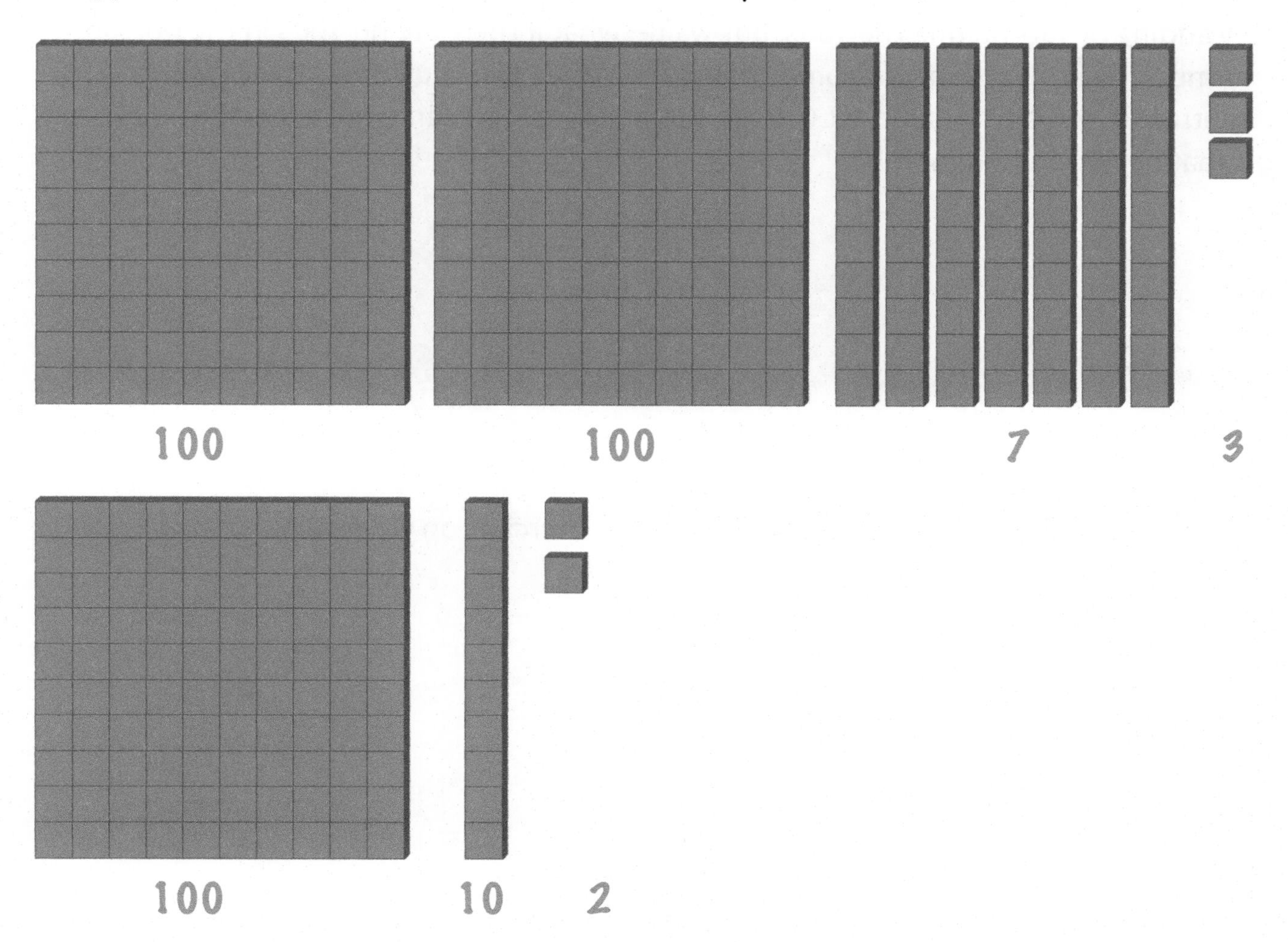

Let's take away 112! Once again, you must start with the 1s first. Take away 2 single blocks from the group of 3. Then, take 1 chunk of 10 away. Finish by taking 1 chunk of 100 away. How many blocks do you have left? That's right! You have 161!

Now let's find the answer to 273 – 112 using long subtraction. Remember to start with the 1s. Then subtract the 10s. Finish with the 100's.

$$\begin{array}{r} 273 \\ -112 \\ \hline 1 \end{array} \qquad \begin{array}{r} 273 \\ -112 \\ \hline 61 \end{array} \qquad \begin{array}{r} 273 \\ -112 \\ \hline 161 \end{array}$$

Let's try another addition exercise and another subtraction exercise. Remember, you must start by adding or subtracting the 1s. Then you will add or subtract the 10s. Finish by adding or subtracting the 100's. But wait—what if there are no 100's in the bottom number? This means the bottom number has zero 100's. You know how to add and subtract 0! Just add or subtract 0 when this happens. You can write a 0 in the 100's place if that helps you to remember.

$$\begin{array}{r} 451 \\ +\ 34 \\ \hline 485 \end{array} \qquad \begin{array}{r} 369 \\ -\ 34 \\ \hline 335 \end{array}$$

Now try the following addition and subtraction exercises on your own:

$$\begin{array}{r} 723 \\ +\ 54 \\ \hline \end{array} \qquad \begin{array}{r} 135 \\ -\ 12 \\ \hline \end{array} \qquad \begin{array}{r} 426 \\ +\ 3 \\ \hline \end{array}$$

Great job! Let's try a few more special exercises: What is 990 + 5? Imagine that a family of 5 is sitting down for dinner. Then, 990 guests joined them. That's a lot of people! How many people would be eating?

Now imagine that you are 5 days away from your birthday. Your parents decide to delay the party for 990 days. Does that sound like a good plan to you?

Imagine diving from a 5-foot tall diving board? Splash! What would it be like to dive from a 990-foot diving board? That would be a long fall! That would be impossible.

The difference between 5 and 990 is very, very big!

You are very small in comparison with trees and mountains. You are about 4 feet tall. God made super tall trees. The tallest tree in the world is 380 feet tall! The tallest mountain in the world is 29,000 feet tall. Mountains are much, much taller than you. The difference is very, very, very big!

Draw a picture comparing you, the tallest tree in the world, and the tallest mountain in the world.

God is bigger than everything He has made. When Solomon made a house for God, Solomon said:

> "But will God indeed dwell on the earth? Behold, heaven and the heaven of heavens cannot contain You. How much less this temple which I have built!" (1 Kings 8:27)

The skies are big—they are very big! But God is bigger than the sky. He is bigger than outer space. Can you see why Solomon asked how God could come to the temple he built?

Student Exercises

Find the distance between the two numbers in each exercise. Subtract the smaller number from the bigger number. You can also use a ruler or a tape measure if you need to.

5 and 12 ________	1 and 13 ________
4 and 18 ________	7 and 1 ________
11 and 7 ________	3 and 9 ________
14 and 6 ________	15 and 2 ________
11 and 17 ________	17 and 19 ________

Let's add three numbers! Find the sum of the first two numbers and cross them out. Write their sum in the first blank below the exercise. Then add that sum to the last number. What is your answer?

4+2+5 = ________ _____+5 = ________	5+2+6 = ________ _____+6 = ________
3+5+7 = ________ _____+7 = ________	2+7+7 = ________ _____+7 = ________
6+2+6 = ________ _____+6 = ________	1+9+3 = ________ _____+3 = ________
7+0+4 = ________ _____+4 = ________	6+5+5 = ________ _____+5 = ________
5+5+5 = ________ _____+5 = ________	2+3+6 =________ _____+6 = ________

DAY 139 Practice

Student Exercises

Try to add and subtract the smaller numbers below by using your memory. You can also use your blocks, coins, or fingers if you need help.

You will have to use long addition and long subtraction for the bigger numbers. Remember, you need to add or subtract the 1s place first. Then you need to add or subtract the 10s place. Finish by adding or subtracting any 100's.

8 +5 ———	13 + 0 ———	11 + 58 ———
26 + 32 ———	642 +143 ———	534 +322 ———
15 – 7 ———	14 – 8 ———	16 – 9 ———
14 – 5 ———	75 – 53 ———	69 – 34 ———

Extra Challenge

Let's have fun with two more sudoku puzzles. Remember, every group of four squares must have a 1, 2, 3, and 4 in it. But each number can only be used once in each group.

Every horizontal (side-by-side) line of four squares and every vertical (up and down) line of four squares must also have 1, 2, 3, and 4 in it.

1		4	
	3		2
2		3	
	1	2	4

	3		
4			
3	1		4
2	4	3	

DAY 140 Rounding to the Nearest 100

This lesson explores estimation and rounding with bigger numbers, and is followed by two pages of review exercises. This will require about 10 minutes of instruction from the parent/teacher.

Prayer

Pray your own prayer of thanksgiving and praise to God. Pray for His help on this lesson.

Memory

Spend a few minutes with both addition and subtraction flash cards (mix and match).

Lesson

Imagine that you dug into a pile of small rocks with a shovel. How many rocks would make up each scoop?

If you guessed 325, that might be pretty close to the right number. It would take a long time to count all those rocks! Guessing is much faster.

Let's say you wanted to round your guess to the nearest 10. If you rounded 325 to the nearest 10, you would say "There are about 330 rocks in the shovel." What if you rounded to the nearest hundred? You would say, "There are about 300 rocks in the shovel."

Sometimes it's best to take a guess and round to the nearest 100. When you're looking at big numbers of rocks, it's too hard to round to the nearest 10. It's definitely too hard to count them all!

Now let's round 325 to the nearest 100. The number 325 is made up of 3 chunks of 100, 2 chunks of 10, and 5 single 1s. When we round the number to the nearest 100, we must only look at the chunks of 10. Since there are only 2 chunks of 10, we will round down.

325 → 300

What about rounding the number 349? We would round down to 300. 349 only has 4 chunks of 10.

Let's try rounding 350. Would we round it up to the next hundred (400)? Or would we round it down to 300? We would round it up to 400, just like we would round 35 up to 40. 350 has 5 chunks of 10, so we round up. Remember, 5 or more chunks of something means we round up.

Practice rounding these numbers up or down to the nearest 100.

331	500	678
353	355	359

331 →	500 →	678 →
353 →	355 →	359 →

Student Exercises

Try to add and subtract the smaller numbers below by using your memory. Hopefully by this time you won't need to use your blocks, coins, or fingers.

You will have to use long addition and long subtraction for the bigger numbers.

$\begin{array}{r} 8 \\ +3 \\ \hline \end{array}$	$\begin{array}{r} 9 \\ +5 \\ \hline \end{array}$	$\begin{array}{r} 58 \\ +\ 31 \\ \hline \end{array}$
$\begin{array}{r} 62 \\ +36 \\ \hline \end{array}$	$\begin{array}{r} 324 \\ +152 \\ \hline \end{array}$	$\begin{array}{r} 634 \\ +154 \\ \hline \end{array}$
$\begin{array}{r} 18 \\ -\ 7 \\ \hline \end{array}$	$\begin{array}{r} 16 \\ -\ 9 \\ \hline \end{array}$	$\begin{array}{r} 18 \\ -\ 9 \\ \hline \end{array}$
$\begin{array}{r} 17 \\ -\ 8 \\ \hline \end{array}$	$\begin{array}{r} 56 \\ -26 \\ \hline \end{array}$	$\begin{array}{r} 68 \\ -41 \\ \hline \end{array}$

Let's put these bigger numbers in order! Sometimes you will have to count down (or backward) to find the right number. Sometimes you will have to count up (or forward) to find the number that comes next. God made many numbers. He put them all in order!

20, ________ , ________

________ , 50, ________

________ , ________ , 80

________ , 250, ________ , ________

________ , ________ , 740, ________

________ , ________ , 600

397, ________ , ________ , ________

________ , ________ , ________ , ________ , 500

DAY 141 Practice

Student Exercises

Round each of these numbers to the nearest 10. Sometimes you might have to round up. Sometimes you might have to round down. Remember, 5 is always rounded up.

92 → ________	51 → ________
86 → ________	47 → ________
75 → ________	35 → ________

Round each of these numbers to the nearest 100. Rounding is one way that we make good guesses! Remember, 5 in the 10s place means we always round up to the next 100.

145 → ________	632 → ________
934 → ________	578 → ________
350 → ________	289 → ________

Extra Challenge

This symmetrical shape is a star with 10 points. That's 10 symmetries! Each point is marked by 10 numbers — 0, 1, 2, 3, 4, 5, 6, 7, 8, and 9. You can turn this star to 10 different positions, and it will look exactly alike!

Fill in the following number patterns. The first exercise provides the answer. For the second exercise, set your pencil on the number 6. Moving your pencil in the direction of the numbers going up (clockwise), count up 7 positions. Or, "spin up by 7." What number is your pencil pointing at now? Write 1 into the first blank line. Complete the exercises.

0 ↷ 2 = 2

6 ↷ 7 = ______

5 ↷ 2 = ______

2 ↷ 5 = ______

8 ↷ 2 = ______

5 ↷ 6 = ______

11 ↷ 3 = ______

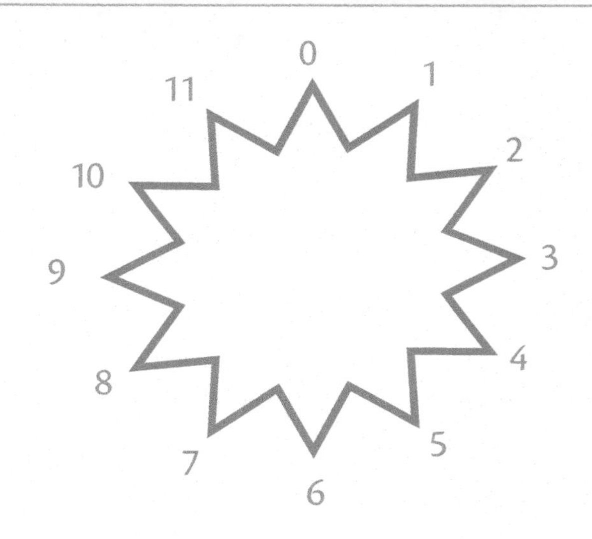

DAY 142 Timelines

This lesson explores timelines using larger numbers, and is followed by one page of review exercises. This will require about 15 minutes of instruction from the parent/teacher.

Prayer

Pray your own prayer of thanksgiving and praise to God. Pray for His help on this lesson.

Memory

Spend a few minutes with both addition and subtraction flash cards (mix and match).

Lesson

> Now Jesus Himself began His ministry at about thirty years of age. (Luke 3:23)

Do you remember how we keep track of the time? We use a clock. What do we use to keep track of days, weeks, or months? A calendar! A calendar helps us keep track of the days and months in a whole year. But is there something we can use to help us keep track of a whole list of years? Yes! We keep track of the years using a group of numbers called a timeline. Jesus was born around the year 0. Then Jesus started His ministry when He was about 30 years old. He died and rose again when He was about 33. It has now been over 2,000 years since Jesus was born. We could show these years on a timeline. Ask your parent or teacher the year you were born. That is the number of years between your birth and Jesus's birth!

Let's look at this timeline and add a few more important things that have happened.

- Jesus died and rose again from the dead around the year 33.
- The Apostle Paul died around the year 66.
- Jerusalem was destroyed in the year 70.
- Patrick of Ireland went to Ireland as a missionary around the year 440.
- The Roman Empire fell in the year 476.
- Augustine of Canterbury took missionaries to England in the year 600.

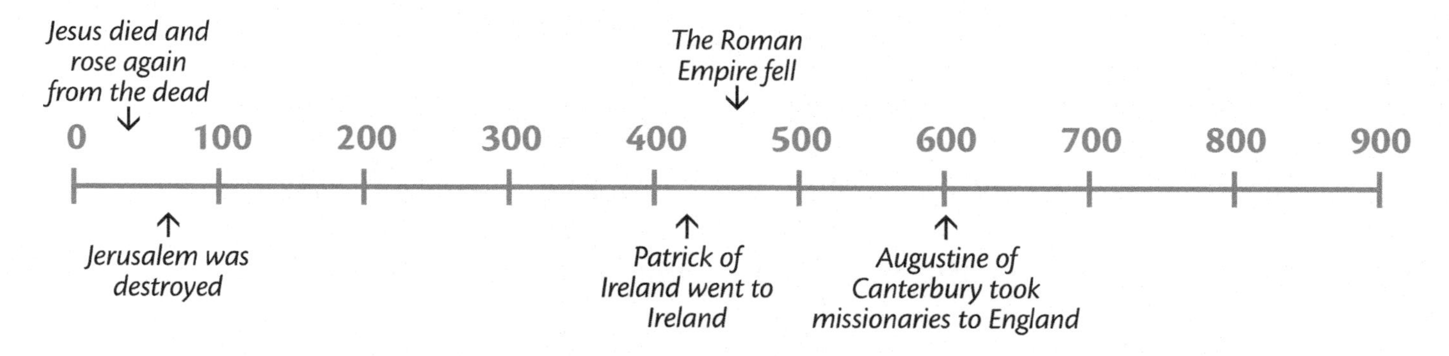

The timeline looks a little bit like a ruler. It measures time over many years.

Patrick's grave in Ireland.

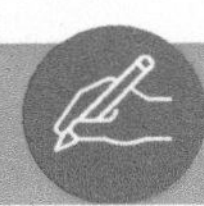

Student Exercises

Try to add and subtract the smaller numbers below by using your memory. Hopefully by this time you won't need to use your blocks, coins, or fingers. You will have to use long addition and long subtraction for the bigger numbers.

8 +7	62 +35	74 +13
354 +231	713 +184	15 − 5
16 − 7	13 − 6	13 − 8
130 −110	144 −121	52 −21

After doing these exercises with God's patterns, what can you thank God for?

Practice DAY 143

Student Exercises

Addition makes bigger numbers. Subtraction makes smaller numbers. God made numbers to work together. Sometimes numbers work together to make bigger numbers. Sometimes numbers work together to make smaller numbers.

For these exercises, you need to decide how these numbers work together. How do 8 and 6 work together to make 14? Of course, they must be added! How do 15 and 7 come together to make 8? Write "+" or "-" in each circle to show how they work together.

8 ◯ 6 = 14	15 ◯ 7 = 8
8 ◯ 5 = 13	8 ◯ 8 = 0
16 ◯ 9 = 7	9 ◯ 7 = 16
6 ◯ 7 = 13	13 ◯ 4 = 9
0 ◯ 19 = 19	8 ◯ 8 = 16

Put these numbers in order from smallest to biggest. Put them in the order God made for them!

What is going on in each picture? Look at each one and try to imagine how each number could be used. God's world gives us lots of things to count and to measure!

The boy hit the baseball...

134, **251**, or **311** feet into the field!

__________, __________, __________

There are...

100, **170**, or **120** people in the parade!

__________, __________, __________

That bridge is...

415, **541**, or **145** feet high!

__________, __________, __________

Go Use God's Big Numbers! DAY 144

This lesson integrates math into everyday life. This is an essential element to learning. The child is encouraged to apply God's patterns and wisdom to life in the home and community. Let's take a break from memory work and academic exercises, and identify ways in which to make math part of everyday life. The following are suggestions or examples, but other ideas may be added to the list.

Depending on the currency used, the parent/teacher may have to adjust the value (and currency symbols) for the items in these exercises.

> Diverse weights and diverse measures,
> They are both alike, an abomination to the LORD. (Proverbs 20:10)

In the last chapter, we talked about buying and selling things. Here is one more reason to learn math: God wants us to be honest when we buy and sell things. He doesn't want us to steal. Math helps us to be honest.

Let's find the length of time between two things that happened a long time ago. Why do you think these lengths are important?

1. Jesus died and rose again around the year 33. Paul was killed by the bad King Nero around the year 66. How many years were there between these two events?

$$66 - 33 = \_\_\_\_\_\_$$

2. Suppose that Paul became a missionary 7 years after Jesus died and rose again, around the year 40. How long was Paul a missionary?

$$66 - 40 = \_\_\_\_\_\_$$

3. Jerusalem was destroyed in the year 70. How many years were there between Jesus' death and the destruction of Jerusalem? Read Matthew 24:1-3, 34. In these verses, Jesus is speaking about Jerusalem's destruction. Use your blocks for this one.

$$70 - 33 = \_\_\_\_\_\_$$

4. The Roman Empire fell apart in the year 476. How many years were there between the fall of the Roman Empire and the year the Romans put Jesus to death?

$$476 - 33 = \_\_\_\_\_\_$$

5. Jesus told His disciples to take the Gospel to the whole world. He told them to do this in the year 33. How many years after this did Patrick go to Ireland as a missionary? Use your blocks for this one.

$$440 - 33 = \_\_\_\_\_\_$$

6. Missionaries took the gospel to Iceland around the year 975. Augustine of Canterbury went to England as a missionary in the year 600. How many years were there between these events?

$$975 - 600 = \_\_\_\_\_\_$$

Activity 2

Choose one of these ways to add or subtract big numbers in your own world.

1. Start a little business and sell something. Add up all the money you made. Compare it to the money you had to spend for your business. How much profit did you make? You can figure out your profit this way:

Money people paid you – Money you spent = Profit

2. Compare ages. Families are made of generations. What are generations? Generations are made when two people have children, who have children, who have children. . . . Each set of children makes a generation.

- Your great-great-grandparents make up one generation of your family. Your great-grandparents make another generation. Your grandparents are another generation. Your grandparents had your parents. Your parents had you.
- Compare the ages of people in your family.

- How old is one of your grandparents? How old are you? How much older is your grandparent than you?
- How much older is your great-grandparent than a parent?
- See how many different ways you can compare the ages of people in your family!

3. Make more cookies. First, find a recipe for cookies. How many cookies will this recipe make? Double the recipe to make twice as many cookies. Add the number to itself to figure out how many cookies you will make if you double the recipe.

4. Go on a long trip and keep track of how far you drive. First, take a guess at how far you will be traveling. Then, ask your parents how many miles it will take to get where you are going. Add that number of miles to itself (double it). That's the total number of miles you will be driving to go and come back again. Did you guess correctly? Each time you make a stop on your trip, write down how many miles the car shows you have driven since you left home.

5. Plant a garden. Before you plant the garden, decide how many tomatoes (or some other vegetable) you would like to harvest. How many packages of seeds will you need to buy if each package will give you 100 tomatoes? What if each package gives you 200 tomatoes?

6. Build something. Before you build something, you want to know how many nails you need. Let's say 100 nails come in each box. If you need 400 nails, how many boxes would you need to buy? Hint: Try adding the number of nails from two boxes. If you buy two boxes, how many nails will you have? Will you have enough nails? Would you need to buy three boxes, or four boxes?

7. Plan for somebody's birthday. Whose birthday is coming up next? Use a calendar to figure out how many days it will be until their birthday comes. What if the birthday is more than a month away? You would need to count how many days are left in this month. Add those to the number of days in the next month. Keep adding the number of days in each month until you come to the person's birthday month. Then add the number of days until the birthday.

Answer Key

DAY 1 EXERCISES

Answers will vary.

DAY 2 EXERCISES

Page 1 — Note: Three figures for each exercise designated as a, b, and c. 1: a, 2: b, 3: c, 4: c, 5: b, 6: c

Page 2 — Answers will vary.

DAY 3 EXERCISES

Page 1 — Answers will vary.

Page 2 — Answers will vary.

DAY 4 EXERCISES

Page 1 — Answers will vary.

Page 2 — Note: Three figures for each exercise designated as a, b, and c. 1: a, 2: b or c, 3: c, 4: a, 5: c, 6: b

DAY 5 EXERCISES

Page 1 — Answers will vary.

Page 2 — Answers will vary.

DAY 6 EXERCISES

Page 1 — Note: Three figures for each exercise designated as a, b, and c. 1: a, 2: b, 3: a, 4: b, 5: c, 6: b,

Page 2 — Snake, Eagle, Owl, Squirrel, Monkey, Raccoon

DAY 7 EXERCISES

Page 1 — Answers will vary.

Page 2 — Note: Three figures for each exercise designated as a, b, and c. 1: b, 2: c, 3: c, 4: c, 5: b, 6: c,

DAY 11 EXERCISES

Page 1 — Note: Rectangle on the top left designated as "a" and following left to right. a: 4, b: 3, c: 6, d: 4, e: 6, f: 12, g: 10, h: 4, i: 3, j: 5, k: 7,

Page 2 — Note: Triangle on the top left designated as "a" and following left to right. a: 3, b: 5, c: 6, d: 4, e: 3, f: 12, g: 10, h: 4, i: 4, j: 3, k: 7

DAY 12 EXERCISES

Page 1 — a: 3, b: 5, c: 1, d: 0, e: 6, f: 6

Page 2 — Arches: 5. Circles: Answers will vary. Triangles: Answers will vary. Squares: 6. Rectangles: Answers will vary. Ovals: 6

DAY 13 EXERCISES

Page 1 — a: 6, 7, 8 b: 9, 10, 11 c: 10, 11, 12 d: 1, 2, 3, 4 e: 5, 6, 7, 8 f: 2, 3, 4 g: 5, 6, 7, 8 h: 5, 6, 7

DAY 14 EXERCISES

Page 1 — Answers will vary.

Page 2 — Review student answers...

DAY 15 LESSON

Page 1 — How many people are here? 2

Page 2 — How many people are here? 2

Page 3 — How many people are here now? 4

Page 4 — 1, 1, 2

Page 5 — 2, 2, 4

Page 6 — 4, 3, 7

DAY 15 EXERCISES

Page 1 — a: 2, 3, 5 b: 3, 4, 7 c: 4, 4, 8 d: 4, 2, 6

DAY 16 EXERCISES

Page 1 — a: 2, 3, 4 b: 4, 5, 6 c: 0, 1, 2, 3 d: 6, 7, 8 e: 9, 10, 11, 12 f: 7, 8, 9, 10 g: 0, 1, 2, 3, 4 h: 1, 2, 3, 4, 5

Page 2 — Note: Three figures for each exercise designated as a, b, and c. 1-c, 2: b or c, 3: b, 4: c, 5: b, 6: c

DAY 17 EXERCISES

Page 1 — Answers will vary.

DAY 18 EXERCISES

Page 1 — Answers will vary.

Page 2 — a: 3, 4, 5 b: 8, 9, 10 c: 1, 2, 3, 4 d: 8, 9, 10, 11 e: 2, 3, 4, 5 f: 2, 3, 4, 5 g: 5, 6, 7, 8, 9 h: 8, 9, 10, 11, 12

DAY 19 EXERCISES

Page 1 — a: 10, 11, 12 b: 2, 3, 4 c: 6, 7, 8 d: 5, 6, 7, 8 e: 6, 7, 8, 9 f: 2, 3, 4, 5 g: 4, 5, 6 h: 8, 9, 10, 11, 12

Page 2 — a: 4, 2, 6 b: 3, 3, 6 c: 2, 3, 5 d: 0, 5, 5

DAY 21 LESSON

Page 1 — 1, 1, 2

DAY 21 ACTIVITY

Page 1 — a: 1, 2, 3 b: 2, 2, 4 c: 1, 1, 2

Page 2 — 2, 3, 5

Page 3 — a: 7, b: 7, c: 3, d: 9, e: 5

DAY 22 EXERCISES

Page 1 — a: 1 + 1= 2, b: 1 + 2 = 3, c: 3 + 3 = 6, d: 4 + 2 = 6, e: 5 + 3 = 8, f: 2 + 7 = 9

Page 2 — a: 7, b: 7, c: 12, d: 12, e: 3, f: 10, g: 12, h: 6, i: 9, j: 11

Page 3 — a: 3, b: 3, c: 7, d: 5, e: 4, f: 7, g: 2, h: 6, i: 10

DAY 23 LESSON

Page 1 — a: 9, b: 12, c: 6, d: 7, e: 8, f: 4

DAY 23 EXERCISES

Page 1 — a: 2 + 3 = 5, b: 2 + 2 = 4, c: 4 + 2 = 6, d: 3 + 1 = 4, e: 6 + 3 = 9, f: 5 + 2 = 7

DAY 24 EXERCISES

Page 1 — a: 1, b: 5, c: 12, d: 7, e: 2, f: 6, g: 0, h: 8, i: 3, j: 5

Page 2 — a: 12, b: 10, c: 5, d: 12, e: 11, f: 6, g: 10, h: 8, i: 6, j: 6, k: 7, l: 12

Page 3 — a: 3, b: 6, c: 4, d: 8, e: 5, f: 7, g: 2, h: 2, i: 1

DAY 25 LESSON

Page 1 — Eight is bigger than five.

DAY 25 EXERCISES

Page 1 — a: 7>2, b: 1<3, c: 6>4, d: 0<12, e: 2<5, f: 8>4

ANSWER KEY

DAY 26 EXERCISES
Page 1 — a: 11, b: 7, c: 2, d: 10, e: 3, f: 9, g: 4, h: 6, i: 12, j: 5
Page 2 — a: <, b: <, c: >, d: >, e: <, f: <, g: >, h: >, i: >, j: <, k: <, l: >
Page 3 — a: 3, b: 5, c: 7, d: 9, e: 11, f: 9, g: 12, h: 10, i: 8, j: 11, k: 12, l: 2

DAY 27 EXERCISES
Page 1 — Answers will vary.

DAY 28 EXERCISES
Page 1 — a: 5, b: 9, c: 10, d: 4, e: 6, f: 3, g: 4, h: 6, i: 2, j: 1
Page 2 — a: <, b: >, c: <, d: <, e: >, f: >, g: <, h: <, i: <, j: <, k: <, l: >
Page 3 — a: >, b: >, c<7, d: <, e: >, f: >
Page 4 — Answers will vary.
Page 5 — a: 1, b: 7, c: 3, d: 9, e: 4, f: 10, g: 6, h: 5, i: 4

DAY 29 ACTIVITY
Page 1 — a: 1 + 4 = 5, b: 1 + 1 + 3 = 5, c: 4 + 3 + 3 = 10 d: 2 + 2 + 2 + 2 = 8

DAY 30 EXERCISES
Page 1 — Answers will vary.
Page 2 — a: 11, b: 2, c: 8, d: 3, e: 9, f: 12, g: 10, h: 4, i: 6, j: 5

DAY 32 EXERCISES
Page 1 — a: <, < b: <, < c: >, > d: >, > e: <, < f: <, < g: >, > h: >, > i: <, < j: <, <
Page 2 — a: even: 2, 6 odd: 3, 7 b: even: odd: 1, 3, 5, 7 c: even: 10, 12 odd: 3, 7 d: even: 4, 6, 8, 10 odd: e: even: 2, 6, 12 odd: 5

DAY 35 LESSON
Page 1 — a:1, b:1, c:2, d:2

DAY 35 EXERCISES
Page 1 — a: 3, b: 4, c: 5, d: 10, e: 9, f: 8, g: 5, h: 4, i: 3, j: 12, k: 12, l: 12

DAY 36 EXERCISES
Page 1 — a: 5, b: 8, c: 11, d: 3, e: 7, f: 12, g: 4, h: 2, i: 0, j: 5
Page 2 — a: 4, b: 2, c: 0, d: 4, e: 3, f: 2, g: 3, h: 5, i: 7
Page 3 — a: 2 - 1 = 1, b: 3 - 1 = 2, c: 7 - 2 = 5

DAY 37 LESSON
Page 1 — 3 - 2 = 1
Page 2 — 4 - 2 = 2
Page 3 — a: 3, b: 5, c: 3, d: 1, e: 5, f: 8

DAY 37 EXERCISES
Page 1 — a: 3 - 2 = 1, b: 5 - 2 = 3, c: 4 - 2 = 2, d: 4 - 3 = 1, e: 6 - 3 = 3, f: 7 - 2 = 5

DAY 38 EXERCISES
Page 1 — a: 1, b: 5, c: 12, d: 7, e: 2, f: 6, g: 0, h: 8, i: 3, j: 9
Page 2 — Answers will vary.
Page 3 — a: 3, b: 3, c: 4, d: 4, e: 4, f: 1, g: 6, h: 8, i: 1, j: 1, k: 1, l: 3

DAY 39 EXERCISES
Page 1 — a: 4, b: 2, c: 9, d: 10, e: 5, f: 12, g: 3, h: 11, i: 6, j: 8
Page 2 — a: 3 - 2 = 1, b: 10 - 1 = 9, c: 2 + 3 = 5, d: 6 - 4 = 2
Page 3 — a: {4}, b: {1}, c: {4, 6}, d: {9}, e: {4, 6, 10}, f: { }

DAY 40 EXERCISES
Page 1 — a: 9, b: 5, c: 0, d: 8, e: 1, f: 7, g: 2, h: 4, i: 11, j: 8
Page 2 — a: 3, b: 4, c: 5, d: 5, e: 4, f: 2, g: 12, h: 11, i: 10, j: 0, k: 1, l: 2

DAY 42 EXERCISES
Page 1 — a: 3, b: 7, c: 8, d: 2, e: 4, f: 1, g: 0, h: 11, i: 5, j: 6
Page 2 — a: +, b: -, c: -, d: +, e: -, f: +, g: +, h: -, i: +, j: -
Page 3 — a: 2, 3, 4, b: 4, 5, 6, c: 6, 7, 8, d: 8, 9, 10, 11, e: 9, 10, 11, 12 f: 2, 3, 4, g: 0, 1, 2, 3, h: 2, 3, 4, 5, 6,

DAY 43 LESSON
Page 1 — 4
Page 2 — 3
Page 3 — 2

DAY 43 EXERCISES
Page 1 — a: 3, b: 7, c: 3, d: 9, e: 7, f: 7, g: 9
Page 2 — a: 5 - 3 = 2, b: 4 - 1 = 3, c: 4 - 1 = 3, d: 3 - 1 = 2, e: 8 - 3 = 5 f: 6 - 2 = 4

DAY 44 EXERCISES
Page 1 — a: 9, b: 0, c: 6, d: 1, e: 7, f: 10, g: 8, h: 2, i: 4, j: 3
Page 2 — a: 1, b: 2, c: 4, d: 3, e: 5, f: 8, g: 9, h: 6, i: 0, j: 3, k: 2, l: 0
Page 3 — a: +, b: -, c: -, d: +, e: -, f: +, g: -, h: +, i: +, j: -

DAY 46 EXERCISES
Page 1 — a: 8, b: 16, c: 18, d: 20, e: 10, f: 4, g: 2, h: 0, i: 12, j: 6,
Page 2 — Answers will vary.
Page 3 — a: {5}, b: {1, 2, 3}, c: {7}, d: {6}, e: {12}, f: { }

DAY 47 ACTIVITY
Pages 1-8 — Answers will vary. There is a boy and a girl; they are on the opposite sides of the page, and the sky is colored while the ground is dark.

DAY 47 EXERCISES
Page 1 — a: 2, b: 6, c: 8, d: 2, e: 12, f: 12, g: 5

DAY 48 EXERCISES
Page 1 — a: 2, b: 4, c: 6, d: 8, e: 10, f: 0, g: 4, h: 8, i: 12, j: 6, k: 10, l: 12
Page 2 — a: 6, b: 8, c: 10, d: 4, e: 6, f: 8, g: 2, h: 4, i: 6, j: 0, k: 2, l: 4

DAY 50 EXERCISES
Page 1 — a: 2, 3, b: 1, 2, c: 3, 3 d: 5, 6
Page 2 — a: 1, 3, 7 b: 1, 2, 5, c: 2, 2, 8 d: 1, 3, 10 e: 3, 2, 11 f: 2, 1, 6

DAY 51 EXERCISES
Page 1 — Total of 5, b: Total of 9, c: Total of 5, d: Total of 8, e:Total of 7, f: Total of 9, g: Total of 10
Page 2 — a: 1, b: 3, c: 4, d: 0, e: 2, f: 4, g: 1, h: 4, i: 2, j: 10, k: 8, l: 5

DAY 52 LESSON
Page 1 — 4, 5
Page 2 — 6, 2
Page 3 — 7, 0

DAY 52 EXERCISES
Page 1 — a: 4, 3 b: 3, 4 c: 2, 5 d: 5, 2 e: 3, 9 f: 9, 3 g: 1, 7 h: 7, 1
Page 2 — a: 24, 25, 26 b: 27, 28, 29 c: 30, 31, 32, d: 29, 30, 31, 32 e: 31, 32, 33, 34 f: 35, 36, 37 g: 37, 38, 39, 40 h: 38, 39, 40, 41, 42

DAY 53 EXERCISES
Page 1 — a: 3, 8 b: 4,4 c: 5, 7 d: 8, 0 e: 0, 9 f: 9, 9 g: 5, 0 h: 0, 2
Page 2 — a: 67, 68, 69 b: 69, 70, 71 c: 72, 73, 74, d: 49, 50, 51, 52 e: 58, 59, 60, 61 f: 70, 71, 72, g: 60, 61, 62, 63 h: 46, 47, 48, 49, 50
Page 3 — a: 10, b: 4, c: 11

DAY 54 EXERCISES
Page 1 — a: 8, b: 18, c: 16, d: 20, e: 10, f: 4, g: 2, h: 0
Page 2 — a: 1, 7, 3, 173 b: 2,4, 6, 246 c: 0, 8, 7, 87 d: 1, 5, 3, 153

DAY 55 EXERCISES
Page 1 — a: 90, 91, 92 b: 79, 80, 81 c: 68, 69, 70 d: 59, 60, 61, 62 e: 48, 49, 50, 51 f: 38, 39, 40, g: 30, 31, 32, 33 h: 16, 17, 18, 19, 20
Page 2 — a: 20, b: 2, c: 14, d: 4, e: 16, f: 0, g: 18, h: 6, i: 10,

DAY 56 LESSON
Page 1 — a: 12, b: 7, c: 12 + 7 = 19

DAY 56 EXERCISES

Page 1 — a: 4, 3 b: 6, 8 c: 2, 5 d-1, 7 e: 3, 9 f: 9, 5 g: 5, 2 h: 7, 5

DAY 57 EXERCISES

Page 1 — a: 13, b: 15, c: 14, d: 9, e: 7, f: 8, g: 8, h: 7, i: 9,
Page 2 — a: {4}, b: {2, 3}, c: {8}, d: {19}, e: {40}, f: {}

DAY 58 EXERCISES

Page 1 — a: 3 + 10 = 13 b: 6 + 10 = 16 c: 8 + 6 = 14 d: 12 + 6 = 18 e: 17 + 6 = 23
Page 2 — a: 16, b: 18, c: 20, d: 11, e: 13, f: 15, g: 7, h: 6, i: 5, j: 8, k: 7, l: 6

DAY 59 EXERCISES

Page 1 — a: 70, b: 72, c: 74, d: 69, e: 68, f: 67, g: 68, h: 70, i: 72
Page 2 — a: 11, b: 12, c: 14, d: 12, e: 14, f: 16, g: 9, h: 7, i: 9, j: 7, k: 7, l: 8

DAY 60 EXERCISES

Page 1 — a: 31, b: 30, c: 29, d: 31, e: 32, f: 27, g: 28, h: 27, i: 29,
Page 2 — Answers will vary.

DAY 61 EXERCISES

Page 1 — Review student answers.
Page 2 — a: +, b: +, c: -, d: -, e: -, f: +, g: +, h: -, i: +, j: +

DAY 62 EXERCISES

Page 1 — a: 13, b: 15, c: 12, d: 15, e: 13, f: 14, g: 6, h: 8, i: 8, j: 10, k: 9, l: 5
Page 2 — a: 56, 57, 58 b: 59, 60, 61 c: 62, 63, 64, d: 55, 56, 57, 58 e: 51, 52, 53, 54 f: 49, 50, 51, 52 g: 57, 58, 59, 60, 61 h: 61, 62, 63, 64, 65

DAY 63 ACTIVITY

Page 6 — Answers will vary.

DAY 64 EXERCISES

Page 1 — a: 7 in., b: 6 in., c: 3 in., d: 2 in., e: 5 in.

DAY 65 EXERCISES

Page 1 — a: 9, b: 3, c: 11, d: 5, e: 7, f: 2, g: 13, h: 16, i: 15, j: 7, k: 10, l: 8
Page 2 — a: 25, 26, 27 b: 34, 35, 36 c: 38, 39, 40 d: 49, 50, 51, 52 e: 53, 54, 55, 56 f: 58, 59, 60 g: 70, 71, 72, 73 h: 76, 77, 78, 79, 80

DAY 66 EXERCISES

Page 1 — Note: Three figures for each exercise designated as a, b, and c. 1: a or b, 2: b, 3: a, 4: c, 5: c, 6: a

DAY 67 EXERCISES

Page 1 — a: +, b: -, c: +, d: -, e: -, f: +, g: -, h: +, i: +, j: -
Page 2 — a: 20, b: 2, c: 12, d: 4, e: 16, f: 8, g: 18, h: 6, i: 10

DAY 68 EXERCISES

Page 1 — a: 32, b: 30, c: 31, d: 29, e: 28, f: 25, g: 27, h: 32, i: 30
Page 2 — a: 5 - 3 = 2 b: 0 + 0 = 0 c: 1 + 0 = 1 d: 7 - 4 = 3 e: 2 + 2 = 4 f: 0 + 5 = 5 g: 5 + 1 = 6 h: 9 - 2 = 7

DAY 69 LESSON

Page 1 — a: 3 + 2 + 4 = 9, 5 + 4 = 9 b: 4 + 2 + 2 = 8, 6 + 2 = 8 c: 5 + 3 + 1 = 9, 8 + 1 = 9

DAY 69 EXERCISES

Page 1 — Answers will vary.

DAY 70 EXERCISES

Page 1 — Answers will vary.
Page 2 — a: 2 + 4 + 1 = 7, 6 + 1 = 7 b: 4 + 4 + 3 = 11, 8 + 3 = 11 c: 1 + 2 + 3 = 6, 3 + 3 = 6 d: 3 + 3 + 3 = 9, 6 + 3 = 9 e: 6 + 2 + 1 = 9, 8 + 1 = 9 f: 7 + 0 + 2 = 9, 7 + 2 = 9 g: 5 + 7 + 1 = 13, 12 + 1 = 13 h: 2 + 4 + 6 = 12, 6 + 6 = 12
Page 3 — Answers will vary.

DAY 71 EXERCISES

Page 1 — a: 12, b: 6, c: 10, d: 4, e: 2, f: 6, g: 14, h: 15, i: 14, j: 11, k: 8, l: 6

DAY 72 EXERCISES

Page 1 — Answers will vary.
Page 2 — a: < b: > c: > d: < e: < f: < g: < h: > i: < j: >

DAY 73 EXERCISES

Page 1 — a: 3, 5 b: 4, 2, c: 5, 8 d: 6, 7 e: 7, 0 f: 8, 3 g: 9, 0 h: 0, 9
Page 2 — a: 5 + 1 + 3 = 9, 6 + 3 = 9 b: 4 + 3 + 2 = 9, 7 + 2 = 9 c: 3 + 0 + 2 = 5, 3 + 2 = 5 d: 5 + 3 + 0 = 8, 8 + 0 = 8 e: 3 + 3 + 3 = 9, 6 + 3 = 9 f: 4 + 4 + 4 = 12, 8 + 4 = 12 g: 2 + 2 + 6 = 10, 4 + 6 = 10 h: 6 + 4 + 1 = 11, 10 + 1 = 11

DAY 75 EXERCISES

Page 1 — a: 4 - 1 = 3, b: 8 - 2 = 6, c: 40 - 10 = 30, d: 150ml, e: 30 - 10 = 20

DAY 76 EXERCISES

Page 1 — a: < b: > c: < d: > e: > f: < g: > h: > i: < j: <
Page 2 — Answers will vary.

DAY 77 EXERCISES

Page 1 — a: 0, 5 b: 1, 3 c: 2, 4 d: 3, 5 e: 4, 0 f: 5, 0 g: 6, 8 h: 7, 1
Page 2 — a: {4, 6}, b: {1, 2}, c: {10}, d: {}, e: {4, 8}, f: {12, 18}

DAY 78 EXERCISES

Page 1 — a: 1, b: 1, c: 1, d: 1, e: 2, f: 2, g: 2, h: 2, i: 4, j: 4
Page 2 — a: 8, b: 10, c: 12, d: 7, e: 2, f: 2, g: 13, h: 15, i: 16, j: 11, k: 9, l: 7

DAY 79 EXERCISES

Page 1 — 1) a: Answers will vary. b: 2. c: 6 - 4 = 2
2) 8 + 3 = 11
3) 5 + 2 = 7
4) 12 - 8 = 4
Page 2 — a: 27, 36, 45, 63, 72 b: 17, 34, 48, 72, 95 c: 15, 25, 45, 55, 65

DAY 80 EXERCISES

Page 1 — a: 20 b: 20 c: 30 d: 50 e: 70 f: 70 g: 70 h: 20 i: 90 j: 80
Page 2 — a: 3 b: 3 c: 2 d: 2 e: 2 f: 2 g: 5 h: 5 i: 0 j: 0

DAY 81 EXERCISES

Page 1 — a: 70 b: 20 c: 80 d: 60 e: 60 f: 60 g: 30 h: 40 i: 30 j: 20

DAY 82 EXERCISES

Page 1 — a: {12} b: {20, 21} c: {} d: {1, 3} e: {2, 8} f: {13, 19, 29, 31}
Page 2 — a: 5 b: 3 c: 0 d: 0 e: 1 f: 4 g: 0 h: 0 i: 2

DAY 83 ACTIVITY

Page 1 — Answers will vary.
Page 2 — Answers will vary.
Page 3 — Answers will vary.
Page 4 — Answers will vary.

DAY 83 EXERCISES

Page 1 — a: 1, 11, 21, 31, 51 b: 13, 23, 31, 32, 33 c: 46, 54, 56, 64, 65

DAY 84 EXERCISES

Page 1 — a: -, b: +, c: -, d: +, e: -, f: -, g: +, h: -, i: +/-, j: +/-, k: +, l: +
Page 2 — Answers will vary.

DAY 85 ACTIVITY

Page 1 — Answers will vary.
Page 2 — Answers will vary.
Page 3 — Answers will vary.

DAY 86 EXERCISES

Page 1 — a: 3, 2, 4, 1 b: 2, 4, 1, 3 c: 2, 4, 1, 3
Page 2 — a: 7 + 1 = 8 b: 5 + 4 = 9 c: 4 + 3 = 7 d: 5 + 1 = 6 e: 7 + 2 = 9 f: 1 + 1 = 2 g: 6 + 5 = 11 h: 6 + 1 = 7

ANSWER KEY

DAY 88 LESSON
Page 1 — a: 28, b: 69, c: 97

DAY 88 EXERCISES
Page 1 — 1: b, 2: c, 3: a, 4: c, 5: b, 6: c
Page 2 — a: 14, b: 15, c: 16, d: 15, e: 16, f: 17, g: 16, h: 17

DAY 89 EXERCISES
Page 1 — a: 27, 36, 62, 63, 72 b: 11, 44, 61, 73, 97 c: 25, 35, 47, 62, 63
Page 2 — a: 70, b: 70, c: 10, d: 50, e: 90, f: 20, g: 60, h: 0, i: 20, j: 30

DAY 90 LESSON
Page 1 — a: 47, b: 78, c: 99

DAY 90 EXERCISES
Page 1 — Answers will vary.
Page 2 — Answers will vary.

DAY 91 EXERCISES
Page 1 — a: 7, b: 12, c: 14, d: 16, e: 25, f: 24, g: 87, h: 58
Page 2 — a: 1, b: 3, c: 9, d: 7, e: 9, f: 6, g: 12

DAY 92 EXERCISES
Page 1 — a: -, b: -, c: -, d: -, e: +, f: +, g: +, h: -, i: -, j: +

DAY 93 EXERCISES
Page 1 — a: 8, b: 6, c: 14, d: 16, e: 20, f: 60, g: 25, h: 46, i: 94,
Page 2 — a: <, <, b: >, >, c: >, >, d: <, <, e: <, <, f: >, > g: <, <, h: <, <,

DAY 94 EXERCISES
Page 1 — a: 11, b: 8, c: 3, d: 9, e: 12, f: 7, g: 10, h: 6

DAY 95 EXERCISES
Page 1 — a: 6, b: 5, c: 4, d: 8, e: 21, f: 40, g: 31, h: 23, i: 22
Page 2 — a: Blue Dot and Purple Dot, b: Blue Dot, c: Green Triangle, d: Green Triangle, e: Green Star, f: Red Triangle

DAY 96 EXERCISES
Page 1 — Note: Three figures for each exercise designated as a, b, and c. 1: b, 2: c, 3: a, 4: b, 5: c, 6: a

DAY 97 EXERCISES
Page 1 — Answers will vary.
Page 2 — a: 50, b: 51, c: 52, d: 50, e: 49, f: 48, g: 51, h: 48, i: 49

DAY 98 ACTIVITY
Page 1 — 36

DAY 99 LESSON
Page 1 — Answers will vary.

DAY 99 EXERCISES
Page 1 — a: 8, b: 9, c: 13, d: 14, e: 86, f: 48, g: 3, h: 7, i: 8, j: 9, k: 43, l: 51

DAY 100 EXERCISES
Page 1 — a: 5, b: 1, c: 3, d: 3, e: 1, f: 5, g: 5, h: 8, i: 4, j: 8, k: 7, l: 10
Page 2 — a: 2, b: 4, c: 0, d: 5, e: 0, f: 1, g: 0, h: 6, i: 3

DAY 101 ACTIVITY
Page 1 — Answers will vary.

DAY 101 EXERCISES
Page 1 — a: 90, b: 50, c: 90, d: 50, e: 80, f: 40, g: 70, h: 20, i: 60, j: 0
Page 2 — a: >, > b: <, > c: >, < d: <, < e: <, > f: >, > g: <, > h: >, <

DAY 102 EXERCISES
Page 1 — a: 3, b: 4, c: 6, d: Answers will vary. e: Answers will vary. f: 7
Page 2 — a: 6, 5 b: 9, 4 c: 2, 2 d: 8, 7 e: 4, 5 f: 0, 9 g: 3, 0 h: 4, 3

DAY 103 EXERCISES
Page 1 — a: 5, b: 9, c: 5 + 9 = 14
Page 2 — Answers will vary.
Page 3 — Note: Three figures for each exercise designated as a or b. 1: a, 2: b, 3: b, 4: b
Page 4 — Review student answers.

DAY 104 EXERCISES
Page 1 — a: 20, b: 2, c: 15, d: 400
Page 2 — a: 10, b: 12, c: 14, d: 16, e: 27, f: 57, g: 2, h: 3, i: 6, j: 8, k: 21, l: 52

DAY 105 ACTIVITY
Page 1 — Answers will vary.
Page 2 — a: {20, 30} b: {35}, c: {4, 18, 46} d: {}, e: {20, 40}, f: {23, 43}

DAY 106 EXERCISES
Page 1 — a: 7, b: 25, c: 33, d: 78
Page 2 — a: 13, b: 15, c: 13, d: 14, e;11, f: 8, g: 10, h: 9, i: 8

DAY 107 EXERCISES
Page 1 — a: 7, b: 13, c: Answer will vary d: Answer will vary e: 7, f: 8
Page 2 — a: 4 + 1 = 5, b: 7 + 5 = 12, c: 5 + 12 = 17

DAY 109 EXERCISES
Page 1 — a: 8, b: 6, c: 5, d: 9, e: 16, f: 13, g: 25, h: 21, i: 10, j: 13, k: 22, l: 15

DAY 110 EXERCISES
Page 1 — a: 7, b: 60 miles, 1hr./60 min., c: Answers will vary.
Page 2 — a: sides = 5, corners = 5 b: sides = 7, corners = 7 c: sides = 10, corners = 10 d: sides = 4, corners = 4 e: sides = 6, corners = 6 f: sides = 3, corners = 3 g-sides = 12, corners = 12

DAY 111 EXERCISES
Page 1 — a: 0, 1, 2, 3, 4, 5, b: 0, 1, 2, 3, 4 c: 0, 1, 2 d: 0, 1, 2, 3 e: 0, 1, 2, 3, 4, f: 0, 1, 2, 3
Page 2 — a: 0, 1, 2 b: 3, 4, 5, c: 6, 7, 8 d: 5, 6, 7, 8 e: 9, 10, 11, 12 f: 3, 4, 5, 6, 7 g: 6, 7, 8, 9, 10

DAY 112 EXERCISES
Page 1 — Answers will vary.
Page 2 — Answers will vary.

DAY 113 ACTIVITY
Page Using Wheel #1 — a: 2, b: 0, c: 2
Page Using Wheel #2 — d: 2, e: 0, f: 4

DAY 113 EXERCISES
Page 1 — a: 15 =15 b: 23 = 23 c: 30 = 30 d: 14 = 14 e: 29 = 29
Page 2 — a: 9, b: 21, c: 33, d: 51

DAY 114 EXERCISES
Page 1 — a: 2 b: 4 c: 0 d: 2 e: 0 f: 0 g: 3 h: 5
Page 2 — a: 1 b: 1 c: 3 d: 2 e: 5 f: 4 g: 1 h: 1 i: 2 j: 3
Page 3 — a: 3, 4, 5, 1 kitten, b: 4, 6, 8, 10, 2 puppies

DAY 115 EXERCISES
Page 1 — Review student answers.

DAY 116 EXERCISES
Page 1 — a: 7, b: 7, c: 5, d: 5, e: 2, f: 4, g: 5, h: 0
Page 2 — Answers will vary.
Page 3 — a: 4, 6, 8, 10, 12, 2 puppies, b: 6, 9, 12, 3 piglets

DAY 117 EXERCISES
Page 1 — a: 10, b: 11, c: 12, d: 7, e: 8 f: 9, g: 7, h: 9, i: 11, j: 10, k: 11, l: 12
Page 1 — a. even: 2, 4, odd: 3, 5 b. even: odd: 1, 3, 5, 7 c. even: 6, 8 odd: 5, 7 d. even: 4, 6, 8 odd: 9 e. even: 6, 8, 12 odd: 7

DAY 118 LESSON

Page 1 — a-Wheel 1: 3, Wheel 2: 3 b-Wheel 1: 0, Wheel 2: 4 c-Wheel 1: 2, Wheel 2: 5 d-Wheel 1: 3, Wheel 2: 4 e-Wheel 1: 2, Wheel 2: 0

DAY 119 PRACTICE

Page 2 — a: 0, 1, 2, 3 b: 0, 1, 2, 3, 4, 5, 6 c: 0, 1, 2, 3, 4, 5, 6, 7, 8, 9 d: 0, 1, 2, 3, 4, e: 0, 1, 2, 3, 4, 5, f: 0, 1, 2, 3, 4 ,5

Page 1 — Answers will vary.

DAY 121 ACTIVITY

Page 1 — Answers will vary.

DAY 121 EXERCISES

Page 1 — a: 9, b: 13, c: 14, d: 76, e: 98 f: 3, g: 3 h: 6 i: 22, j: 13, k: 42, l: 37

DAY 122 EXERCISES

Page 1 — a: 2, b: 11, c: 8, d: Answers will vary. e: Answers will vary. f: 5, g: 16

Page 2 — a: 13, b: 14, c: 13, d: 15, e: 7 f: 15, g: 10, h: 9, i: 9, j: 8, k: 8, l: 7

DAY 123 LESSON

Page 1 — Answers will vary.

DAY 123 EXERCISES

Page 1 — a: 24, 25, 26 b: 27, 28, 29 c: 30, 31, 32 d: 29, 30, 31, 32 e: 31, 32, 33, 34 f: 35, 36, 37 g: 53, 54, 55, 56 h: 67, 68, 69, 70, 71

Page 2 — a: 10, b: 0, c: 50, d: 30, e: 80 f: 40, g: 70, h: 50, i: 90, j: 90

DAY 124 EXERCISES

Page 1 — a. 3: 00 b. 2: 40 c. 10: 30 d. 7: 25 e. 6: 05 f. 1: 50

Page 2 — Review student answers.

Page 3 — a: 11 - 9 = 2, b: 38 - 33 = 5 c: Answers will vary.

DAY 125 EXERCISES

Page 1 — a: 1, b: 3, c: 9, d: 7, e: 9 f: 6, g: 12

Page 2 — a. 5:00, b. 6:50, c. 10:00, d. 3:30, e. 10:10 f. 10:45

Page 3 — Review student answers.

DAY 126 EXERCISES

Page 1 — Review student answers.

Page 2 — a: 11, b: 13, c: 12, d: 14, e: 66 f: 88, g: 9, h: 8, i: 8, j: 7, k: 32, l: 42

DAY 127 EXERCISES

Page 1 — a: >, > b: <, > c: >, < d: <, < e: <, > f: >, > g: >, > h: <, >

Page 2 — a: 8, b: 8, c: 10, d: 10, e: 11 f: 65, g: 5, h: 7, i: 23, j: 9, k: 12, l: 72

DAY 128 EXERCISES

Page 1 — a. 10: 00, b. 1: 00, c. 3: 00, d. 6: 00, e. 6: 00 f. 12: 00

Page 2 — a: Sunday b: Saturday c: Friday d: Tuesday e: one week f: two weeks g: zero weeks

DAY 129 LESSON

Page 1 — a: 3 days, b: 4 days , c: 12 years d: 75 years old, e: 100 years old f: 25 years

DAY 129 EXERCISES

Page 1 — a: June b: April, c: April d: July e: September f: January

DAY 130 EXERCISES

Page 1 — a: {1, 5} b: {2, 4, 6, 8} c: {} d: {2, 3, 4} e: {8} f: {1}

Page 2 — Review student answers.

DAY 131 ACTIVITY

Page 1 — a: 50 + 30 = 80, 95 - 80 = 15 b: 30 + 30 = 60, 85 - 60 = 25 c: 25 + 25 + 25 = 75 75 - 40 = 35

Page 2 — a: 80, b: 60, c: 50, d: Answers will vary.

DAY 132 LESSON

Page 1 — a: 431 people b: 624 people c: 200 people d: 303 people e: 412 people

Page 2 — a: 1, 2, 3 b: 4, 5, 3, c: 7, 4, 5 d: 4, 5, 6 e: 3, 6, 6 f: 1, 4, 5 g: 0, 4, 5, h: 0, 0, 7

DAY 133 EXERCISES

Page 1 — Answers may vary.

Page 2 — No need to be in particular order: 4+4, 5+3, 6+2, 7+1, 8+0

Page 3 — a: 173, b: 246, c: 87

DAY 134 LESSON

Page 1 — a: < b: > c: > d: > e: < f: < g: < h: > i: <

DAY 134 EXERCISES

Page 1 — a: 7, b: Answers will vary. c: 6 d: 5 e: Answers will vary. f: Answers will vary. g: Answers will vary.

Page 2 — Review student answers.

DAY 135 EXERCISES

Page 1 — a: < b: > c: > d: < e: < f: < g: < h: > i: > j: >

Page 2 — a. 1: 50, b. 12: 15, c. 10: 15, d. 12: 50, e. 8: 00 f. 5: 05

Page 3 — Review student answers.

DAY 136 LESSON

Page 1 — Answers will vary.

DAY 136 EXERCISES

Page 1 — Note: Three figures for each exercise designated as a, b, and c. 1: c, 2: b, 3: a, 4: b, 5: b, 6: a

DAY 137 EXERCISES

Page 1 — a: 2, b: 2, c: 6, d: 11, e: 4, f: 2, g: 10

Page 2 — a: 0, 0, 6 b: 3, 6, 5 c: 1, 2, 5, d: 1, 4, 4, e: 2, 5, 6 f: 0, 4, 9 g: 5, 5, 5 h: 9, 8,7

DAY 138 LESSON

Page 1 — a: 777, b: 123, c: 429,

DAY 138 EXERCISES

Page 1 — a: 7, b: 12, c: 14, d: 6, e: 4, f: 6, g: 8, h: 13, i: 6, j: 2

Page 2 — a: 6 + 5 = 11 b: 7 + 6 = 13 c: 8 + 7 = 15 d: 9 + 7 = 16 e: 8 + 6 = 14 f: 10 + 3 = 13 g: 7 + 4 = 11, h: 11 + 5 = 16, i: 10 + 5 = 15, j: 5 + 6 = 11

DAY 139 EXERCISES

Page 1 — a: 13, b: 13, c: 69, d: 58, e: 785, f: 856, g: 8, h: 6, i: 7, j: 9, k: 22, l: 35

Page 2 — Review student answers.

DAY 140 LESSON

Page 1 — a: 300, b: 500, c: 700, d: 400, e: 400 f: 400

DAY 140 EXERCISES

Page 1 — a: 11, b: 14, c: 89, d: 98, e: 476 f: 788, g: 11, h: 7, i: 9, j: 9, k: 30, l: 27

Page 2 — a: 20, 21, 22 b: 49, 50, 51 c: 78, 79, 80 d: 249, 250, 251, 252 e: 738, 739, 740, 741 f: 397, 398, 399, 400 g: 496, 497, 498, 499, 500

DAY 141 EXERCISES

Page 1 — a: 90 b: 50, c: 90, d: 50, e: 80 f: 40

Page 2 — a: 100, b: 600 c: 900 d: 600 e: 400 f: 300

Page 3 — a: 2, b: 1, c: 7, d: 7, e: 10 f: 11, g: 2

DAY 142 EXERCISES

Page 1 — a: 15, b: 97, c: 87, d: 585, e: 897 f: 10, g: 9, h: 7, i: 5, j: 20, k: 23, l: 31

DAY 143 EXERCISES

Page 1 — a: +, b: -, c: +, d: -, e: -, f: +, g: + h: -, i: +, j: +

Page #2 — a: 134, 251, 311 b: 100, 120, 170 c: 145, 415, 541

DAY 144 ACTIVITY 1

Page 1 — a: 33, b: 26, c: 47, d: 443, e: 407, f: 375